Training Methods to Improve Sports Performance and Health

Training Methods to Improve Sports Performance and Health

Editors

Bruno Gonçalves
Jorge Bravo
Hugo Folgado

MDPI • Basel • Beijing • Wuhan • Barcelona • Belgrade • Manchester • Tokyo • Cluj • Tianjin

Editors
Bruno Gonçalves
Departamento de Desporto e
Saúde, Escola de Saúde e
Desenvolvimento Humano
Universidade de Évora
Évora
Portugal

Jorge Bravo
Departamento de Desporto e
Saúde, Escola de Saúde e
Desenvolvimento Humano
Universidade de Évora
Évora
Portugal

Hugo Folgado
Departamento de Desporto e
Saúde, Escola de Saúde e
Desenvolvimento Humano
Universidade de Évora
Évora
Portugal

Editorial Office
MDPI
St. Alban-Anlage 66
4052 Basel, Switzerland

This is a reprint of articles from the Special Issue published online in the open access journal *International Journal of Environmental Research and Public Health* (ISSN 1660-4601) (available at: www.mdpi.com/journal/ijerph/special_issues/Training_Methods).

For citation purposes, cite each article independently as indicated on the article page online and as indicated below:

LastName, A.A.; LastName, B.B.; LastName, C.C. Article Title. *Journal Name* **Year**, *Volume Number*, Page Range.

ISBN 978-3-0365-2967-7 (Hbk)
ISBN 978-3-0365-2966-0 (PDF)

Contents

About the Editors

Bruno Gonçalves

Bruno Gonçalves is an associate professor at the University of Évora and researcher at the Comprehensive Health Research Centre (CHRC). He graduated with a PhD in collective movement behaviour in association football, where he developed expertise in capturing and filtering data using high-tech instruments and in advanced processing computations. His research focuses on the identifcation of new variables that capture collective, exploratory, adaptive, and emergent behaviour principles in team-sport settings.

Jorge Bravo

Jorge Bravo is an assistant professor at the University of Évora and researcher at the Comprehensive Health Research Centre (CHRC). He graduated with a PhD in Sports Performance at the University of Extremadura, developing expertise in the management of physiological biomarkers in children and adolescents. His research interests are the conception and development of interactive exercise programs for health management, particularly in cardiac patients and older people.

Hugo Folgado

Hugo Folgado is an assistant professor at the Department of Sport and Health of the University of Évora. His main research interests are performance analysis and technology development and application in team sports, particularly in football associations. His works explore players' movement coordination and performance during competitive matches and training situations. He also studies context manipulation for the development of players, particularly considering multidimensional performance indicators.

International Journal of
*Environmental Research
and Public Health*

MDPI

Article

External Focus or Differential Learning: Is There an Additive Effect on Learning a Futsal Goal Kick?

Sara Oftadeh [1,*], Abbas Bahram [1], Rasoul Yaali [1], Farhad Ghadiri [1] and Wolfgang I. Schöllhorn [2,*]

[1] Faculty of Physical Education and Sport Sciences, Kharazmi University of Tehran, Tehran 14155-6455, Iran; bahram@khu.ac.ir (A.B.); r.yaali@khu.ac.ir (R.Y.); ghadiri@khu.ac.ir (F.G.)
[2] Department for Training and Movement Science, Johannes Gutenberg-University Mainz, 55128 Mainz, Germany
* Correspondence: s.oftadeh@khu.ac.ir (S.O.); schoellw@uni-mainz.de (W.I.S.)

Abstract: (1) Background: How to optimally promote the process of acquiring and learning a new motor skill is still one of the fundamental questions often raised in training and movement science, rehabilitation, and physical education. This study is aimed at investigating the effects of differential learning (DL) and the elements of OPTIMAL theory on learning a goal-kicking skill in futsal, especially under the conditions of external and internal foci. (2) Methods: A total of 40 female beginners were randomly assigned to, and equally distributed among, five different interventions. Within a pretest and post-test design, with retention and transfer tests, participants practiced for 12 weeks, involving two 20-min sessions per week. The tests involved a kicking skill test. Data were analyzed with a one-way ANOVA. (3) Results: Statistically significant differences with large effect sizes were found between differential learning (DL) with an external focus, DL with an internal focus, DL with no focus, traditional training with an external focus, and traditional training with control groups in the post-, retention, and transfer tests. (4) Conclusions: The results indicate the clear advantages of DL. It is well worth putting further efforts into investigating a more differentiated application of instructions combined with exercises for DL.

Keywords: differential learning; OPTIMAL theory; external focus; motor learning; futsal

 check for **updates**

Citation: Oftadeh, S.; Bahram, A.; Yaali, R.; Ghadiri, F.; Schöllhorn, W.I. External Focus or Differential Learning: Is There an Additive Effect on Learning a Futsal Goal Kick? *Int. J. Environ. Res. Public Health* **2022**, *19*, 317. https://doi.org/10.3390/ijerph19010317

Academic Editors: Bruno Gonçalves, Jorge Bravo and Hugo Folgado

Received: 30 October 2021
Accepted: 24 December 2021
Published: 29 December 2021

Publisher's Note: MDPI stays neutral with regard to jurisdictional claims in published maps and institutional affiliations.

1. Introduction

How to optimally promote the process of acquiring and learning a new motor skill more effectively is still reckoned as one of the core questions in movement science, therapy, and training science [1]. Classically, the traditional approach to skill acquisition is based on reductionist-atomistic thinking, accompanied by an excessive emphasis on cognition since the "cognitive turn" in the 1960s [2]. This is a way of thinking that is mainly based on linear models, explicit verbalization, a lot of imitations, and the internalization of knowledge about a correct prototype that is shown by a lot of repetition and corrective instructions to avoid making mistakes [3]. In the same context, learners are expected to copy the role models of motor patterns [4], mostly derived from the averages of the best athletes in their disciplines. To support the learners during this process, manual [5], mechanical [6–8], verbal [9], and visual [10] guidance is also suggested in the form of augmented feedback [11–13]. Following this traditional logic of a control loop with external feedback, the main theoretically assumed key factors influencing skill learning [14] would first need to be identified, and their interactions would need to be appropriately understood by practitioners and coaches to establish a strong theoretical basis for an appropriate pedagogical approach [15].

More recently, the assumptions proposed for this traditional repetitive and model-oriented approach have been increasingly challenged by the upcoming system dynamics and complexity theory in general [16], and football in particular [17]. From a practical perspective, the repetitive approach became questioned. The process of acquiring more difficult gross motor movements, in which all learners typically begin with the same exercise and then progress to the final learning goal with the identical sequence of exercises, is increasingly being challenged. In this process, each exercise must be repeated until only small differences are made apparent from the partial target model. When the purported target movement has been successfully achieved, the approach stagnates in the exclusive repetition of the movements, and only the hope for additional conditional progress provides further motivation. By reliably identifying individuals not only through their biomechanical movement patterns [18,19], but also through a recognition of the emotions [20] and the fatigue phases [21] within an individual, as well as of individuality, even across disciplines [22], the main premises of the traditional approach have been deconstructed. As a result, the question arises about the necessity of repetition and prototypes, since individuals, situations, and ongoing changes must be dealt with. Given the extremely low probability of executing two identical movements [21,23,24], particularly during a futsal match or similar games with an infinite number of player constellations, traditional learning methods rely primarily on repetitions to realize an ideal movement, and sustainability in error corrections should be reconsidered [25].

Although differential learning (DL) theory is explicitly silent about psychological aspects and applies instructions with both internal (joint angles) [26] and external foci (variable targets) [27], as well as with a metaphoric (moving in animal styles) [28] focus in its experiments, DL is occasionally associated with the internal focus [29]. In the case of learning a tennis forehand stroke [30], the focus should be on performing the forehand with an extended elbow, and then with a flexed elbow, then stiff knees, and so on [31]. In a volleyball experiment, the DL idea of increased noise was realized by means of moving similar various animals (metaphorically) that were creatively announced by each server [28]. To achieve maximum variety in field hockey, participants either changed their bodily positions in response to some internally connected instructions, or they had to change the target for which they were externally attempting [27]. Nonetheless, systematic investigations concerning the interdependencies of variable movements and the foci of instruction are missing.

A quite different and more recently suggested approach to support motor learning has been developed with the OPTIMAL approach [32]. While the DL approach, in the beginning, focused more on the variety of external stimuli that result from an interaction of active and passive forces, as well as on the conditions applied on the organization of the central nervous structures through variable movements via mechanoreceptors [16,33] before the increased fluctuations were adapted to an individual's movement noise [26] the OPTIMAL approach is more rooted in sports psychology, as it addresses the specific mental aspects of movements, such as motivation and attention [34]. Lewthwaite and Wulf proposed the OPTIMAL theory of motor learning, involving two motivational variables (enhanced expectancies and support for autonomy) and an attention variable (e.g., the focus of external attention) affecting optimal performance. The enhanced expectancies, mainly initiated by corresponding instructions, refer to an increase in individual expectations for positive experiences or success [34], while support for autonomy is said to be a condition that supports the need of people to be in control or to be autonomous over their actions [35]. As for the attention variable, studies on the OPTIMAL theory indicate that having an instructed external focus of attention on the intended movement effect (e.g., implement trajectory, hitting the target, exerting force against the ground) typically results in more effective and efficient performances or learning, as compared with the internal focus on body landmarks [36]. In fact, OPTIMAL theory predicts that an external focus of attention is more beneficial for motor learning than an internal one [37].

Therefore, the general aim of this study was to evaluate the effects of the DL approach and the elements of the OPTIMAL theory on the goal-kicking performance in futsal, in comparison to a purely repetitive approach. More specifically, the hypotheses to be tested were that the DL intervention would not only outperform the external-focus intervention, but also that both interventions would have significantly larger effects compared to repetition-only learning, and, finally, that the combination of both interventions would outperform the DL-only intervention, with or without an internal attentional focus [29,36].

2. Materials and Methods

2.1. Participants

A total of 40 school girls, aged 16 to 18 years and recruited from the city of Baghmalek, Khuzestan Province, Iran, voluntarily participated in this study, and were randomly assigned to five equal groups with different instructions during the interventions: (a) DL with an external focus (DL/EF n = 8); (b) DL with an internal focus (DL/IF n = 8); (c) DL without a specific focus (DL/C n = 8); (d) Traditional training with an external focus (T/EF n = 8); and (e) Traditional training without a specific focus (T/C n = 8). The inclusion criteria were as follows: no history of formal futsal training, and no reported disease or injury. The exclusion criterion is specified as follows: absenteeism in any of the training sessions. The study was approved by the university's Institutional Review Board, with the ethics code: "IR.KHU.REC.1399.04". Using G-Power (version 3.1), the sample size was estimated at the significance level of 0.05, with a 0.80 statistical power, and an effect size of 0.6 (medium to large effect size), using the statistical one-way ANOVA method [37,38]. Accordingly, 40 people were assigned to five groups with different interventions.

The demographic information for all the groups is presented in Table 1. Specifically, the average age of all the participants was 16.97 ± 2.28 years, their mean body mass was 58.62 ± 4.27 kg, and their mean body height was 159.36 ± 3.73 cm.

Table 1. Demographic characteristics of the five intervention groups.

Variable	Total Mean ± SD	DL/EF Mean ± SD	DL/IF Mean ± SD	DL/C Mean ± SD	T/EF Mean ± SD	T/C Mean ± SD
Age (year)	16.97 ± 2.28	17.3 ± 2.27	17.07 ± 2.16	16.87 ± 2.11	16.79 ± 2.33	16.82 ± 2.53
Mass (kg)	58.62± 4.27	59.74 ± 4.09	58.83 ± 4.47	58.74 ± 3.96	58.4 ± 4.16	57.41 ± 4.67
Height (cm)	159.36 ± 3.73	159.28 ± 3.22	159.99 ± 3.92	158.33 ± 3.72	160.12 ± 4.09	159.11 ± 3.74
Number	40	8	8	8	8	8

Having chosen the participants according to the inclusion criteria, the consent form was completed, and the COVID-19 protocols were observed for all the people who played a role in the study, including the subjects, their parents, and the examiners. All the health protocols were applied to all of the individuals, according to the COVID-19 requirements. All procedures and measurements complied with the Declaration of Helsinki (54th Revision 2008, Korea) regarding human subjects.

2.2. Design

A pretest and post-test design, with retention and transfer tests, was chosen for this study. Then, a futsal shooting skill was evaluated as an example of learning a gross motor skill. The chronological schedule of the design is shown in Table 2. After the pretest, all the subjects participated in a 12-week intervention period, with two 20-min sessions per week. Ten minutes after the last intervention, a post-test was conducted. One week after the post-test, all the subjects participated in the retention and transfer tests.

Table 2. Study design.

Groups	Pretest	Training	Post-Test	Retention Test	Transfer Test
DL/EF	Before the start of the training period, a futsal shoot pretest had been taken by all the groups.	12-week DL exercises (two 20-min sessions per week) with external focus	Ten minutes after the end of the last training session, the futsal shooting test was taken.	One week after the post-test, all the groups performed a futsal shooting test.	One week after the post-test, the futsal shoot transfer test (in the presence of spectators), immediately after the retention test was performed.
DL/IF		12-week DL exercises (two 20-min sessions per week) with internal focus			
DL/C		12-week DL exercises (two 20-min sessions per week) with no attention instruction			
T/EF		12-week traditional exercises (two 20-min sessions per week) with external focus			
T/C		12-week traditional exercises (two 20-min sessions per week) with no attention instruction			

2.3. Tests

All tests and interventions were executed applying a standard futsal ball and a standard futsal goal in a standard indoor futsal court.

In the goal-shooting test, participants had to shoot the ball from a 6-m line towards the goal, without a goalkeeper, in seven different situations, with each one repeated five times. Overall, each participant performed 35 shooting movements in a blocked order (7 situations × 5 times = 35 trials). We used the average scores obtained in the 35 trials as the final score of each person in the shooting test. The goal-shooting test was adopted from Schöllhorn [26,39]. The seven different goal-shooting situations were as follows:

1. Five immobile futsal balls were shot towards the goal after a short approach from Position 1 (Figure 1);
2. Five futsal balls were shot towards the goal after 10 m of dribbling from Position 1 (Figure 1);
3. Five futsal balls were shot towards the goal after 5 m of dribbling from Position 2 (Figure 1);
4. Five futsal balls were shot towards the goal from Position 1 after a pass from the right (Figure 1);
5. Five futsal balls were shot towards the goal after 5 m of dribbling from Position 3 (Figure 1);
6. Five futsal balls were shot towards the goal from Position 1 after a pass from the left (Figure 1);
7. Five futsal balls were shot towards the goal from Position 1 after crossing a 40-cm height obstacle with a vertical jump (Figure 1).

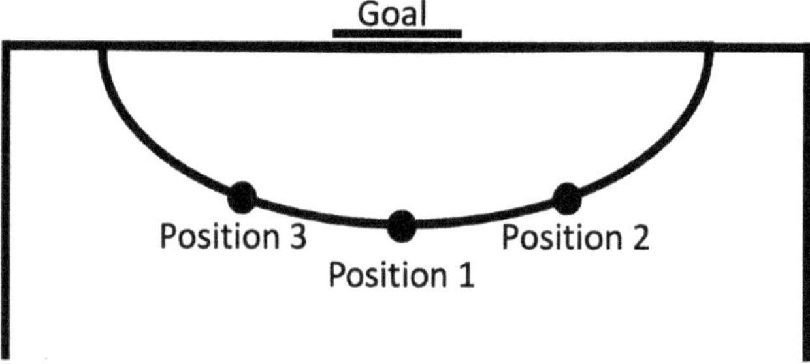

Figure 1. Participant positions for goal-shooting.

The participants' shooting positions, and the ways in which they scored their shots, are shown in Figure 1.

The precision of the shots was measured by dividing the goal into scoring zones. The zones were chosen based on the likelihood that a goal would be scored in each one of them. Meanwhile, because of this, the regions that were more difficult to reach by a goalkeeper were scored higher, and vice versa. Shots that closely missed the goal still scored 1 point (Figure 2) [39]. Additional spectators were asked to produce a more stressful environment, and to monitor the learned content for stability against psychological disturbances during the transfer test. According to the findings by Henz et al. [31,40], differential learning is accompanied by a downregulation of the frontal lobe that is associated with higher stress resistance, and it should consequently lead to better test performances in the DL intervention groups.

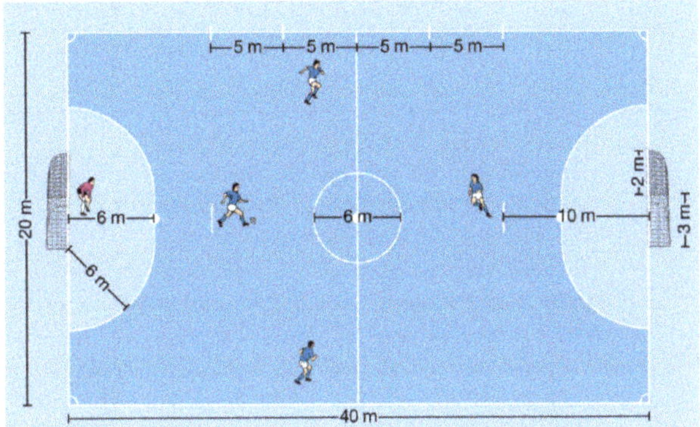

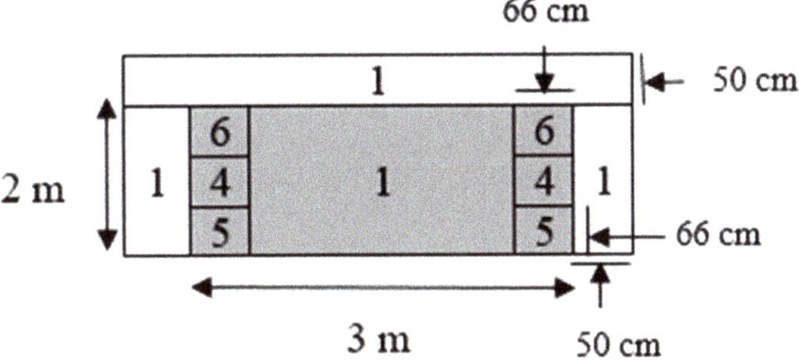

Figure 2. The goal-shooting test scores (Schöllhorn et al., 2012).

2.4. Interventions

The DL exercises are described in Table 3. Each exercise was assigned a number. The numbers were written on slips of paper and were all put in a box. The slips of paper with the numbers were then randomly drawn from the box during each intervention session. The assigned exercises were then performed by the participants. Up to three of the exercises listed in Table 3 were combined.

Table 3. DL Exercises [41,42].

Standing position	Smooth and vertical, bend forward and bend backward
Joint position	The maximal fixed, the middle position, and the maximal stretched
Hand of the standing foot side	Overhead, under the hip, in front, back, and lateral
Joint movement	Flexion, extention, abduction, adduction, internal and external rotation
Standing foot	Standing on toes, standing on heels, and standing on the whole feet
Shooting direction	Shooting to the center, the left, the right, the center, the top, and the bottom
Movement velocity	Slow, submaximal, and maximal
Hand of the shooting leg side	Overhead, under the hip, in front, back, and lateral
Eyes	Both eyes open, both eyes closed, left eye closed, and right eye closed
Foot position	Left foot front, right foot front, and feet parallel
Ball	Large, small, heavy, light balls, and other sports balls
Muscles	Maximally tensed, activated, and relaxed

Depending on the focus of each group, the participants received the following instructions, with respect to their focus of attention during the kicking movement. The specific instructions were specified as follows:

DL/EF Practices: Firstly, the DL/EF practice group received verbal instructions for the movement to be executed according to the combinations chosen from Table 3, and secondly, they received a verbal instruction for the external focus, where they had to focus on a target related to the area of the goal they intended to shoot at (e.g., shooting the ball with the middle of the foot while focusing on the upper left corner of the goal).

DL/IF Practices: The DL/IF practice group received, firstly, verbal instructions for the movement that had to be executed according to the combinations chosen from Table 3, and secondly, a verbal instruction for the internal focus, aimed at focusing on a body-related zone (e.g., shooting the ball with the midfoot, and focusing on the moment the foot hits the ball.)

DL/C Practices: The DL/C practice group received only verbal instructions for the movement that had to be executed according to the combinations chosen from Table 3 (e.g., shooting the ball with the middle of the foot).

T/EF Practices: The T/EF practice group received a visual nonverbal demonstration for the most common futsal shooting pattern provided by the teacher. Then, the participants were asked to follow the prototype pattern provided, with an additional instruction for an external focus of attention (e.g., shooting the ball with the toes and focusing on the upper left corner of the goal).

T/C Practices: The T/C practice group only received a visual nonverbal demonstration for the most common futsal shooting pattern, provided by the teacher to compellingly be imitated (e.g., shooting at the goal with your toes).

All the practice sessions and instructions were administered by one of the authors of this article, who is also a futsal expert.

2.5. Statistical Analysis

The acquisition and learning rates were assessed using a one-way ANOVA test. The variables were tested to assess the normality (Shapiro–Wilk Test), homoscedasticity (Levene's test), and sphericity (Mauchly's test) ($p > 0.05$). The effect sizes (r or η^2) were also estimated for all of the comparisons [43], and they were comparatively classified as a small effect (r = 0.10 ~ d = 0.2 or η^2 = 0.01), a medium effect (r = 0.25 ~ d = 0.5 or η^2 = 0.06), or a large effect ((r = 0.37 ~ d = 0.8 or η^2 = 0.14) [44]. The level of significance was set to 0.05, and the statistical analyses were conducted using SPSS 23.0 (IBM Corp., Armonk, NY, USA).

3. Results

The descriptive results obtained after all four tests were conducted on all five groups are graphically displayed in Figure 3. Specifically, the highest acquisition rates were achieved by the differential training groups. Within the DL groups, the group with an external focus achieved the highest scores in the post-test and in the retention and transfer tests.

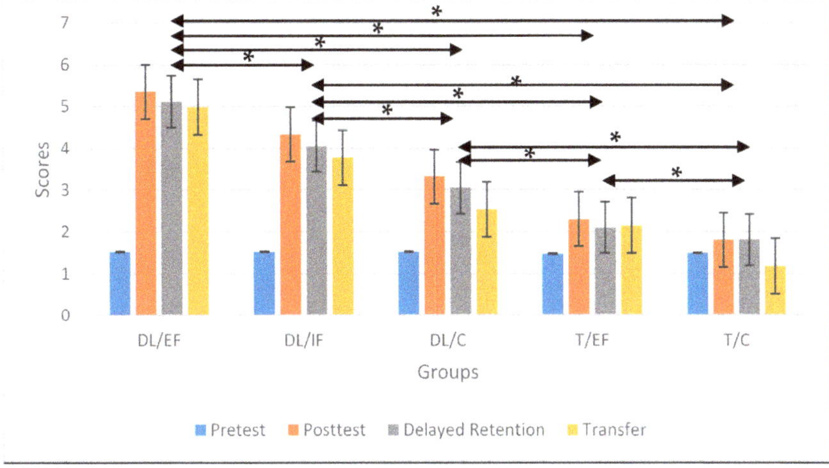

Figure 3. Average scores and standard deviations of pre-, post-, retention and transfer test results for the five intervention groups. DL/EF: differential learning with an external focus; DL/IF: differential learning with an internal focus; DL/C: differential learning without a specific focus; T/EF: traditional repetitive training with an external focus; T/C: traditional repetitive training without a specific focus. * Major statistically significant differences between groups (in all tests, a significant difference was observed between all groups).

The one-way ANOVA test showed that there was no difference between the subjects in the pretest, but for the post-test, the data showed a statistically significant difference between the groups with a large effect size ($F_{(35.4)}$ = 698.55, p = 0.001, η^2 = 0.98). Moreover, the results of the LSD post hoc test revealed a statistically significant difference between the DL/EF, DL/IF, and DL/C groups compared to the T/EF and T/C groups (p < 0.05; r > 0.83), with the external, internal, control differential, external, and control traditional groups accounting for the highest scores, followed by the control traditional group. Moreover, an ANOVA of the retention test data indicated a statistically significant difference between the groups ($F_{(35.4)}$ = 2104.9, p = 0.001, η^2 = 0.99). The LSD post hoc test results show a statistically significant difference between the DL/EF, DL/IF, and DL/C groups and the T/EF and T/C groups (p < 0.05; r > 0.82), and the highest scores indicate the DL groups with external and internal foci, as well as the control group with no focus instructions, followed by the traditionally repetitive external and control groups (Table 4).

Table 4. Fisher's least significant differences (LSDs) of averages comparing changes in pre- and post-retention and transfer tests. * Deemed to be statistically highly significant, as observed.

(I) Group	(J) Group	Prepost Test			Preretention Test			Pretransfer Test		
		Mean Difference (I-J)	SD	ES (r)	Mean Difference (I-J)	SD	ES (r)	Mean Difference (I-J)	SD	ES (r)
DL/EF	DL/IF	1.01 *	0.07	0.95	1.05 *	0.04	0.98	1.21 *	0.07	0.96
	DL/C	2.02 *	0.07	0.98	2.06 *	0.04	0.99	2.44 *	0.07	0.99
	T/EF	3.05 *	0.07	0.98	3.01 *	0.04	0.99	2.83 *	0.07	0.99
	T/C	3.54 *	0.07	0.99	3.3 *	0.04	0.99	3.8 *	0.07	0.99
DL/IF	DL/C	1 *	0.07	0.99	1 *	0.04	0.99	1.23 *	0.07	0.97
	T/EF	2.03 *	0.07	0.98	1.95 *	0.04	0.99	1.62 *	0.07	0.98
	T/C	2.52 *	0.07	0.99	2.24 *	0.04	0.99	2.58 *	0.07	0.99
DL/C	T/EF	1.02 *	0.07	0.95	0.95 *	0.04	0.98	0.38 *	0.07	0.74
	T/C	1.52 *	0.07	0.99	1.23 *	0.04	0.99	1.35 *	0.07	0.98
T/EF	T/C	0.49 *	0.07	0.83	0.28 *	0.04	0.82	0.96 *	0.07	0.96

The ANOVA of the transfer test revealed a statistically significant difference between the groups ($F_{(35.4)}$ = 739.28, p = 0.001, η^2 = 0.98). Furthermore, the LSD post hoc test results show a statistically significant difference between the DL/EF, DL/IF, and DL/C groups and the T/EF and T/C groups (p < 0.05; r > 0.74), and the highest scores related to the external, internal, control differential, external, and control traditional groups (Table 4). Most intriguingly, all the effect sizes were large in both the ANOVA and in all of the post hoc tests, on the basis of Cohen's classification [44].

4. Discussion

The purpose of this study was to explore the distinguished effects of various forms of the differential learning approach [16], combined with elements of OPTIMAL theory [32], on learning the futsal goal-shooting skill. The statistical analysis revealed significant differences between the interventions DL/EF, DL/IF, DL/C, T/EF, and T/C training in the post-, retention, and transfer tests. The present test results were consistent with previous research conducted on DL (e.g., [45–47]), which show the advantages of DL in comparison to the rather repetitive approaches in the acquisition and learning phases. However, gender must be kept in mind as a possible moderator, as other studies have examined students of the same age, but only male students.

A 12-week intervention was used in this study, as opposed to the 4-to-6-week interventions used in earlier DL-related football studies [26,39]. Compared to repetition-based training, it appears that the benefits of differential learning increase over time. This has also been observed in research on tennis, with varying intervention durations, where the rising advantage of DL increased over time [48–50]. As the intervention duration increases, the DL advantages are mainly gained over traditional training in the acquisition phase, and they remain for the learning phase. These results are also consistent with the assumption that, in the DL approach, fluctuations in learning subsystems are advantageously used for learning by destabilizing the system to prepare a self-organized phase transition [45]. By amplifying such fluctuations, the system can potentially experience new solutions, involving new combinations of given activations [19]. As a result, a self-organizing process is initiated and exploited that stimulates the system to develop a new coordination strategy that typically results in more effective or stable patterns of movement [45]. The amplified fluctuations and intermittencies tend to increase the variances in other anatomical regions of the body, and lead to a highly unpredictable adaptation process [45]. The whole process finally corresponds to the somehow unspecific formulated cybernetic law of requisite variety [51], or to the "bliss of abundance" [52,53]. The first was formulated with "only variety can destroy variety", in general, as a law in cybernetics for the regulation of processes in nature

be of interest to perform the same study with boys or adults. Moreover, in this research, long-term follow-ups were not conducted, which should be used in future research. Given that the present research was performed on beginners in a certain culture, future research should be performed on skilled individuals with different cultural backgrounds as well. In future research, other sports skills, including team behavior, should also be considered. In this study, the highly variable intervention method of differential learning, with assumingly chaotic characteristics, was used to teach a single futsal skill. In the search for optimal variation structures, future studies should also differentiate gradual and chaotic differential learning approaches by adapting the noise of the exercises to the noise inherent to the learner to find the optimal stochastic resonance [49], as well as their interactions with internal and external foci. Another interesting question is the influence of the test sequence. Future research would have to clarify to what extent the interventions have different influences on blocked or random test orders.

5. Conclusions

Since all the DL interventions, whether with external or internal foci, showed better performance in the post-, retention, and transfer tests than all combinations of traditional training with the foci, the results suggest the more beneficial influence of the DL training method compared to the traditional training methods, and compared to the focus interventions. Nonetheless, attentional foci seem to provide a kind of positive fine-tuning in addition to DL intervention. It would be of great interest to know what additional influences the other aspects of OPTIMAL theory have in connection with DL, or whether these are possibly formed and promoted precisely by DL.

The DL and the OPTIMAL approaches are both still in the beginning stages of their research course development. Thus, what has been obtained from the present study can be added as further evidence to the body of theoretical basics in these theories. To enhance and create more effective exercises for the acquisition of motor skills, research findings should be made available to instructional and learning designers [64].

Author Contributions: Conceptualization, S.O., A.B., R.Y. and F.G.; methodology, S.O., A.B., R.Y., F.G. and W.I.S.; formal analysis, S.O.; validation, S.O., A.B., R.Y., F.G. and W.I.S., investigation, S.O.; resources, S.O.; data curation, S.O.; writing—original draft preparation, S.O.; writing—review and editing, S.O., A.B., R.Y., F.G. and W.I.S.; visualization, S.O., A.B., R.Y., F.G. and W.I.S.; supervision, R.Y.; project administration, S.O. All authors have read and agreed to the published version of the manuscript.

Funding: This research received no external funding.

Institutional Review Board Statement: The study was conducted according to the guidelines of the Declaration of Helsinki and approved by the Institutional Review Board of Kharazmi University, Teheran, Iran, with the ethics code of "IR.KHU.REC.1399.04".

Informed Consent Statement: Written informed consent was obtained from all the parents of the participants involved in the study.

Data Availability Statement: The data that support the findings of this study are available from the corresponding author, S.O., upon reasonable request.

Acknowledgments: This paper is based on a research project that was conducted by the spiritual aids of research affairs at the Kharazmi University of Tehran.

Conflicts of Interest: The authors declare no conflict of interest.

References

1. Schöllhorn, W.I. Individualität-ein vernachlässigter Parameter. *Leistungssport* **1999**, *29*, 7–11.
2. Miller, G.A.; Galanter, E.; Pribram, K.H. *Plans and the Structure of Behavior*; Holt, Rhinehart, & Winston: New York, NY, USA, 1960.
3. Abernethy, B.; Maxwell, J.P.; Masters, R.S.; Van Der Kamp, J.; Jackson, R.C. *Attentional Processes in Skill Learning and Expert Performance. In Handbook of Sport Psychology*, 3rd ed.; John Wiley & Sons, Inc.: Hoboken, NJ, USA, 2007; pp. 245–263.

4.   Farrow, D.; Reid, M. The effect of equipment scaling on the skill acquisition of beginning tennis players. *J. Sports Sci.* **2010**, *28*, 723–732. [CrossRef]
5.   Ludgate, K.E. The effect of manual guidance upon maze learning. *Psychol. Monogr.* **1923**, *33*. [CrossRef]
6.   Koch, H.L. The influence of mechanical guidance upon maze learning. *Psychol. Monogr.* **1923**, *32*. [CrossRef]
7.   Mendoza, L.; Schöllhorn, W.I. Training of the sprint start technique with biomechanical feedback. *J. Sports Sci.* **1993**, *11*, 25–29. [CrossRef] [PubMed]
8.   Gray, R. Comparing the constraints led approach, differential learning and prescriptive instruction for training opposite-field hitting in baseball. *Psychol. Sport Exerc.* **2020**, *51*, 101797. [CrossRef]
9.   Wang, T.L. The influence of tuition in the acquisition of skill. *Psychol. Monogr.* **1925**, *34*, 51. [CrossRef]
10. Melcher, R.T. Children's Motor Learning with and without Vision. *Child. Dev.* **1934**, *5*, 315–350. [CrossRef]
11. Von Wright, J.M. A Note on Role of 'Guidance' in Learning. *Br. J. Psychol.* **1957**, *48*, 133–137. [CrossRef] [PubMed]
12. Holding, D.H.; Macrae, A.W. Guidance, Restriction and Knowledge of Results. *Ergonomics* **1964**, *7*, 289–295. [CrossRef]
13. Adams, J.A. Historical review and appraisal of research on the learning, retention, and transfer of human motor skills. *Psychol. Bull.* **1987**, *101*, 41–74. [CrossRef]
14. Newell, K.M. Physical Education in Higher Education: Chaos Out of Order. *Quest* **1990**, *42*, 227–242. [CrossRef]
15. Lee, M.C.Y.; Chow, J.Y.; Komar, J.; Tan, C.W.K.; Button, C. Nonlinear Pedagogy: An Effective Approach to Cater for Individual Differences in Learning a Sports Skill. *PLoS ONE* **2014**, *9*, e104744. [CrossRef] [PubMed]
16. Schöllhorn, W.I. Practical consequences of systems dynamic approach to technique and strength training. *Acta Acad. Olymp. Est.* **2000**, *8*, 25–37.
17. Williams, A.M.; Hodges, N.J. Practice, instruction, and skill acquisition in soccer: Challenging tradition. *J. Sports Sci.* **2005**, *23*, 637–650. [CrossRef]
18. Schöllhorn, W.I.; Bauer, H.U. Identifying Individual Movement Styles in High Performance Sports by Means of Self-Organizing Kohonen Maps. In Proceedings of the XVI International Symposium on Biomechanics in Sports, Konstanz, Germany, 21–25 July 1998; pp. 574–577.
19. Schöllhorn, W.I.; Nigg, B.M.; Stefanyshyn, D.J.; Liu, W. Identification of individual walking patterns using time discrete and time continuous data sets. *Gait Posture* **2002**, *15*, 180–186. [CrossRef]
20. Janssen, D.; Schöllhorn, W.I.; Lubienetzki, J.; Fölling, K.; Kokenge, H.; Davids, K. Recognition of Emotions in Gait Patterns by Means of Artificial Neural Nets. *J. Nonverb. Behav.* **2008**, *32*, 79–92. [CrossRef]
21. Burdack, J.; Horst, F.; Aragonés, D.; Eekhoff, A.; Schöllhorn, W.I. Fatigue-Related and Timescale-Dependent Changes in Individual Movement Patterns Identified Using Support Vector Machine. *Front. Psychol.* **2020**, *11*, 2273. [CrossRef]
22. Horst, F.; Janssen, D.; Beckmann, H.; Schöllhorn, W.I. Can Individual Movement Characteristics Across Different Throwing Disciplines Be Identified in High-Performance Decathletes? *Front. Psychol.* **2020**, *11*, 2262. [CrossRef]
23. Bernstein, N.A. *The Co-Ordination and Regulation of Movements*; PergamonPress Ltd.: Oxford, UK, 1967.
24. Hatze, H. Motion Variability—its Definition, Quantification, and Origin. *J. Mot. Behav.* **1986**, *18*, 5–16. [CrossRef]
25. Schöllhorn, W.I.; Michelbrink, M.; Welminsiki, D.; Davids, K. Increasing stochastic perturbations enhances acquisition and learning of complex sport movements. In *Perspectives on Cognition and Action in Sport*; Raab, M., Araujo, D., Ripoll, H., Eds.; Nova Science Publishers: Hauppauge, NY, USA, 2009; pp. 59–73.
26. Schöllhorn, W.I.; Beckmann, H.; Michelbrink, M.; Sechelmann, M.; Trockel, M.; Davids, K. Does Noise Provide a Basis for the Unification of Motor Learning Theories? *Int. J. Sport Psychol.* **2006**, *37*, 186–206.
27. Beckmann, H.; Winkel, C.; Schöllhorn, W.I. Optimal range of variation in hockey technique training. *Int. J. Sport Psychol.* **2010**, *41*, 5–45.
28. Römer, J.; Schöllhorn, W.I.; Jaitner, T.; Preiss, R. Differenzielles lernen im Volleyball. *Sportunterricht* **2009**, *58*, 41–45.
29. Serrien, B.; Tassignon, B.; Baeyens, J.-P.; Clijsen, R. A critical review on the theoretical framework of differential motor learning and meta-analytic review on the empirical evidence of differential motor learning. *SportRxiv* **2018**. [CrossRef]
30. Schöllhorn, W.I.; Humpert, V.; Oelenberg, M.; Michelbrink, M.; Beckmann, H. Differenzielles und mentales Training im Tennis. *Leistungssport* **2008**, *38*, 10–14.
31. Henz, D.; John, A.; Merz, C.; Schöllhorn, W.I. Post-task effects on EEG brain activity differ for various differential learning and contextual interference protocols. *Front. Hum. Neurosci.* **2018**, *12*, 19. [CrossRef] [PubMed]
32. Wulf, G.; Lewthwaite, R. Optimizing performance through intrinsic motivation and attention for learning: The OPTIMAL theory of motor learning. *Psychon. Bull. Rev.* **2016**, *23*, 1382–1414. [CrossRef] [PubMed]
33. Bauer, H.U.; Schöllhorn, W.I. Self-Organizing Maps for the Analysis of Complex Movement Patterns. *Neural Process. Lett.* **1997**, *5*, 193–199. [CrossRef]
34. Lewthwaite, R.; Wulf, G. Optimizing motivation and attention for motor performance and learning. *Curr. Opin. Psychol.* **2017**, *16*, 38–42. [CrossRef] [PubMed]
35. Elias, L.J.; Bryden, M.P.; Bulman-Fleming, M.B. Footedness is a better predictor than is handedness of emotional lateralization. *Neuropsychologia* **1998**, *36*, 37–43. [CrossRef]
36. Chua, L.K.; Wulf, G.; Lewthwaite, R. Onward and upward: Optimizing motor performance. *Hum. Mov. Sci.* **2018**, *60*, 107–114. [CrossRef] [PubMed]

37. An, J.; Wulf, G.; Kim, S. Increased carry distance and X-factor stretch in golf through an external focus of attention. *J. Motor. Learn Dev.* **2013**, *1*, 2–11. [CrossRef]
38. Schöllhorn, W.I.; Sechelmann, M.; Trockel, M.; Westers, R. Nie das Richtige trainieren, um richtig zu spielen. *Leistungssport* **2004**, *5*, 13–17.
39. Schöllhorn, W.I.; Hegen, P.; Davids, K. The nonlinear nature of learning-A differential learning approach. *Open Sports Sci. J.* **2012**, *5*, 100–112. [CrossRef]
40. Henz, D.; Schöllhorn, W.I. Differential training facilitates early consolidation in motor learning. *Front. Behav. Neurosci.* **2016**, *10*, 199. [CrossRef]
41. Wagner, H. *Optimierung Komplexer Bewegungsmuster Bei Wurfbewegungen*; Meyer & Meyer: Osnabrück, Germany, 2005.
42. Wagner, H.; Müller, E. The effects of differential and variable training on the quality parameters of a handball throw. *Sports Biomech.* **2008**, *7*, 54–71. [CrossRef]
43. Field, A. *Discovering Statistics Using SPSS*; Sage Publications: Thousand Oaks, CA, USA, 2009.
44. Cohen, J. *Statistical Power Analysis in the Behavioral*, 2nd ed.; Lawrence Erlbaum Associates: Hillsdale, NJ, USA, 1988; pp. 10–74.
45. Bozkurt, S. The Effects of Differential Learning and Traditional Learning Trainings on Technical Development of Football Players. *Educ. Train.* **2018**, *6*, 25–29. [CrossRef]
46. Yaali, R.; Faezi, G.; Oftadeh, S. The effect of type practice of Differential Learning and Contextual Interference on the Retention and Transfer of Badminton backhand short serve. *J. Res. Sports Manag. Mot. Behav.* 2010. Available online: https://jrsm.khu.ac.ir/article-1-2935-fa.html (accessed on 7 October 2021). (In Persian)
47. Tassignon, B.; Verschueren, J.; Baeyens, J.P.; Benjaminse, A.; Gokeler, A.; Serrien, B.; Clijsen, R. An Exploratory Meta-Analytic Review on the Empirical Evidence of Differential Learning as an Enhanced Motor Learning Method. *Front. Psych.* **2021**, *12*, 1186. [CrossRef]
48. Schöllhorn, W.I.; Oelenberg, M.; Michelbrink, M. Can Mental Training Enhance the Learning Effect After Differencial Training? A Tennis Serve Task. In Proceedings of the 12th European Congress of Sport Psychology, Kallithea, Halkidiki, Greece, 4–9 September 2007; p. 241.
49. Schöllhorn, W.I.; Mayer-Kress, G.; Newell, K.; Michelbrink, M. Time scales of adaptive behavior and motor learning in the presence of stochastic perturbations. *Hum. Mov. Sci.* **2009**, *28*, 319–333. [CrossRef]
50. Yildirim, Y.; Kizilet, A. The Effects of Differential Learning Method on the Tennis Ground Stroke Accuracy and Mobility. *Educ. Learn.* **2020**, *9*, 146–154. [CrossRef]
51. Ashby, W.R. *An Introduction to Cybernetics*; Chapman & Hall Ltd.: London, UK, 1961.
52. Gelfand, I.M.; Latash, M.L. On the problem of adequate language in motor control. *Mot. Contr.* **1998**, *2*, 306–313. [CrossRef] [PubMed]
53. Latash, M.L. The bliss (not the problem) of motor abundance (not redundancy). *Exp. Brain Res.* **2012**, *217*, 1–5. [CrossRef]
54. Frank, T.D.; Michelbrink, M.; Beckmann, H.; Schöllhorn, W.I. A quantitative dynamical systems approach to differential learning: Self-organization principle and order parameter equations. *Biol. Cybern.* **2008**, *98*, 19–31. [CrossRef]
55. Lohse, K.R.; Sherwood, D.E.; Healy, A.F. How changing the focus of attention affects performance, kinematics, and electromyography in dart throwing. *Hum. Mov. Sci.* **2010**, *29*, 542–555. [CrossRef]
56. Wulf, G.; Chiviacowsky, S.; Drews, R. External focus and autonomy support: Two important factors in motor learning have additive benefits. *Hum. Mov. Sci.* **2015**, *40*, 176–184. [CrossRef]
57. Simpson, T.; Cronin, L.; Ellison, P.; Carnegie, E.; Marchant, D. A test of optimal theory on young adolescents' standing long jump performance and motivation. *Hum. Mov. Sci.* **2020**, *72*, 102651. [CrossRef]
58. Magosso, E.; De Crescenzio, F.; Ricci, G.; Piastra, S.; Ursino, M. EEG alpha power is modulated by attentional changes during cognitive tasks and virtual reality immersion. *Comput. Intell. Neurosci.* **2019**, *2019*, 7051079. [CrossRef]
59. Hatami, F.; Tahmasbi, F.; Hadi, H. The effect of internal and external focus of attention on EEG changes in darts throwing skill. *Neuropsychology* **2018**, *3*, 115–130.
60. Sauseng, P.; Klimesch, W.; Schabus, M.; Doppelmayr, M. Fronto-parietal EEG coherence in theta and upper alpha reflect central executive functions of working memory. *Int. J. Psychophysiol.* **2005**, *57*, 97–103. [CrossRef] [PubMed]
61. Ray, W.J.; Cole, H.W. EEG alpha activity reflects attentional demands, and beta activity reflects emotional and cognitive processes. *Science* **1985**, *228*, 750–752. [CrossRef] [PubMed]
62. Perkins-Ceccato, N.; Passmore, S.R.; Lee, T.D. Effects of focus of attention depend on golfers' skill. *J. Sports Sci.* **2003**, *21*, 593–600. [CrossRef]
63. John, A.; Schöllhorn, W.I. Acute effects of instructed and self-created variable rope skipping on EEG brain activity and heart rate variability. *Front. Behav. Neurosci.* **2018**, *12*, 311. [CrossRef]
64. Conole, G.; Fill, K. A learning design toolkit to create pedagogically effective learning activities. *J. Interact. Media Educ.* **2005**, *1*. [CrossRef]

International Journal of
Environmental Research and Public Health

Article

How the Number of Players and Floaters' Positioning Changes the Offensive Performance during Futsal Small-Sided and Conditioned Games

David Pizarro [1,*], Alba Práxedes [1], Bruno Travassos [2,3], Bruno Gonçalves [3,4,5] and Alberto Moreno [6]

1   Faculty of Life Sciences and Nature, University of Nebrija, 28015 Madrid, Spain; apraxedes@nebrija.es
2   Research Center in Sport Sciences, Health and Human Development (CIDESD), Department of Sport Sciences, University of Beira Interior, 6201-001 Covilhã, Portugal; BrunoTravassos@hotmail.com
3   Portugal Football School, Portuguese Football Federation, 5001-801 Oeiras, Portugal; bgoncalves@uevora.pt
4   Departamento de Desporto e Saúde, Escola de Saúde e Desenvolvimento Humano, Universidade de Évora, 7004-516 Évora, Portugal
5   Comprehensive Health Research Centre (CHRC), Universidade de Évora, 7004-516 Évora, Portugal
6   Faculty of Sport Sciences, University of Extremadura, 06006 Cáceres, Spain; amorenodfcd@gmail.com
*   Correspondence: dpizarro@nebrija.es

Abstract: This study aims to analyse the effects of floater positioning within futsal Gk + 3vs3 + Gk and Gk + 2vs2 + Gk small-sided and conditioned games (SSCG) on youth offensive performance on an action per minute per player basis. Three experimental conditions were carried out through the manipulation of floater positioning: floaters off (FO), final line floaters (FLF) and lateral floaters (LF). Thirty male futsal players (U19 age category) participated in the study and played once within each situation in a random order on different days. Offensive performance based on "action per minute per player" was analysed through indirect and external systematic observation. Results showed significant differences between both SSCGs (2vs2 and 3vs3). Specifically, according to the game principles analysed, 3vs3 is associated with higher values of passing and dribbling action to progress towards the goal without beating a defensive line (moderate to large effect size), while 2vs2 is associated with higher values of passing and dribbling actions that beating a defensive line (moderate to very large effect size). In addition, 2vs2 is associated with dribbling and shooting actions to shoot at goal with the lowest level of opposition (moderate effect size). Indeed, whilst the 2vs2 game format seems to promote more 1vs1 situations, the 3vs3 game format encourages more ball possession and collective tactical behaviours. Thus, training tasks intended to improve dribbling and shooting actions should use a smaller number of players whereas tasks intended to improve passing actions for ball possession should include a higher number of players with or without floaters. It seems that the number of players can influence the tactical behaviour of the team. These findings should be considered for the design of futsal training tasks, according to the main objective of the training session. For example, if the coach aims to promote the number of dribbles and shots within a SSCG, 2vs2 SSCG situations should be prioritised.

Keywords: ecological dynamics; training tasks; technical–tactical training; game principles

Citation: Pizarro, D.; Práxedes, A.; Travassos, B.; Gonçalves, B.; Moreno, A. How the Number of Players and Floaters' Positioning Changes the Offensive Performance during Futsal Small-Sided and Conditioned Games. *Int. J. Environ. Res. Public Health* **2021**, *18*, 7557. https://doi.org/10.3390/ijerph18147557

Academic Editor: Jorge Pérez-Gómez

Received: 15 June 2021
Accepted: 14 July 2021
Published: 15 July 2021

Publisher's Note: MDPI stays neutral with regard to jurisdictional claims in published maps and institutional affiliations.

1. Introduction

In team sports such as futsal, in which open motor skills predominate, it is required that players continuously co-adapt their actions such as positioning, passes, dribbles, or shots to the movements of opponents, teammates, and the surrounding environment, leading to the emergence of opportunities for action [1–3] and to ensure functional collective behaviour [4–6]. In the last few decades, based on the ecological dynamics approach, non-linear pedagogy has emerged as a new teaching–learning perspective to promote a holistic approach that highlights the need to maintain the perception–action couple on the design of practice tasks [5]. For example, manipulating small-sided and conditioned games (SSCG), coaches can highlight the actions and the information that will support players' performance. SSCG (commonly used modified games that take place in tight spaces, involving small numbers of players and with modified rules) are modified games that optimize the physical, physiological, technical, and tactical demands of sports instead of replicating a real match [7]. However, the advantages of playing SSCG are dependent on the task's goals and design [8] that guides players to explore the functional behaviours of each task according to the coaches' primary purposes [9].

The manipulation of task constraints in SSCG seems to be an effective approach to skill acquisition [10] that allows coaches to optimize specific offensive or defensive behaviours of players by breaking the game into specific game subunits, i.e., Gk + 1vs1 + Gk until Gk + 3vs3 + Gk [11]. This is likely to maintain the perceptual-motor demands of the match and the spatial-temporal relations between teammates and opponents, instead of replicating the technical and tactical demands of sports [7]. Indeed, coaches should go from simplified units with a low number of players to highlight the informational constraints that promote the development of offensive or defensive foundations of players to more complex units until the numerical relation of the game to develop the game principles and strategic requirements that support collective behaviour of teams according to the perceptual and action demands of competition [11].

Previous studies have attempted to provide a broader comprehension of the impact of altering SSCG characteristics (task constraints), such as the number of players per team [12,13], the court size [1], number of targets [14] or even the manipulation of the numerical relation between teams using floaters (jokers in other studies) [15–21]. Interestingly, one of the task constraints that have been studied recently is the accomplishing of tactical principles of attack performance [22]. These are referred to as keep possession, progress towards the goal (with or without beating a defensive line), or shoot at goal with the lowest level of opposition [23]. For example, regarding the manipulation of floaters, previous studies suggested that the use of on-field floaters increased players' decision-making efficiency due to their distribution over the breadth of the field [24]. Moreover, on-field floaters might have afforded more opportunities for passing the ball, allowing the team to maintain ball possession [15]. Hence, manipulation of the relevant task constraints (e.g., presence of floaters and their positioning) for each goal can guide players to explore the environment of play, improving their tactical and creative behaviour [25].

The effects of such manipulations to design appropriate learning environments that help players' development of adaptative technical–tactical behaviours to changes in the game environment [8] must be well understood by coaches, particularly in futsal. This perspective explains the interest of researchers and practitioners in this topic and the growing number of studies over the last few years [7,26,27]. However, no information exists regarding the technical–tactical changes promoted by the manipulation of the number of players and floater positioning. Thus, the main purpose of this study was to analyse the manipulation on the number of players (Gk + 3vs3 + Gk and Gk + 2vs2 + Gk) and floater positioning on youth players' technical–tactical offensive actions according to game principles.

2. Materials and Methods

2.1. Participants

Thirty male futsal players from the under-19 (U19) category (age, M = 17.714 and SD = 0.713) of four different Spanish clubs agreed to participate in this study. All the participants had the same level of expertise (i.e., average skill level) and participated in the same competition (the first regional league). All teams had the same amount of training (i.e., players perform two training sessions of 60 min per week with an official match played on weekends).

2.2. Design and Procedures

The study designed consisted of an independent measure approach under three experimental conditions (three SSCG) that manipulated the floater positioning. These SSCG (Gk + 3vs3 + Gk; Gk + 2vs2 + Gk) were designed using the presence and absence of "Floaters" (2 Floaters; one per team) as key task constraints: (a) "Floaters Off" (FO); (b) "Final Line Floaters" (FLF) (associated with the 3-1 offensive system) and (c) "Lateral Floaters" (LF) (associated with the 4-0 offensive system). In 3vs3 situations, tests were conducted on a 30 m long by 15 m wide field. In 2vs2 situations, tests were conducted on a 20 m long by 10 m wide field (see Figure 1). These measures respected the player-space ratio used by futsal players according to the maximum length and width dimensions (40 m × 20 m) of the real game (for each team player, 10 m large and 5 m regular, excluding goalkeepers).

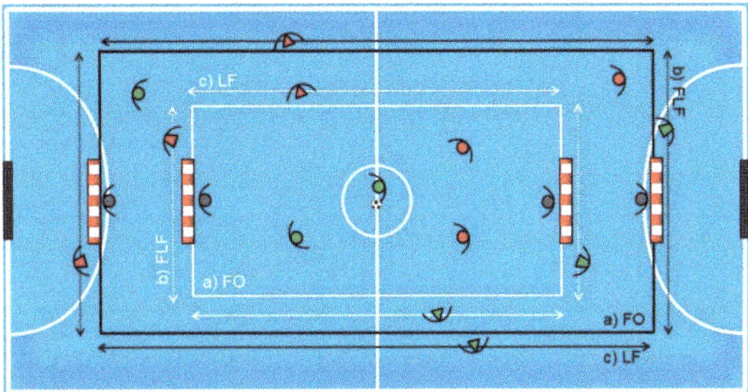

Figure 1. 3vs3 and 2vs2 experimental conditions. (**a**) FO: Floaters Off; (**b**) FLF: Final Line Floaters; (**c**) LF: Lateral Floaters.

Players were randomly distributed into five groups of six individuals for 3vs3 SSCG (G1 to G5); seven groups of four individuals for 2vs2 SSCG (G1 to G7, two players were randomly excluded for 2vs2; goalkeepers and floaters with goalkeepers and floaters no measurements were taken). All participants played once in each situation in random order and on a different day. Each 3vs3 testing had the following organization: warm-up (12′) + SSCG of 12′: 3′-1′-3′-1′-3′-1′ (3′ period = playing; 1′ period = resting); and 2vs2 testing: warm-up (10′) + SSCG of 10′: 2′-1′-2′-1′-2′-1′ (2′ period = playing; 1′ period = resting). During the rest intervals between bouts, players could drink water.

Game situations were explained, and participants were asked to play at their best level to succeed in the SSCG (score in the opposite goal). Coaches and experimenters did not provide any verbal feedback during the SSCG. Floater players were only allowed to perform offensive actions, with a maximum of two touches, and their actions were limited to the space between two marks parallel to each line (side or final) and could not score a goal. In addition, goalkeepers could not leave the goal line, and a throw-in was granted after the ball crossed the lines delimiting the floaters' area. During the test, players were asked not to go inside the floaters' area and balls were placed around the field to allow a quick restart of the game if the ball went out of play. In between bouts, players were allowed to drink water.

2.3. Data Collection

All game actions within SSCGs were recorded using a video camera, recording angle conversion lens (×0.75): VCL-HGA07B and a Hama Gamma tripod Series. The camera was placed in the corner of the playing field, at the height of 4 m, guaranteeing an optimal view of all the game actions (see Figure 2). Videos were transferred to a computer (Acer Aspire E15). Subsequently, data were recorded on a Microsoft Office Excel 2010 sheet and exported to SPSS Inc., Released 2009 (PASW Statistics for Windows, Version 18.0, Page: 4

SPSS Inc., Chicago, IL, USA). Offensive performance measured as "action per minute per player" was analysed through indirect and external systematic observation, a methodology used in previous studies to measure players' behaviour in real game situations [28].

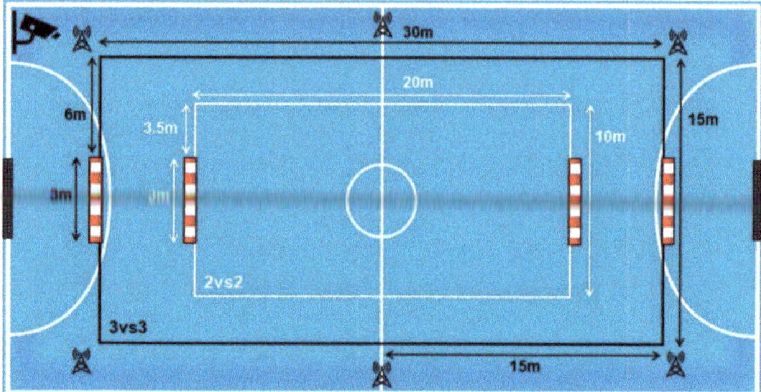

Figure 2. Pitch size and camera positioning.

Two external researchers conducted the observations. As a preparatory stage to the observations, the expert met with the observer to clarify possible doubts about the observation instrument and the coding criteria of the dependent variable on the actions mentioned. Then, the observations were carried out using more than 10% of the sample (n = xxx) [29]. Interobserver reliability was calculated using the following formula: agreements/(agreements + disagreements) × 100. Once this value was calculated, the Cohen kappa index was used. Values above 0.90 were obtained for all training sessions, surpassing the value of 0.81 from which adequate concordance is considered [30], thus achieving the necessary reliability for the subsequent coding of the dependent variables. To guarantee the time reliability of the measurement, the same coding was performed at two different moments, with a time difference of 10 days. Cohen kappa values were found to be higher than 0.92, which reflected a reliable concordance.

All the passing, dribbling, and shooting actions of each player in the team were analysed according to the following game principles: 1st principle—to keep possession (BP) (for passing and dribbling, when the action was developed horizontally or backwards); 2nd A principle—to progress towards the goal without beating a defensive line (P) (for passing and dribbling, when the action developed was forward, not beating a defensive line); 2nd B principle—to progress towards the goal beating a defensive line (PDL) (for passing and dribbling, when the action developed was forward, beating a defensive line); 3rd principle—to shoot at goal with the lowest level of opposition (S) (for passing and dribbling, when the action ended at the first touch after it, regardless of its direction; and for any shooting action).

2.4. Statistical Analysis

The statistical analysis was completed using The Jamovi Project (Jamovi). A descriptive analysis is presented on Table 1, with mean and standard deviation (Mean ± SD). An independent sample t-test was performed to identify differences in considered variables between the game formats 2vs2 vs. 3vs3. Statistical significance was set at $p < 0.05$. Additionally, to overcome the shortcomings associated with traditional N-P null hypothesis significance testing, the standardized Cohen's d, with 95% confidence intervals were used as the effect size (ES) of the differences [31–33]. Thresholds for effect size statistics were: 0.0–0.19, trivial; 0.20–0.59, small; 0.6–1.19, moderate; 1.2–1.99, large; and ≥2.0, very large [33].

Table 1. Descriptive (Mean ± SD) and inferential analysis of the considered variables according to the SSCG formats.

Game Principle	Actions	Constraints	SSCG		Mean Difference with 95% CI	Effect Size
			2vs2	3vs3		
1st	Passing	Floaters Off	4.0 ± 2.5	5.1 ± 2.7	1.1 [−0.3, 2.5]	Unclear
		Lateral Floaters	4.3 ± 3.4	4.9 ± 2.2	0.6 [−0.9, 2.1]	Unclear
		Final Lines Floaters	3.0 ± 1.6	3.2 ± 1.7	0.2 [−0.6, 1.1]	Unclear
	Dribbling	Floaters Off	3.6 ± 2.0	2.6 ± 2	−1.0 [−2.0, 0.1]	Small
		Lateral Floaters	2.6 ± 1.7	2.0 ± 1.5	−0.7 [−1.6, 0.2]	Small
		Final Lines Floaters	3.1 ± 2.0	2.2 ± 2.3	−0.8 [−1.9, 0.3]	Unclear
2nd A	Passing	Floaters Off	0.8 ± 0.8	4.0 ± 2.3	3.1 [2.2, 4.1] *	Large
		Lateral Floaters	2.2 ± 2.0	4.7 ± 2.9	2.4 [1.1, 3.7] *	Moderate
		Final Lines Floaters	0.8 ± 0.9	5.9 ± 2.9	4.9 [3.8, 6.1] *	Very Large
	Dribbling	Floaters Off	2.9 ± 1.6	3.9 ± 2.9	1.0 [−0.3, 2.2]	Unclear
		Lateral Floaters	1.7 ± 1.4	3.1 ± 2.5	1.3 [0.2, 2.4] *	Moderate
		Final Lines Floaters	2.1 ± 1.3	2.5 ± 2.3	0.4 [−0.5, 1.4]	Unclear
2nd B	Passing	Floaters Off	2.7 ± 1.6	1.1 ± 1.1	−1.5 [−2.2, −0.7] *	Moderate
		Lateral Floaters	4.3 ± 2.8	1.1 ± 1.0	−3.3 [−4.4, −2.2] *	Large
		Final Lines Floaters	4.7 ± 2.0	1.2 ± 1.2	−3.5 [−4.4, −2.7] *	Very Large
	Dribbling	Floaters Off	3.5 ± 1.8	1.9 ± 1.5	−1.6 [−2.5, −0.7] *	Moderate
		Lateral Floaters	1.8 ± 1.5	1.9 ± 1.7	0.1 [−0.7, 1.0]	Unclear
		Final Lines Floaters	2.2 ± 2.1	1.8 ± 1.3	−0.4 [−1.4, 0.5]	Unclear
3rd	Passing	Floaters Off	0.7 ± 0.9	0.9 ± 1.1	0.2 [−0.3, 0.7]	Unclear
		Lateral Floaters	0.6 ± 0.8	1.0 ± 0.9	0.4 [−0.1, 0.8]	Small
		Final Lines Floaters	0.3 ± 0.5	0.6 ± 0.9	0.4 [−0.1, 0.8]	Small
	Dribbling	Floaters Off	2.1 ± 1.4	0.6 ± 0.8	−1.4 [−2.0, −0.8] *	Large
		Lateral Floaters	2.2 ± 1.5	0.8 ± 1.0	−1.4 [−2.1, −0.7] *	Moderate
		Final Lines Floaters	1.3 ± 0.9	0.4 ± 0.6	−0.8 [−1.2, −0.4] *	Moderate
	Shooting	Floaters Off	4.0 ± 1.9	2.7 ± 1.8	−1.2 [−2.2, −0.2] *	Moderate
		Lateral Floaters	3.8 ± 1.5	2.8 ± 1.9	−1.1 [−2.1, −0.1] *	Moderate
		Final Lines Floaters	3.7 ± 1.6	2.9 ± 1.7	−0.7 [−1.6, 0.2]	Small

* $p < 0.05$. Abbreviations: 1st = to keep possession; 2nd A = to progress towards the goal without beating a defensive line; 2nd B = to progress towards the goal beating a defensive line; 3rd = to shoot at goal with the lowest level of opposition.

3. Results

For 3vs3 SSCG, a total of 1352 passing (1st principle, n = 573; 2nd A principle, n = 548, 2nd B principle, n = 127; 3rd principle, n = 104); 920 dribbling (1st principle, n = 256; 2nd A principle = 371, 2nd B principle, n = 215; 3rd principle, n = 78); and 342 shooting (3rd principle, n = 342) actions occurred. For 2vs2 SSCG, a total of 1087 passing (1st principle, n = 418; 2nd A principle = 155, 2nd B principle, n = 396; 3rd principle, n = 55); 1044 dribbling (1st principle, n = 318; 2nd A principle = 235, 2nd B principle, n = 277; 3rd principle, n = 214); and 421 shooting (3rd principle, n = 421) actions occurred.

The descriptive and inferential analysis between actions per minute per player developed in two small-sided and conditioned games (2vs2–3vs3) according to the floater positioning (task constraint) and the game principle (GP) presented in Table 1. Additionally, Figure 3 shows the standardized (Cohen) differences for the pairwise comparations.

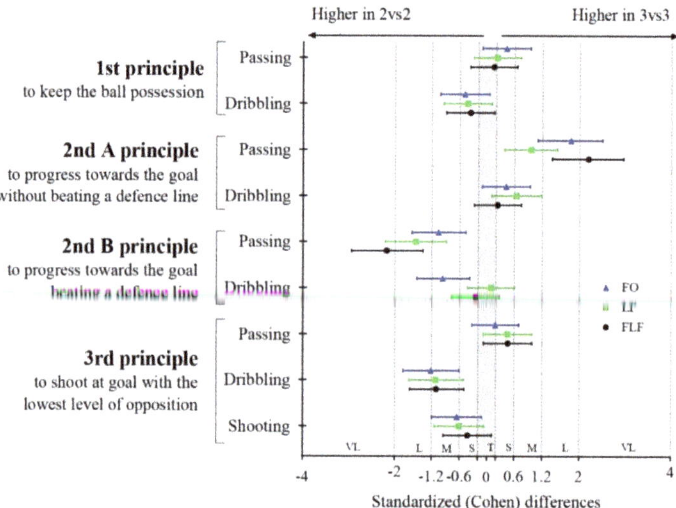

Figure 3. Standardized Cohen differences for the pairwise comparations between 2vs2 and 3vs3 for each action and game principle considered.

Non-significant differences were identified for passing and dribbling actions in the 1st principle (BP) for any task constraints between both SSCG. According to passing actions in 2nd A principle (P), the results showed significantly higher values in 3vs3 than in 2vs2 SSCG in FO (mean differences [95% confidence interval]; 3.1 [2.2, 4.1], $p < 0.01$, large ES), LF (2.4 [1.1, 3.7], $p < 0.01$, moderate ES) and FLF (4.9 [3.8, 6.1], $p < 0.01$, large ES). Regarding dribbling actions in 2nd A principle (P), results showed significantly higher values in 3vs3 than in 2vs2 SSCG in LF (1.3 [0.2, 2.4], $p < 0.05$, moderate ES).

When considering the passing actions in 2nd B principle (PDL), results showed significantly higher values in 2vs2 than in 3vs3 SSCG in FO (-1.5 [-2.2, -0.7], $p < 0.01$, moderate ES), LF (-3.3 [-4.4, -2.2], $p < 0.01$, large ES), and FLF (-3.5 [-4.4, -2.7], $p < 0.01$, very large ES). Regarding dribbling actions in 2nd B principle (PDL), results showed significantly higher values in 2vs2 than in 3vs3 SSCG in FO SSCG (-1.6 [-2.5, -0.7], $p < 0.01$, moderate ES).

For passing actions in the 3rd principle (S), no significant difference was identified. For dribbling actions performed in 3rd principle (S), results showed significantly higher values in 2vs2 than in 3vs3 SSCG in FO (-1.4 [-2.0, -0.8], $p < 0.01$, large ES), LF (-1.4 [-2.1, -0.7], $p < 0.01$, moderate ES) and FLF (-0.8 [-1.2, -0.4], $p < 0.01$, moderate ES). Finally, for the shooting actions in 3rd principle (S), results showed significantly higher values in 2vs2 than in 3vs3 SSCG in FO (-1.2 [-2.2, -0.2], $p < 0.05$, moderate ES) and LF (-1.1 [-2.1, -0.1], $p < 0.05$, moderate ES).

4. Discussion

This study aimed to analyse the manipulation of the number of players (Gk + 3vs3 + Gk and Gk + 2vs2 + Gk) and floater positioning on youth players' technical–tactical offensive actions according to game principles. The highest values of passing were observed in the 3vs3 SSCG, where most dribbles and shots occurred in the 2vs2 SSCG. These results seem to indicate that the number of players per team as a task constraint can influence players and teams' possibilities for action, and consequently their tactical behaviour. One of the first constraints that coaches need to consider when designing the practice tasks is the number of players involved [34]. When the goal is to create passing lines and maintain ball possession through passing, the 3vs3 SSCG should be used whereas if the focus is on dribbling and shooting, the 2vs2 situation can ensure a greater number of these actions.

Furthermore, the manipulation of the number of players constrains not only the actions per se but the emergence of each action in relation to the game principles that support different purposes of the teams [35]. However, further studies are required with more participants from different levels of practice to generalize our results.

4.1. First Game Principle (1st = to Keep Possession)

With regard to the first game principle (BP), no significant difference was observed between the 2vs2 and the 3vs3 SSCG or the addition of floaters in the side or final line of the field. Contrary to previous research [15], a different number of players or floaters seems not to influence the number of passing or dribbling actions by players to maintain ball possession. Thus, a link between the goal and the manipulations promoted should be considered to understand the impacts of such manipulations on players' and teams' tactical behaviour [36].

4.2. Second Game Principle (2nd = to Progress towards the Goal)

Regarding the second game principle, two different categories were considered: 2nd A principle—to progress towards the goal without beating a defensive line (P) and 2nd B principle—to progress towards the goal by beating a defensive line (PDL).

Results of the 2nd A principle (P) revealed significantly higher values of passing in for the 3vs3 compared to the 2vs2 situation when players try to progress towards the goal without beating a defensive line in all experimental conditions (FO, FLF and LF). Thus, the number of players per team might be more determining for the emergence of progression without breaking a defensive line compared to the presence or absence of floaters. In agreement with previous research, the use of 3vs3 could be considered a more balanced defensive structure of play, defined by two defensive lines, not allowing an easy effective progression. As Gonçalves et al. (2016) and Vilar et al. (2014) pointed out, manipulating the number of players per team stimulates the emergence of new play patterns that support the emergence of different individual action possibilities for both attacking and defending players. Thus, it could be that in the 2vs2 the number of passing possibilities of the attacking team is limited (specifically, only one), so the defending team increases the pressure on the attacking players and the possibilities to do successful penetration (i.e., beating a defensive line) increases too. On the contrary, in the 3vs3 the defending team could retreat its position on the field by decreasing the distance between teammates and their own goal. As Pizarro et al. (2021) pointed out, when the defending team retreats its position, the distance between attacking and defending players increases and consequently, the probability of developing passing actions without beating the line increases. Furthermore, in 2vs2 teams, there is only one defensive line, which affords more advantage to progress, compared to the two existing defensive lines of the 3vs3 conditions that allows a better space equilibrium. Indeed, when a team has more players, the game is more positional and less variable, increasing the balance between teams [36].

No significant differences were observed for dribbling except for the condition LF, which revealed a higher number of dribbling actions favouring 3vs3. For the other side, the floater in the side-line allows more opportunities for dribbling in 2vs2. In line with previous research, the addition of the floater probably promoted a retreat of defenders on the field to guarantee the protection of space near the goal. Usually, when playing against unfavourable numerical relationships, the defender tends to decrease the space for action [27], maintaining the space equilibrium between defensive lines, not allowing passing actions, but inviting more 1vs1 dribbling situations [20,37]. Due to the 3vs3 structure allowing more than one defensive line, usually, such dribbling actions also do not afford the possibility to beat defensive lines.

Conversely, regarding the 2nd B principle (PDL), results revealed significantly higher values of passing in favour of 2vs2 when players try to progress towards the goal beating a defensive line in all experimental conditions (FO, FLF and LF). Interestingly, the effect tends to increase with the addition of floaters. With the increase in floaters, the number of passing actions that beat defensive lines in the 2vs2 conditions tends to increase in comparison with the 3vs3 conditions. In line with previous assumptions, the use of fewer defensive players decreased the number of defensive lines, increasing the need for each player to mark the opponent to maintain the spatial-temporal relations not to allow progression. It opens new possibilities to increase the mobility of attacking players to create passing lines for progression [38]. The addition of floaters promoted a numerical unbalance between teams, giving an advantage to attacking teams and allowing them to progress on the field, and consequently putting less pressure from defenders on the ball carrier, opening up more passing lines to the floaters [38]. The use of floaters in the final line in particular increases the number of passing lines and defenders attracted to the ball and seems to promote higher spatial unbalance for the emergence of passing opportunities.

Regarding dribbling actions, higher values were obtained in favour of 2vs2 when players try to progress towards the goal, beating a defensive line without the presence of floaters. In line with previous research, the absence of floaters and the small number of players (2vs2) seems to promote the emergence of 1vs1 situations, thus enabling the attacking players to perform more dribbling actions towards the opposite goal and beating a defensive line [20,37]. As previously stated, the addition of floaters tends to decrease the pressure of defenders on a ball carrier's possibilities for passing actions instead of possibilities for dribbling [39].

4.3. Third Game Principle (3rd = to Shoot at Goal with the Lowest Level of Opposition)

Concerning the third game principle, only the dribbling and shooting revealed significant differences between conditions. No significant differences were observed for passing actions. The emergence of passing actions that support the shoot is quite similar for both conditions used, revealing the lower values of actions to support shooting.

The analysis of dribbling actions revealed significant differences in all the experimental conditions. Specifically, significantly higher values were obtained in favour of 2vs2 in comparison with 3vs3. Despite defenders in both seeking to maintain their position between the ball and the goal, not allowing a misalignment between the ball and the goal [40], variability in the attacking players' relations with opponents and the ball is attributed to their constant explorative performances as they seek to break the symmetry with the defending players because of creating opportunities for scoring goals [41]. However, the explorative behaviours of the attacking team take place under the constraints imposed by the defending team. As noted, the defensive team tries to maintain spatiotemporal relations with the offensive team. In contrast, the offensive team attempts to disrupt the status quo at opportune times by advancing their position on the field, reaching the free attacking player, and finding chances for goal-scoring possibilities [42]. Therefore, the relevant issue is how players change their exploratory behaviours that disrupt the status quo: in 3vs3 through passing actions and in 2vs2 through dribbling and shooting.

4.4. Study Limitations and Future Research

As this research only involved male futsal players under 19, the generalization of findings to more diverse samples is limited. Additionally, the small sample size may refrain from achieving more robust inferences. Future research should overcome these issues and utilize players of varying ages, ability, and gender. On the other hand, this intervention was carried out in a natural context, where some contextual variables are challenging to control.

5. Conclusions and Practical Implications

This study has shown that manipulating the number of players (Gk + 3vs3 + Gk and Gk + 2vs2 + Gk) and floater positioning influence players´ technical–tactical behaviours in 3vs3 and 2vs2 SSCG. In the 2vs2, players perform more dribbling and shooting actions than in the 3vs3, where players developed more passing actions. However, these results vary depending on the game principle analysed. Specifically, 3vs3 is associated with passing and dribbling action to progress towards the goal without beating a defensive line, while 2vs2 is associated with passing and dribbling actions aimed at beating a defensive line. It probably happens because the defending team in 3vs3 form a zonal defence, necessitating the prioritization of avoiding creating penetrative passing lines and shooting at goal and increasing the pressure on the attacking players. Thus, within 2vs2 there seems to be more opportunities for 1vs1. According to the development steps, the overall results stress that the 2vs2 seems to highlight individual actions even with the presence of floaters, while the 3vs3 highlights more relational actions and collective tactical behaviours. However, as results have shown, there are differences between the individual actions developed according to the SSCG and the game principle. According to the main objective of training sessions, such information may support coaches in designing training tasks by manipulating task constraints (number of players and floaters that should be assigned to each goal).

Author Contributions: Conceptualization, all authors; methodology, D.P., A.P. and A.M.; software, B.G. and B.T.; validation, all authors; formal analysis, D.P. and B.G.; investigation, D.P., A.P., B.T. and A.M.; data curation, D.P. and B.G.; writing—original draft preparation, D.P., A.P., B.T., and A.M.; writing—review and editing, D.P., A.P., B.G. and B.T.; visualization, all authors; supervision, all authors. All authors have read and agreed to the published version of the manuscript.

Funding: This research received no external funding.

Institutional Review Board Statement: The study was conducted according to the guidelines of the Declaration of Helsinki and approved by the Ethics Committee of Universidad de Extremadura (239/2019).

Informed Consent Statement: Informed consent was obtained from all subjects involved in the study.

Acknowledgments: This study has been carried out thanks to the contribution of the Junta de Extremadura through the European Regional Development Fund. A way to make Europe (GR18129).

Conflicts of Interest: The authors declare no conflict of interest.

References

1. Coutinho, D.; Gonçalves, B.; Wong, D.P.; Travassos, B.; Coutts, A.J.; Sampaio, J. Exploring the effects of mental and muscular fatigue in soccer players' performance. *Hum. Mov. Sci.* **2018**, *58*, 287–296. [CrossRef] [PubMed]
2. Travassos, B.; Araújo, D.; Duarte, R.; McGarry, T. Spatiotemporal coordination behaviors in futsal (indoor football) are guided by informational game constraints. *Hum. Mov. Sci.* **2012**, *31*, 932–945. [CrossRef]
3. Travassos, B.; Duarte, R.; Vilar, L.; Davids, K.; Araújo, D. Practice task design in team sports: Representativeness enhanced by increasing opportunities for action. *J. Sports Sci.* **2012**, *30*, 1447–1454. [CrossRef]
4. Araújo, D.; Davids, K. Team Synergies in Sport: Theory and Measures. *Front. Psychol.* **2016**, *7*, 1449. [CrossRef]
5. Chow, J.Y.; Davids, K.; Button, C.; Renshaw, I. *Nonlinear Pedagogy in Skills Acquisition: An Introduction*; Routletge: New York, NY, USA, 2016.
6. Ric, A.; Hristovski, R.; Gonçalves, B.; Torres, L.; Sampaio, J.; Torrents, C. Timescales for exploratory tactical behaviour in football small-sided games. *J. Sports Sci.* **2016**, *34*, 1723–1730. [CrossRef] [PubMed]
7. Sarmento, H.; Clemente, F.M.; Harper, L.D.; Da Costa, I.T.; Owen, A.; Figueiredo, A.J. Small-sided games in soccer. A systematic review. *Int. J. Perform. Anal. Sport* **2018**, *18*, 693–749. [CrossRef]
8. Davids, K.; Araújo, D.; Correia, V.; Vilar, L. How Small-Sided and Conditioned Games Enhance Acquisition of Movement and Decision-Making Skills. *Exerc. Sport Sci. Rev.* **2013**, *41*, 154–161. [CrossRef]
9. Passos, P.; Araújo, D.; Davids, K.; Shuttleworth, R. Manipulating Constraints to Train Decision Making in Rugby Union. *Int. J. Sports Sci. Coach.* **2008**, *3*, 125–140. [CrossRef]
10. Sgrò, F.; Bracco, S.; Pignato, S.; Lipoma, M. Small-Sided Games and Technical Skills in Soccer Training: Systematic Review and Implications for Sport and Physical Education Practitioners. *J. Sports Sci.* **2018**, *6*, 9–19. [CrossRef]
11. Sampaio, J.; Macas, V. Measuring Tactical Behaviour in Football. *Int. J. Sports Med.* **2012**, *33*, 395–401. [CrossRef]
12. Clemente, F.M.; Wong, D.P.; Martins, F.M.L.; Mendes, R.S. Acute Effects of the Number of Players and Scoring Method on Physiological, Physical, and Technical Performance in Small-sided Soccer Games. *Res. Sports Med.* **2014**, *22*, 380–397. [CrossRef]
13. Práxedes, A.; Moreno, A.; Gil-Arias, A.; Claver, F.; Del Villar, F. The effect of small-sided games with different levels of oppo-sition on the tactical behaviour of young footballers with different levels of sport expertise. *PLoS ONE* **2018**, *13*, e0190157. [CrossRef]
14. Travassos, B.; Coutinho, D.; Gonçalves, B.; Pedroso, P.; Sampaio, J. Effects of manipulating the number of targets in U9, U11, U15 and U17 futsal players' tactical behaviour. *Hum. Mov. Sci.* **2018**, *61*, 19–26. [CrossRef]

15. Castellano, J.; Silva, P.; Usabiaga, O.; Barreira, D. The influence of scoring targets and outer-floaters on attacking and defending team dispersion, shape and creation of space during small-sided soccer games. *J. Hum. Kinet.* **2016**, *51*, 153–163. [CrossRef]
16. Clemente, F.; Dellal, A.; Wong, D.; Martins, F.L.; Mendes, R. Heart rate responses and distance coverage during 1 vs. 1 duel in soccer: Effects of neutral player and different task conditions. *Sci. Sports* **2016**, *31*, e155–e161. [CrossRef]
17. Clemente, F.M.; Wong, D.P.; Martins, F.M.L.; Mendes, R. Differences in U14 football players' performance between different small-sided conditioned games. *RICYDE. Rev. Int. Cienc. Deporte* **2015**, *11*, 376–386. [CrossRef]
18. Clemente, F.M.; Martins, F.M.L.; Mendes, R.S.; Campos, F. Inspecting the performance of neutral players in different small-sided games. *Mot. Rev. Educ. Física* **2015**, *21*, 45–53. [CrossRef]
19. Hill-Haas, S.V.; Dowson, B.T.; Couts, A.J.; Rowsell, G.J. Time-motion characteristics and physiological responses of small-sided games in elite youth players: The influence of player number and rule changes. *J. Strength Cond. Res.* **2010**, *24*, 2149–2156. [CrossRef] [PubMed]
20. Padilha, M.B.; Guilherme, J.; Serra-Olivares, J.; Roca, A.; Teoldo, I. The influence of floaters on players' tactical behaviour in small-sided and conditioned soccer games. *Int. J. Perform. Anal. Sport* **2017**, *17*, 721–736. [CrossRef]
21. Pizarro, D.; Práxedes, A.; Travassos, B.; Gonçalves, B.; Domínguez, A.M. Floaters as coach's joker? Effects of the floaters positioning in 3vs3 small-sided games in futsal. *Int. J. Perform. Anal. Sport* **2021**, *21*, 197–214. [CrossRef]
22. Serra-Olivares, J.; González-Víllora, S.; López, L.M.G.; Araújo, D. Game-Based Approaches' Pedagogical Principles: Exploring Task Constraints in Youth Soccer. *J. Hum. Kinet.* **2015**, *46*, 251–261. [CrossRef]
23. Bayer, C. *The Teaching of Collective Sports Games*; Hispano Europea: Barcelona, Spain, 1992.
24. Ric, Á.; Hristovski, R.; Torrents, C. Can joker players favor the exploratory behavior in football small-sided games? *Res. Phys. Educ. Sport Health* **2015**, *4*, 35–39.
25. Gonçalves, B.; Marcelino, R.; Torres-Ronda, L.; Torrents, C.; Sampaio, J. Effects of emphasising opposition and cooperation on collective movement behaviour during football small-sided games. *J. Sports Sci.* **2016**, *34*, 1346–1354. [CrossRef]
26. Gonçalves, B.; Esteves, P.T.; Folgado, H.; Ric, A.; Torrents, C.; Sampaio, J. Effects of Pitch Area-Restrictions on Tactical Behavior, Physical, and Physiological Performances in Soccer Large-Sided Games. *J. Strength Cond. Res.* **2017**, *31*, 2398–2408. [CrossRef] [PubMed]
27. Travassos, B.; Gonçalves, B.; Marcelino, R.; Monteiro, R.; Sampaio, J. How perceiving additional targets modifies teams' tactical behavior during football small-sided games. *Hum. Mov. Sci.* **2014**, *38*, 241–250. [CrossRef] [PubMed]
28. Travassos, B.; Araújo, D.; Davids, K.; O'Hara, K.; Leitão, J.; Cortinhas, A. Expertise effects on decision-making in sport are constrained by requisite response behaviours—A meta-analysis. *Psychol. Sport Exerc.* **2013**, *14*, 211–219. [CrossRef]
29. Tabachnick, B.G.; Fidell, L.S. *Using Multivariate Statistics*; Pearson Education Inc: New York, NY, USA, 2007.
30. Fleiss, J.L.; Levi, B.; Cho Paik, M. *Statistical Methods for Rates and Proportions*; Wiley: New York, NY, USA, 2003.
31. Cumming, G. *Understanding the New Statistics: Effect Sizes, Confidence Intervals, and Meta-Analysis*; Routledge: New York, NY, USA, 2012.
32. Ho, J.; Tumkaya, T.; Aryal, S.; Choi, H.; Claridge-Chang, A. Moving beyond P values: Data analysis with estimation graphics. *Nat. Methods* **2019**, *16*, 565–566. [CrossRef] [PubMed]
33. Hopkins, W.G.; Marshall, S.W.; Batterham, A.M.; Hanin, J. Progressive Statistics for Studies in Sports Medicine and Exercise Science. *Med. Sci. Sports Exerc.* **2009**, *41*, 3–12. [CrossRef]
34. Sullivan, M.O.; Woods, C.T.; Vaughan, J.; Davids, K. Towards a contemporary player learning in development framework for sports practitioners. *Int. J. Sports Sci. Coach.* **2021**, 17479541211002335. [CrossRef]
35. Pizarro, A.P.; Práxedes, A.; Travassos, B.; del Villar, F.; Moreno, A. The effects of a nonlinear pedagogy training program in the technical-tactical behaviour of youth futsal players. *Int. J. Sports Sci. Coach.* **2019**, *14*, 15–23. [CrossRef]
36. Travassos, B. *Manipulação de Exercícios de Treino no Futsal. Da Conceptualização à Práctica [Manipulating Training Exercises in Futsal. From Conceptualization to Practice]*; Prime Books: Estoril, Portugal, 2020.
37. Duarte, R.; Araújo, D.; Davids, K.; Travassos, B.; Gazimba, V.; Sampaio, J. Interpersonal coordination tendencies shape 1-vs-1 sub-phase performance outcomes in youth soccer. *J. Sports Sci.* **2012**, *30*, 871–877. [CrossRef]
38. Vilar, L.; Esteves, P.T.; Travassos, B.; Passos, P.; Lago-Peñas, C.; Davids, K. Varying numbers of players in small-sided soccer games modifies action opportunities during training. *Int. J. Sports Sci. Coach.* **2014**, *9*, 1007–1018. [CrossRef]
39. Travassos, B.; Araújo, D.; Davids, K.; Vilar, L.; Esteves, P.T.; Correia, V. Informational constraints shape emergent functional behaviors during performance of interceptive actions in team sports. *Psychol. Sport Exerc.* **2012**, *13*, 216–223. [CrossRef]
40. Vilar, L.; Araújo, D.; Davids, K.; Button, C. The role of ecological dynamics in analysing performance in team sports. *Sports Med.* **2012**, *42*, 1–10. [CrossRef] [PubMed]
41. Corrêa, U.C.; Alegre, F.A.M.; Freudenheim, A.; Dos Santos, S.; Tani, G. The game of futsal as an adaptive process. *Nonlinear Dyn. Psychol. Life Sci.* **2012**, *16*, 185–204.
42. Travassos, B.; Vilar, L.; Araújo, D.; McGarry, T. Tactical performance changes with equal vs unequal numbers of players in small-sided football games. *Int. J. Perform. Anal. Sport* **2014**, *14*. [CrossRef]

International Journal of
Environmental Research and Public Health

Article

The Effect of Differential Repeated Sprint Training on Physical Performance in Female Basketball Players: A Pilot Study

Jorge Arede [1,2,3,*], Sogand Poureghbali [4], Tomás Freitas [5,6,7], John Fernandes [8], Wolfgang I. Schöllhorn [9] and Nuno Leite [1,10]

1   Department of Sports Sciences, Exercise and Health, University of Trás-os-Montes and Alto Douro, 5001-801 Vila Real, Portugal; nleite@utad.pt
2   School of Education, Polytechnic Institute of Viseu, 3504-501 Viseu, Portugal
3   Department of Sports, Higher Institute of Educational Sciences of the Douro, 4560-708 Penafiel, Portugal
4   Institute of Sport Science, Otto-von-Guericke-Universität Magdeburg, 39104 Magdeburg, Germany; sogandpoureghbali@gmail.com
5   UCAM Research Center for High Performance Sport, Catholic University of Murcia (UCAM), 30107 Murcia, Spain; tfreitas@ucam.edu
6   NAR-Nucleus of High Performance in Sport, São Paulo 04753-060, Brazil
7   Faculty of Sport Sciences, Catholic University of Murcia (UCAM), 30107 Murcia, Spain
8   School of Sport and Health Sciences, Cardiff Metropolitan University, Cardiff CF23 6XD, UK; jfmtfernandes@hotmail.co.uk
9   Institute of Sport Science, Training and Movement Science, University of Mainz, 55122 Mainz, Germany; schoellw@uni-mainz.de
10 Research Center in Sports Sciences, Health Sciences and Human Development, CIDESD, University of Trás-os-Montes and Alto Douro, 5001-801 Vila Real, Portugal
*   Correspondence: jorge_arede@hotmail.com

check for updates

Citation: Arede, J.; Poureghbali, S.; Freitas, T.; Fernandes, J.; Schöllhorn, W.I.; Leite, N. The Effect of Differential Repeated Sprint Training on Physical Performance in Female Basketball Players: A Pilot Study. *Int. J. Environ. Res. Public Health* **2021**, *18*, 12616. https://doi.org/10.3390/ijerph182312616

Academic Editors: António Carlos Sousa and Paul B. Tchounwou

Received: 18 October 2021
Accepted: 25 November 2021
Published: 30 November 2021

Publisher's Note: MDPI stays neutral with regard to jurisdictional claims in published maps and institutional affiliations.

Abstract: This pilot study aimed to determine the effects of differential learning in sprint running with and without changes of direction (COD) on physical performance parameters in female basketball players and to determine the feasibility of the training protocol. Nine female basketball players completed 4 weeks of repeated sprint training (RST) with (COD, $n = 4$) or without (NCOD, $n = 5$) changes of direction. A battery of sprints (0–10 and 0–25 m), vertical jumps (counter movement jump (CMJ), drop jump, and single-leg CMJs), and COD tests were conducted before and after intervention. NCOD completed two sets of ten sprints of 20 m, whereas COD performed 20 m sprints with a 180 degree turn at 10 m, returning to the starting line. Before each sprint, participants were instructed to provide different fluctuations (i.e., differential learning) in terms of varying the sprint. Both groups had 30 s of passive recovery between two sprints and 3 min between sets. A significant effect of time for the 0–10 m sprint, CMJ, and single leg-CMJ asymmetries were observed. Adding "erroneous" fluctuation during RST seems to be a suitable and feasible strategy for coaches to enhance physical performance in young female basketball players. However, further studies including larger samples and controlled designs are recommended to strengthen present findings.

Keywords: jumping; sprinting; change-of-direction; fluctuations; bilateral asymmetry

1. Introduction

The success of team sports depends, to a large extent, on the physical abilities of a player but also on the technical and tactical skills [1]. In team sports (e.g., basketball), repeated bouts of high-intensity activities (i.e., sprinting, jumping) are interspersed with periods of low-to-moderate activity or passive recovery [1]. Moreover, the physical demands are complex and challenge athletes to have highly and simultaneously developed speed, agility, strength, power, and endurance qualities [1]. Therefore, it seems plausible that practitioners working within team sports design training regimes that adequately match the specific team sport requirements.

Players' ability to perform repeated sprints is one of the critical determinants of team sports performance [2], including in basketball [3]. In team sports, sprinting activities correspond to a small percentage of total distance covered or number of activities [4,5]; however, short sprint activities are intercepted by a short period of time (i.e., every 21 to 39 s), for example, in basketball [6]. In this regard, although the total distance covered during competition has not changed over the years, the requirement for high-intensity running and longer sprint distances has increased [7]. Consequently, it is astute that practitioners develop methods of enhancing sprint and repeat sprint ability (RSA) in team sports athletes. Moreover, since sprints in these contexts are not exclusively straight-line, it is considered beneficial to prepare athletes to sprint in different directions, challenging the technical model.

Various methods are used to develop sprint and change of direction (COD) performance, including assisted and resisted sprinting techniques, resistance training, plyometrics, and rotational flywheel training [8–10]. A comparison between RSA- and COD-training revealed comparable performance improvements in soccer-relevant physical performance factors [11]. However, evidence suggests that sprints with COD and short shuttle runs are more demanding than linear sprints [12], resulting in higher cardiorespiratory involvement and blood lactate accumulation [12]. Moreover, sprints with COD include deceleration and acceleration efforts, creating heightened metabolic and mechanical demands needed to overcome inertia and rapidly generate propulsive forces in a new direction [13]. This might evoke more significant stimuli in performance components associated with neuromuscular factors such as jumps, sprints, and repeated sprint performance [14]. However, sprints and sprints with COD are performed during the game in unpredictable situations and in different contexts, so it is warranted to employ training methods that promote both the ability to adapt and the development of the necessary physical capabilities.

From a dynamical systems perspective, fluctuations play a key role in the adaptation of living systems when transiting from one stable state to another [15]. On the basis of bio-mechanically identified, different levels of fluctuations in several highly automatized movements [16,17], the differential learning (DL) model proposed a concrete practical transfer of the system dynamics approach to gross motor movements by utilising and actively increasing fluctuations in an athlete's movement learning process [18]. In contrast to the traditional training strategy, where fluctuations are considered errors that must be minimized, the DL approach considers the fluctuations in moving systems as potential sources and necessary for learning. In an analogy to artificial neural nets that are modelling phenomena of real neurons [19], an increase of noise during the learning phase fosters the process of interpolation and, as a consequence, allows the system a better performance in the subsequent application phase [15]. Evidence for better performance by interpolation compared to extrapolation in natural neuro-motor systems was provided by a study by Catalano and Kleiner [20]. The experiment shows significantly better performance when testing within the trained range than when testing outside. Accordingly, the DL theory recommends increasing the range already during training to increase the probability of being able to interpolate in case of emergency or already during the next movement alone and thus to be able to react more adequately to the new elements that will surely come. Fluctuations can be increased along a neuromechanical strategy [18] in which the angles, angular velocities, and angular accelerations of all joints or their rhythms are varied, since each biomechanical variable can be associated with a physiological proprioceptor that provides the information necessary for training to the central nervous system for reorganization. Fluctuations also can be increased by all kinds of instructions (e.g., or metaphors) or modification of the surrounding in form of restrictions or enriched environments. Thereby, a major intention of DL is to increase the possibilities of the athlete rather than constraining them. Meanwhile, evidence is provided that fluctuations that can be quantified by the amount or structure of noise can also be increased or modified by means of emotions [21] or fatigue [22,23].

and can be observed impressively in this study and the studies connected with DL: The additional variation in exercise caused the initial large variation in the hit performance to "destroy" it, resulting in a greater hit performance. Gelfand and Latash [52,53] formulated something similar, but more poetically, with the "bliss of abundance" as an alternative view of movement variety that occurs even in repetitive motor learning processes, and thus plead for a rethinking of the traditional motor control theories that are mainly based on singular solutions.

In contrast to traditional prototypical training, DL exercises encourage learners to actively perform movement errors (e.g., throwing to the left instead of forwarding), rather than avoid them [54]. Participants performed shooting to the left, to the right, upwards, and shooting straight. Errors are considered as neutral fluctuations ("you never know, what they are good for") and are not only experienced to know what is correct, which was already included in contrast learning and would give the learner convergent guidance in contrast to divergent self-organization. In DL the fluctuations are essential elements for a more stable training of the neural nets involved.

Our findings are consistent with those of other researchers [55–57] who have examined the typology of attention focus, which, according to the OPTIMAL theory, suggests that an external focus of attention results in increased learning skills when compared to an internal focus of attention. Nevertheless, based on the results, the focus of external attention combined with DL, as in the DL/EF condition, resulted in a higher, statistically significant improvement than the T/EF. These results provide an indication of the stronger influence of the physically dominated variations of the DL approach, compared to the mentally supportive traditional training method. Regarding the futsal goal-shooting transfer test, statistically, significant differences were observed between all five intervention groups.

A neurophysiological explanation for these phenomena would be appealing but has, so far, been problematic. This is because the available EEG studies on both DL and external foci always refer to acute effects, but not to medium-term effects, after multiple interventions. This is complicated by the fact that the EEG studies on DL focus on the effects immediately after a whole series of movements, whereas the studies on an external focus tend to focus on the EEG activation immediately before and during a single movement. The EEG studies on DL show an increase in power primarily in the frontal brain areas in the alpha and theta frequency bands after the intervention [31,40] whereas a reduction in these frequency bands in the central brain areas is observed during external focus tasks [58]. A decrease in alpha power was also observed during a dart-throwing task with an external focus, in comparison to one with an internal focus [59], but, unfortunately, the theta band was not analyzed, which is assumed to be of major interest for high-concentration tasks [60]. The degree to which these phenomena depend on the complexity of the movement task, the performance level, or the age of the subjects requires further investigation [61,62]. The same applies to the question of the extent to which a decrease in the alpha frequencies during training leads to an increase in the alpha frequencies after training, or the extent to which a sustained increase during sleep leads to a permanent change. Nonetheless, the practical interventions derived from OPTIMAL theory seem to be a complement to DL that could be of great interest when it comes to the absolute limits of performance.

The current study findings were based on two assumptions. First, that DL exercises lead to detecting adaptive solutions through increased fluctuations in the individual, and they appear to support learning ability [46]. Second, that DL exercise, using an optimal relation of the internal and external noise frequencies [63], moves the brain to a more stress-resilient state by downregulating frontal lobe activation [31]. The participants in all the DL-related groups were able to establish an adaptive response to the task modification, even though we included spectators in the test and created a more stressful environment.

Epistemologically, the applied statistics and methodology do not claim to be generalizable. Therefore, what is often interpreted as a limitation turns out, upon closer inspection, to be an aspect to be considered in future studies. One such aspect is related to the influence of the sex of the subjects. While this study was conducted on pubescent girls, it would

Increasing the noise serves to destabilize the learning system and to launch a genuine self-organizing process. In its most extreme form, DL leads to movement variations without repetition and without correction [24]. Movement corrections in DL are avoided to enable the athlete to find their own optimal solution (which would not be the case if the athlete would be guided by information about "errors"). Moreover, external feedback can be overestimated and redundant in case of enough variation [25]. Here, it is important to differentiate augmented (or external) feedback from receiving emotional support, which could have a different influence on the training process but has not been the subject of DL research so far. The continuous confrontation with new movement challenges in DL training results in flexible and adaptable movement patterns [26,27]. According to DL theory, increased fluctuations result in better skill acquisition and better learning rates than traditional models [28,29]. Hereby, the noise should be optimized rather than maximized, which is covered by the stochastic resonance principle of the DL-theory [24,27]. Thereby, in the ideal case, the noise provided by the given exercises should be adapted to the individual and momentary noise provided by athlete. Epistemologically, the majority of studies on DL follow the strategy of conceptual replication instead of direct replication or reproduction, which logically fall too short because of the Duhem–Quine thesis, and so far provide corroboration (not verification) of the DL theory [30]. However, a differentiation of effects comparing models that were influenced and inspired by DL theory, such as the rather eclectic constraints-led approach [31] or the gradual and stochastic DL [26], are pending. Indeed, the benefits of the training programs based on DL in both technical and physical skills have been reported in team sports [32–35].

Based on the previous, including the DL approach, fluctuation in repeated sprints and sprints with COD in a training program is assumed to have the potential for eliciting physical performance improvements. Therefore, the aim of this study was to examine the effect of a four-week training intervention involving repeated differential sprint training with COD (COD) vs. without COD (NCOD) on a series of physical tests (i.e., jumping, landing, sprinting, and cutting). A better understanding of the effects of differential repeated sprint training on various aspects of physical performance may help practitioners to better schedule and design training tasks to improve these aspects. Due to the lack of comparable findings, we propose the null hypothesis, i.e., that there will be no difference in the efficacy of the repeated differential sprint training with or without COD. Furthermore, we hypothesised that repeated differential sprint training is a feasible training strategy and beneficial for physical parameters.

2. Materials and Methods

2.1. Participants

Sixteen female basketball players from the under-19 age group up to the amateur senior level volunteered in this study. All participants completed, in sum, about 270 min of basketball training (three basketball sessions/week, 90 min/session) and one to two competitive matches per week. Only participants who participated in at least 90% of the workouts were considered for data analysis, which resulted in the exclusion of seven players from post-testing analysis (NCOD, $n = 5$; COD, $n = 4$). Nine players were finally assessed. Post hoc observed power calculations (G*Power, version 3.1.9.7; University of Düsseldorf; Düsseldorf, Germany) for repeated measures ANOVA, including two groups and two measurements ($\alpha = 0.05$, $d = 0.25$), revealed power (β) between 0.11 and 0.25. Written and informed consent was obtained from all participants' parents, and player approval was obtained before the beginning of this investigation. The present study was approved by the Institutional Research Ethics Committee and conformed to the recommendations of the Declaration of Helsinki.

2.2. Procedures

This pilot study incorporated a parallel-groups, repeated measures design, whereby participants were randomly divided into two groups with repeated sprinting training with

(COD, n = 8) and without COD (NCOD, n = 8). The training period lasted 4 weeks and was carried out within the regular training sessions. The tests were performed one and two weeks before the commencement of the training period and one week after the intervention. Physical performance tests (PPT) were performed under the same experimental conditions (training session time and indoor basketball court). A 10 min standardized warm-up was performed (5 min jogging, dynamic stretching, 10 bilateral squats, core exercises, 10 unilateral squats, and three vertical unilateral jumps) before all testing. PPTs were conducted in the following order, respecting the principles of National Strength and Conditioning Association for testing order [36]: anthropometrical measurements, jumping tests (countermovement jump (CMJ), single-leg countermovement jumps (SLCMJs)), drop jumps (DJ), single leg drop jumps (SLDJ), the 505 test, and straight sprinting tests (0–10 and 0–25 m splits time).

2.3. Training Program

The participants included in both training groups participated in two weekly training sessions during in-court practice in a four-week period. All the intervention drills were performed at the beginning of the training session, after the warm-up period. The differential repeated sprint training was comprised by two sets of ten sprints of 20 m with 30 s of passive recovery between sprints and 3 min of passive recovery between sets. The NCOD group performed all repetitions straight, while the COD group ran to a mark situated 10 m from the starting line, performed a 180° COD using alternatively the right or left leg to push off, before returning to the starting line (total of 20 m) (Figure 1). Before each repetition, all participants were verbally instructed by the main researcher to perform a different fluctuation (Table 1; Figure 2) or a combination of fluctuations. No instructed movement fluctuation was repeated more than once in each training session. These fluctuations were selected based on previous studies involving the DL approach exercises for motor skills [28,37].

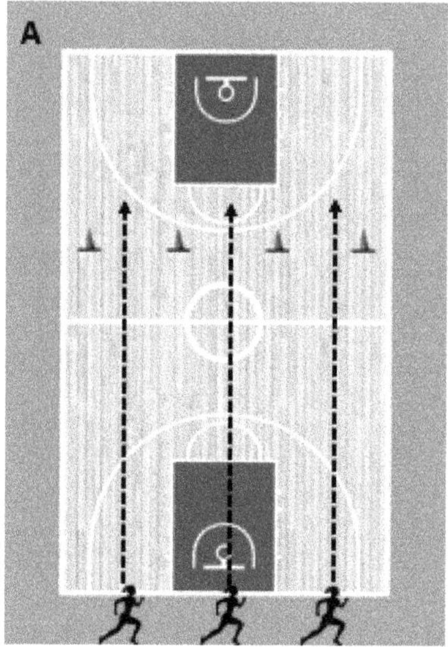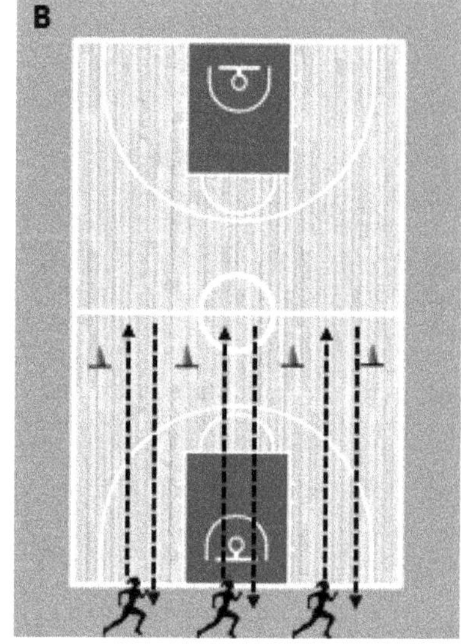

Figure 1. Examples of different experimental setups. (**A**) NCOD group and (**B**) COD group.

Table 1. Examples of the fluctuations performed during differential repeated sprint training interventions.

#	Body Part	Fluctuations	#	Body Part	Fluctuations
1		Head back	29		Trunk rotation to the left
2		Head forward	30		Trunk rotation to the right
3		Head back and forth	31	Trunk	Trunk tilted laterally to the left
4	Head	Head rotated to the left	32		Trunk angled laterally to the right
5		Head rotated to the right	33		Torso tilted back
6		Head rotated left and right	34		Trunk tilted forward
7		Head tilted to the left	35		Hands on hip
8		Head tilted to the right	36		Right hand on hip
9		Closed right eye	37		Left hands on hip
10		Left eye closed	38		Hands behind head
11	Eyes	Blinking eyes	39		Hands on forehead
12		Look to the right	40	Hands	Right hand behind head
13		Look left	41		Left hand behind head
14		Two arms up	42		Hands behind back
15		Two arms close to the torso	43		Right hand behind back
16		Arm rotation forward	44		Left hand behind back
17		Arm rotation back	45		Clapping ahead
18		Alternate forward arm rotation	46		Clap behind back
19		Alternate backward arm rotation	47		Clap front and back
20	Arms	Arms open to the side			
21		Arms open down			
22		Crossed arms			
23		Arms stretched forward			
24		Arms stretched back			
25		Right arm up + left arm down			
26		Left arm up + right arm down			
27		Left arm up + right arm to the side			
28		Right arm up + left arm to the side			

Figure 2. Examples of differential repeated sprint training fluctuations.

2.4. Measurements

Bilateral and Unilateral Countermovement Jumps (CMJ). CMJs were assessed according to the Bosco Protocol. Participants performed three successful SLCMJs with each leg in the vertical and horizontal directions. Participants began standing on one leg, then descended into a countermovement before extending the stance leg to jump as far or as high as possible in the vertical and horizontal directions. The landing was performed on both feet simultaneously. A successful trial included hands remaining on the hips throughout the movement and balance being maintained for at least 3 s after landing. If the trial was considered unsuccessful, a new trial was performed. In the horizontal direction, the participants started with the selected leg positioned just behind a starting line. The jump height was recorded using an infrared optical system (OptoJump Next—Microgate, Bolzano, Italy).

Bilateral and Unilateral Rebound Drop Jumps (DJ). Participants stood on top of a 30 cm high box with hands placed on the hips. Then, they dropped down, landed on both legs, and jumped vertically as high as possible with the shortest ground contact time possible. In unilateral rebound jumps (SLRJ), they hopped down diagonally (45° anterolateral), landed on the same leg within the infrared optical system, and then jumped vertically as high as possible with the shortest contact time possible [38]. The reactive strength index (RSI) was automatically calculated using Optojump Next software, version 1.12.1.0, through the following formula: jump height/contact time [38].

The 505 test (COD). Each participant was instructed to run to a mark situated 10 m from the starting line, perform a 180° COD using the right or left leg to push off, and return to a mark located 5 m away, covering a total of 15 m [39]. The participants were asked to pass the line indicated on the ground with their entire foot at each turn. The 505 test total time was recorded with 90 cm height photoelectric cells separated by 1.5 m (Witty, Microgate, Bolzano, Italy). Each participant performed three sprints with COD for each side with 2 min of rest between them. Players began each trial in standing positions with their feet 0.5 m behind the first timing gate. The lower limb asymmetry index (ASI) was determined by adhering to the procedures described by Bishop and colleagues [40] using the following formula: ASI = 100/Max Value (right and left)*Min Value (right and left)* − 1 + 100. The COD deficit (CODD) for the double 180° COD test for each leg was calculated via the following formula: mean double 180° COD time—mean 10 m time [39].

Speed tests. The average running speeds were evaluated by 10 m (0–10 m) and 25 m (0–25 m) split times. Running times were recorded with 90 cm high photoelectric cells separated by 1.5 m. Each participant performed three trials with 2 min of rest between each of the trials. Players began each trial in an upright standing position with their feet 0.5 m behind the first timing gate.

2.5. Statistical Analyses

Data are presented as mean ± standard deviation (SD). Reliability of test measures computed using an average measures two-way random intraclass correlation coefficient (ICC) with absolute agreement, inclusive of 90% confidence intervals, and the coefficient of variation (CV). The ICC was interpreted as poor (<0.5), moderate (0.5–0.74), good (0.75–0.9), and excellent (>0.9) [41]. Coefficient of variation values were considered acceptable if <10% [42]. A paired-samples t-test was used to analyse within-group changes [43]. The threshold values for Cohen's d for within-group effect sizes (ES) statistics were 0–0.2 trivial, >0.2–0.6 small, >0.6–1.2 moderate, >1.2–2.0 large, and >2.0 very large [44]. A 2 × 2 repeated-measure analysis of variance (ANOVA) was performed based on the absolute values of all parameters to determine the main effects between groups (NCOD and COD group) and time (pre- and post-test) [43]. Effect size was evaluated with partial eta squared (η^2_p), and the threshold values were no effect ($\eta^2_p < 0.04$), minimum effect ($0.04 < \eta^2_p < 0.25$), moderate effect ($0.25 < \eta^2_p < 0.64$), and strong effect ($\eta^2_p > 0.64$) [45]. This measure has been widely cited as a measure of effect size and predominantly provided by statistical software [46]. Tukey's post-hoc analysis was used to examine the differences between times according to THE group. All statistical analyses were performed using the SPSS software (version 24 for Windows; SPSS Inc., Chicago, IL, USA).

3. Results

3.1. Sample

The mean age of the included subjects was 19.0 years (SD: 2.4). The mean height of the subjects was 169.8 cm (SD: 5.3). The mean body mass of the subjects in this study was 62.3 kg (SD: 3.9).

3.2. Tests Reliability

Although the concept of systems dynamics with its essential role of fluctuations especially in phase transitions conflicts with the reliability criterium in test theory, the reliability of the chosen test diagnosis was determined for reasons of comparison and evaluation.

All ICC values ranged from moderate to excellent (ICC range = 0.54–0.93), and most (6 of the 10) of CV values were acceptable (CV range = 1.28–16.70%) (Table 2).

Table 2. Reliability data for test variables. Data are presented as value with lower and upper confidence limits.

Test Variables	ICC (95% CL)	CV (%) (95% CL)
CMJ (cm)	0.89 (0.66; 0.97)	5.13 (2.87; 7.64)
DJ (m/s)	0.93 (0.76; 0.98)	16.70 (12.48; 20.84)
0–10 m (s)	0.81 (−0.07; 0.96)	2.63 (1.14; 4.32)
0–25 m (s)	0.84 (0.06; 0.97)	1.28 (0.46; 2.46)
CMJ_R (cm)	0.77 (0.19; 0.96)	10.67 (6.76; 15.07)
CMJ_L (cm)	0.95 (0.72; 0.99)	7.23 (4.30; 10.23)
$SLRJ_R$ (m/s)	0.93 (0.63; 0.99)	14.89 (9.44; 19.37)
$SLRJ_L$ (m/s)	0.95 (0.82; 0.99)	15.04 (9.43; 21.64)
$COD180_R$ (s)	0.54 (−0.24; 0.89)	2.47 (1.39; 3.63)
$COD180_L$ (s)	0.86 (−0.12; 0.97)	2.71 (1.35; 4.18)

Abbreviations: ICC = intraclass correlation coefficient; CV = coefficient of variation; CL = confidence limits; CMJ = countermovement jump height; DJ = drop rebound jump; 0–10 m = 0–10 m sprint time; 0–25 m = 0–25 m sprint time; SLRJ = diagonal single leg rebound jump; COD180 = change of direction test; R = right; L = left.

3.3. Tests Outcomes

Data from all PPTs were comparable between the two groups at baseline (all $p > 0.05$; see Table 3).

A repeated measures ANOVA indicated a significant main effect of time on CMJ (F = 12.64; $p \leq 0.01$; η^2_p = 0.64), 0–10m (F = 13.17; $p \leq 0.01$; η^2_p = 0.65), CMJ_R (F = 6.43; $p \leq 0.05$; η^2_p = 0.48), and CMJ_{ASI} (F = 5.85; $p \leq 0.05$; η^2_p = 0.46). Additionally, a significant main effect of the group on $SLRJ_R$ (F = 6.03; $p \leq 0.05$; η^2_p = 0.46), and $CODD_R$ (F = 13.79; $p \leq 0.01$; η^2_p = 0.66) was observed. Finally, repeated measures ANOVA indicated a significant interaction effect (group x time) on CMJ_L (F = 12.39; $p \leq 0.01$; η^2_p = 0.64) only. The post hoc *t*-test revealed a significant difference between pre-test and post-test on CMJ_L for the COD training group ($p = 0.047$).

Within-group changes for both training groups are described in Table 3. The NCOD training group showed significant improvements in 0–10 m (ES = −1.25, $p \leq 0.05$), CMJ_R (ES = 1.30, $p \leq 0.05$), CMJ_{ASI} (ES = −1.86, $p \leq 0.05$), and $SLRJ_{ASI}$ (ES = −1.50, $p \leq 0.05$). Figures 3 and 4 display the individual changes in performance from pre- to post-test for both groups. Interestingly, the majority of athletes are responding in the same direction. Nevertheless, in both groups, single athletes can be identified who react contrary to the group trend.

Table 3. Inferences of the training programs intervention on player's performance measures.

Variables		Pretest, Mean ± SD	Postest, Mean ± SD	p	Within-Group Effect Size	Between-Groups Pretest Differences (p)	Between-Group Effect Size
CMJ (cm)	NCOD	23.74 ± 3.47	25.17 ± 2.58	0.061	1.16	0.530	0.44
	COD	22.55 ± 0.90	24.53 ± 1.19	0.095	1.21		
DJ (m/s)	NCOD	0.79 ± 0.32	0.86 ± 0.27	0.418	0.40	0.340	0.67
	COD	0.61 ± 0.18	0.63 ± 0.14	0.686	0.22		
0–10 m (s)	NCOD	2.09 ± 0.09	2.02 ± 0.07	0.035 *	−1.25	0.730	−0.24
	COD	2.12 ± 0.10	2.06 ± 0.07	0.132	−1.03		
0–25 m (s)	NCOD	4.34 ± 0.17	4.29 ± 0.13	0.322	−0.61	0.266	−0.81
	COD	4.46 ± 0.10	4.37 ± 0.14	0.183	−0.86		
CMJ_R (cm)	NCOD	15.10 ± 2.47	16.05 ± 2.45	0.044 *	1.30	0.110	1.22
	COD	12.70 ± 0.87	14.90 ± 2.04	0.202	0.82		
CMJ_L (cm)	NCOD	15.96 ± 3.28	14.84 ± 2.91	0.122	−0.87	0.074	1.41
	COD	11.83 ± 2.38	14.61 ± 2.53	0.073	1.36		
CMJ_{ASI} (%)	NCOD	21.86 ± 6.93	11.68 ± 2.87	0.014 *	−1.86	0.881	−0.10
	COD	22.43 ± 2.50	20.30 ± 7.25	0.691	0.22		
$SLRJ_R$ (m/s)	NCOD	0.43 ± 0.12	0.42 ± 0.10	0.672	−0.20	0.123	1.18
	COD	0.32 ± 0.06	0.31 ± 0.03	0.847	0.11		
$SLRJ_L$ (m/s)	NCOD	0.42 ± 0.12	0.43 ± 0.10	0.859	0.09	0.063	1.48
	COD	0.32 ± 0.06	0.31 ± 0.03	0.073	−0.10		
$SLRJ_{ASI}$ (%)	NCOD	27.67 ± 15.53	22.01 ± 14.10	0.028 *	−1.50	0.262	−0.82
	COD	39.19 ± 11.92	31.80 ± 8.03	0.405	−0.48		
$COD180_R$ (s)	NCOD	4.78 ± 0.11	4.76 ± 0.17	0.703	−0.18	0.361	−1.00
	COD	4.86 ± 0.13	5.00 ± 0.11	0.131	1.03		
$COD180_L$ (s)	NCOD	4.78 ± 0.36	4.80 ± 0.18	0.824	0.10	0.550	−0.42
	COD	4.91 ± 0.21	4.93 ± 0.18	0.690	0.22		
COD_{ASY} (%)	NCOD	7.16 ± 2.19	4.44 ± 2.43	0.196	−0.69	0.336	0.69
	COD	5.38 ± 3.02	2.77 ± 0.92	0.190	−0.85		
$CODD_R$ (s)	NCOD	2.74 ± 0.09	2.73 ± 0.14	0.895	−0.06	0.176	−1.01
	COD	2.82 ± 0.05	2.98 ± 0.06	0.009 **	3.06		
$CODD_L$ (s)	NCOD	2.74 ± 0.24	2.79 ± 0.18	0.533	0.30	0.515	−0.46
	COD	2.84 ± 0.16	2.90 ± 0.14	0.074	1.35		

Abbreviations: CMJ = countermovement jump height; DJ = drop rebound jump; 0–10 m = 0–10 m sprint time; 0–25 m = 0–25 m sprint time; SLRJ = diagonal single leg rebound jump; COD180 = change of direction test; CODD = COD deficit; R = right; L = left; ASI = bilateral asymmetry; * significant differences at $p < 0.05$; ** significant differences at $p < 0.01$.

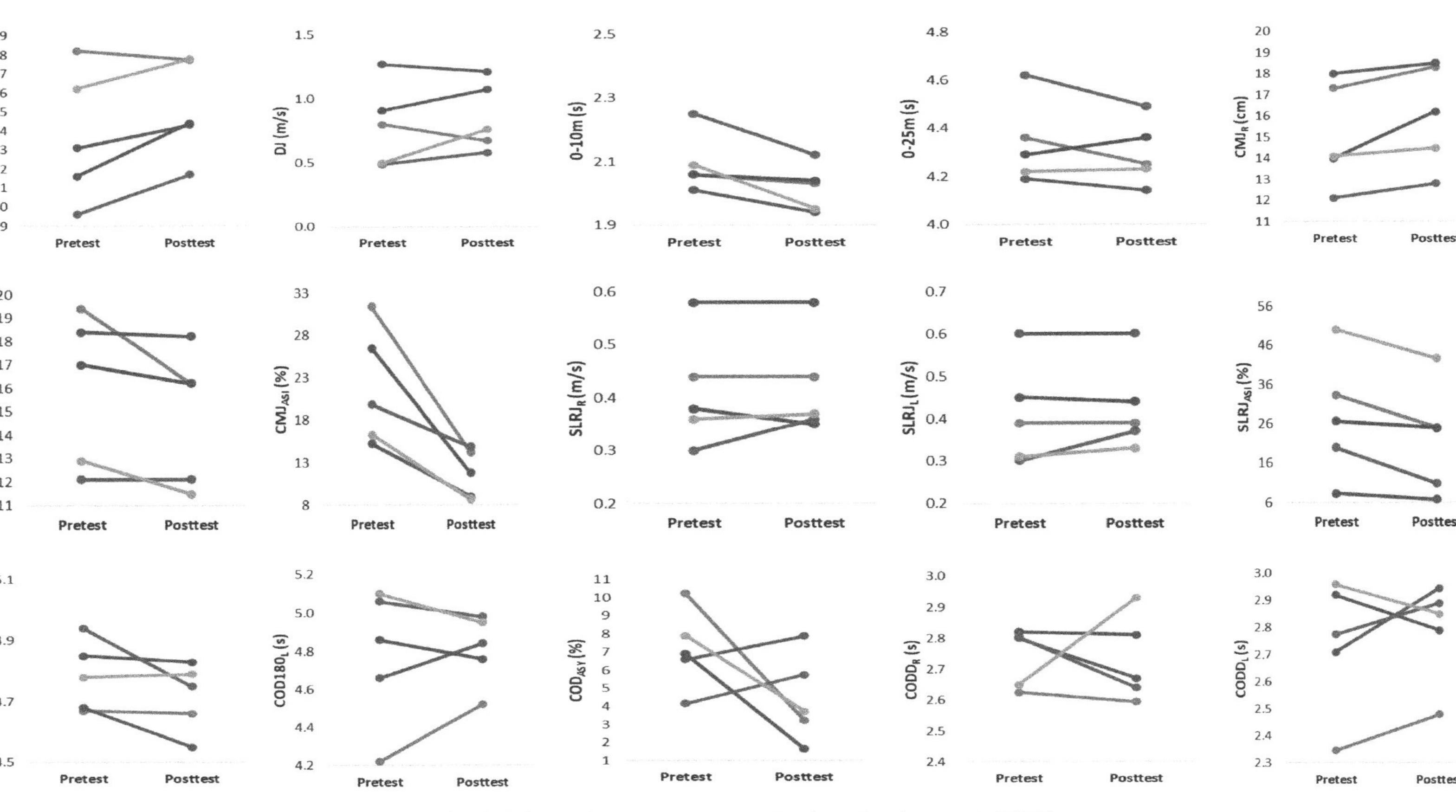

Figure 3. Individual changes from pretest to posttest for physical performance in NCOD group.

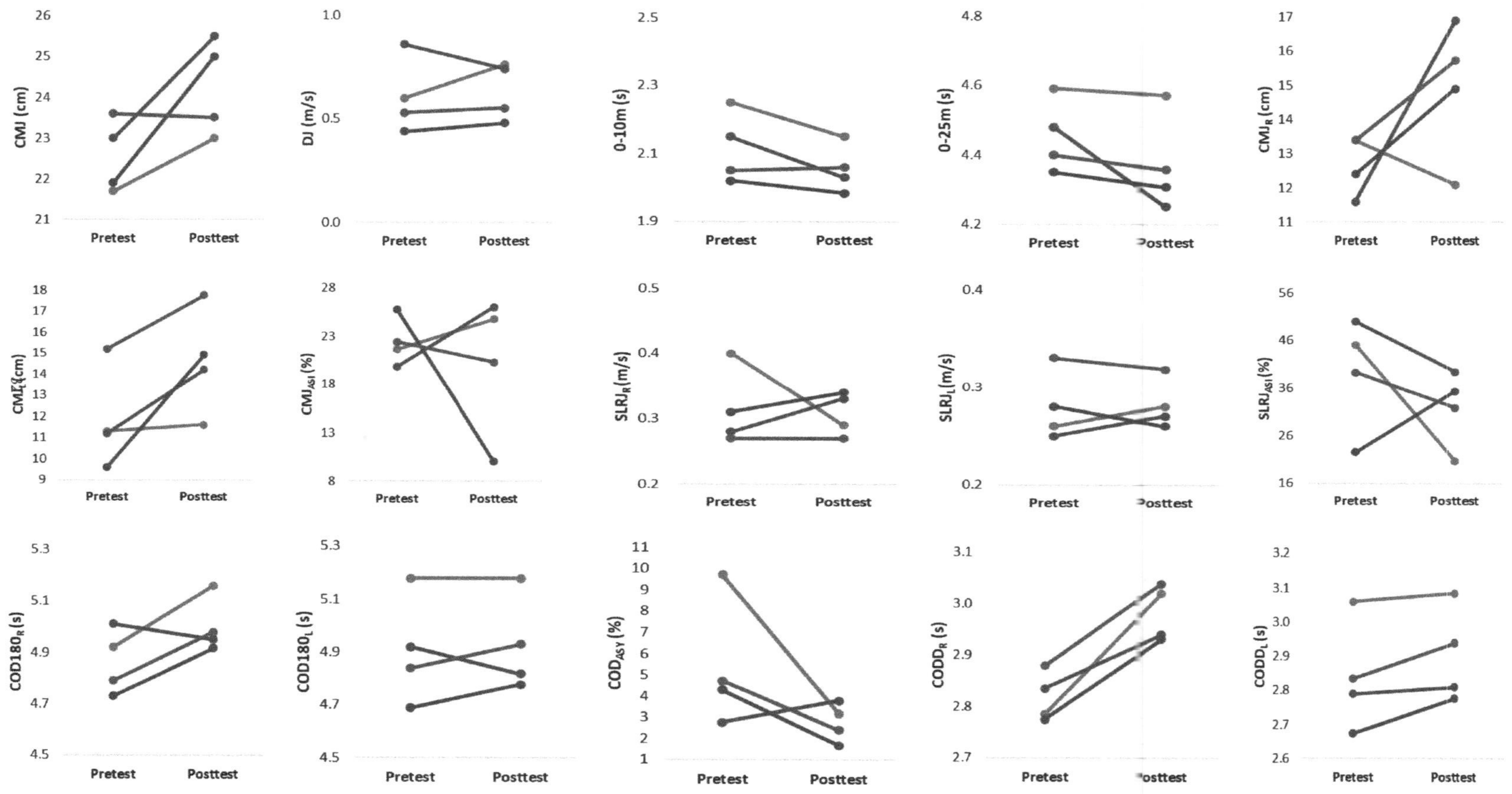

Figure 4. Individual changes from pretest to posttest for physical performance in COD group.

4. Discussion

This pilot study aimed to investigate the effects of the DL approach applied to sprint training with and without COD on physical tests in female basketball players. Methodically, in addition to the classical mean statistics, the presentation of individual results was chosen, since, on the one hand, epistemologically, from mean values only to other mean values can be concluded, but not to the individual athlete, and, on the other hand, for the effectiveness of training measures, individual reactions to interventions are increasingly of interest [47–49]. Considering findings and the absence of injuries and complaints, this study showed that it is feasible to implement DL principles in a repeated sprint training to improve physical performance. Moreover, the study provided indications that differential repeated sprint training has a beneficial effect on CMJ, 0–10 m, CMJ_R, and CMJ_{ASI}. Performing differential repeated sprint training without COD resulted in improved 0–10 m, CMJ_R, CMJ_{ASI}, and $SLRJ_{ASI}$ for all group participants. Furthermore, including COD during differential repeated sprint training resulted in improved 0–25 m sprint time and CMJ_L for all COD group participants.

Differential repeated sprinting training resulted in increased CMJ performance after 4 weeks. This beneficial transfer effect is similar to previous training protocols exploring DL in team sports [32,35], which suggests common underlying mechanisms explaining the improved jumping performance. Recent studies using electroencephalography to analyse brain activation patterns demonstrated neural activation after DL training in frequency bands and brain areas that are assumed to be supportive, especially for motor learning [50]. Improved motor learning and coordination (i.e., neural adaptations) dominate in the early phase of training [36]. That said, most participants may have benefited from the increased experiences of the various combinations of movement fluctuations and repeated highly intensive neuromuscular activations to improve jumping ability on the short-term scale. Specifically, increased neural drive is associated with enhanced agonist muscle recruitment, improved neuronal firing rates, and greater synchronization in the timing of neural discharge [36]. Likewise, central adaptations arising from higher neural activation can lead to increased motor unit activation, resulting in increased high-intensity muscular contraction and increased stiffness (i.e., jumping) [36]. Nevertheless, two of the participants had contrary responses (one per group) in CMJ, indicating that combining movement fluctuations and high-intensity muscular contractions places large stress either on their body or mind, inhibiting their performance. Thus, the present results need to be analysed with caution, with the need to examine the training response on an individual basis. Furthermore, practitioners should be aware that training volume, intensity, duration, or even the frequency of changes can be adjusted for negative or non-responders [51]. More studies with emphasis on individual responses are needed to better understand the effects of manipulating these variables on differential repeated sprint training [51].

The present findings suggest that an improvement in CMJ accompanied an increase in sprint performance. This is in accordance with the relationship between CMJ height and 0–10 m sprint time ($r = -0.51$) previously reported in female athletic populations [52]. The underlying commonality is seen in the fact that both activities take advantage of the stretch-shortening cycle (SSC), where an eccentric action (i.e., stretching) precludes a concentric action (i.e., shortening) [36]. The use of variable movements generates greater neuromuscular [53] and neurophysiological adaptations than movement repetitions [50] and increases the storage of elastic energy during the eccentric phase, leading to larger release of kinetic energy during the concentric phase. This better exploitation of the SSC may have allowed a greater training stimulus to occur over time, which, in turn, resulted in improved sprinting and jumping performance. In fact, after differential repeated sprint training, participants improved their 0–10 m sprint time. Notably, all participants of NCOD were positive responders in this test, which can be related to the specificity versus variability of the practice paradigm [54]. Possibly, the NCOD group may have benefited from better dynamic similarities between the movement patterns in differential repeated sprint training and the 0–10 m sprint test [36]. Additionally, the straight sprint and sprint with COD require specific running techniques [36], since during the COD training protocol, participants not only reduced their velocity prior to the 180° directional change at 10 m but also, due to the slowing down, caused increased eccentric contractions, while NCOD increased their velocity until 20 m.

After the differential repeated sprint training program, participants displayed higher values of CMJ_R compared to the pre-test values. This improvement in unilateral jumping may be indicative of increased strength of ankle- and hip-joint muscles and both static and dynamic postural balance [9]. However, to what extent these specific reactions depend on the individual preferences of the jump and play leg or laterality in general should be clarified by future research. A study analysing the effect of differential jump training on balance performance and postural control of female volleyball players during single leg stance observed a decreased sway area and anterior–posterior and mediolateral sway, indicative of improvements on the aforementioned qualities [34]. Thus, the implementation of movement variations through DL principles during training generates neuromuscular stimuli that may result in improved balance performance and ankle stabilization [55], and consequently, in improved unilateral jumping. Furthermore, previous results demonstrated impaired postural control (i.e., increased body sway) after high-intensity repeated sprints (i.e., ten 30 m sprints with two 180° COD (10 + 10 + 10 m), interspaced by 30 s of passive recovery between sprints) [56]. These findings suggest that repeated sprints result in alterations in the sensory information from the proprioceptive and exteroceptive systems [56]. Thus, continuous exposure to the combination of increased movement fluctuations accompanied by high-intensity muscular contractions can generate beneficial adaptations at the proprioceptive and exteroceptive levels that can result in positive transfer to the unilateral jumping performance in controlled settings. Notwithstanding, apart from one subject, all participants in the COD group improved CMJ_R and CMJ_L after training, including superior improvements in CMJ_L when compared to the NCOD group. Although the figures with individual results are no proof of individuality [48], the different effects on performance indicate that not all athletes achieved similar adaptations.

Following the principle of stochastic resonance idea within DL-theory [28], future research should match the noise of the coach's fluctuations to that of the athletes. Thereby, the extent to which this involves the amount or structure of noise will also need to be clarified [57]. This is particularly important in team sports since improving and strengthening all team members is crucial. The present findings suggest that the inclusion of COD during differential repeated sprint training can yet be more beneficial in unilateral jumping. Optimal movement variability during 180° COD could increase the need for lower-limb stabilisation to maintain balance, therefore requiring input from muscular involvement during triple flexion [58]. Moreover, the 180° COD involves the application of considerable braking and propulsive forces (particularly horizontal) within the final foot contact [59,60], demanding higher muscle activity of the ankle, knee, and hip-joint stabilizers. That said, improvements in unilateral jumping are expected based on the neuromuscular stimuli from simultaneously performing 180° COD with movement variation.

The overall adherence to the training programs is highly associated with the injury rate, enjoyment, motivation, and satisfaction with progress [61]. In this regard, the absence of injuries or complaints and the improvements in physical performance observed during this pilot study, in conjunction with previous adherence in experimental studies [32,35,62], bring increased expectations for a good adherence to the training program in further studies. These expectations are also sustained in brain activation patterns associated with enhanced attentional processes and improved learning processes from the use of differential learning in different motor tasks [50,63,64]. That said, continuous exposure to fluctuations can result in favorable neurophysiological adaptations, which may induce subjects to be more likely to be engaged in a training program, including differential learning approach.

We consider this investigation a pilot study due to the small sample size arising from the dropout rate, frequently observed in youth studies [34,37]. Since we mainly rely on the original Fisher statistics, extended by the effect sizes according to Neyman–Pearson [65], there is no claim for generalization [66]. Rather, we show further supportive evidence for the effectivity of an alternative training approach that encourages self-organization within a divergent learner-oriented teaching/coaching style [67]. In agreement with the Fisher's statistics, we conclude, based on the $p < 0.05$ results, that further research on differential training is promising. Further investigations in less-experienced, semi-professional and professional players is recommended. Furthermore, it may be worth pursuing longitudinal studies wherein the same participants are tested over months in order to fathom the complexity of the individual's continuous changes even more [49]. The original publication on DL in 1999 was especially focused on "Individuality—a neglected parameter" to emphasize the need for alternative theoretical and practical approaches for the dominant group and average oriented streams. Longitudinal changes in each variable could help better understand the long-term training effects and further support our results.

5. Conclusions

Counterintuitively, the findings of the present study displayed positive effects by adding "erroneous" fluctuation during sprint training not only in sprint but also in several jumping performances. Although it is a pilot study with no claim to generalisation, the significant changes from pre- to post-test indicate, according to Fisher's original interpretation, that pursuing this research field is worthwhile. Whether the results due to increased noise are side or main effects demands further research. The higher variability of stimuli during the training also suggests looking for additional effects on prevention of injuries or choking in basketball of female players. This training type is recommended for further experiments in female basketball, where coaches and fitness trainers might go beyond their learned tools and switch from convergent and teacher-oriented training to more divergent thinking approaches with athlete-oriented training. Fluctuations within repeated sprint training may influence the individual's movement patterns towards more effective and stabilized skills. From an epistemological point of view, together with all the other studies on DL, the results can be considered as a further corroboration of the DL theory. Nonetheless, much more research on the structure and amount of noise with respect to the individuality of the athletes and their momentary physiological and emotional states is necessary.

Author Contributions: Data curation, J.A.; formal analysis, J.A.; funding acquisition, N.L.; investigation, J.A. and N.L.; methodology, J.A. and N.L.; project administration, J.A.; validation, J.A.; visualization, J.A.; writing—original draft, J.A., S.P., T.F., J.F., W.I.S. and N.L.; writing—review and editing, J.A., T.F., J.F., W.I.S. and N.L. All authors have read and agreed to the published version of the manuscript.

Funding: This work was supported by the Foundation for Science and Technology (FCT, Portugal), under the project: SFRH/BD/122259/2016 and UID04045/2020.

Institutional Review Board Statement: The present study was approved by the Institutional Research Ethics Committee and conformed to the recommendations of the Declaration of Helsinki.University of Trás-os-Montes and Alto Douro, UID04045/2020.

Informed Consent Statement: Informed consent was obtained from all subjects involved in the study.

Data Availability Statement: The data that support the findings of this study are available from the corresponding author, J.A., upon reasonable request.

Conflicts of Interest: The authors declare no conflict of interest.

References

1. Ziv, G.; Lidor, R. Physical Attributes, Physiological Characteristics, On-Court Performances and Nutritional Strategies of Female and Male Basketball Players. *Sport Med.* **2016**, *39*, 547–568. [CrossRef] [PubMed]
2. Rampinini, E.; Bishop, D.; Marcora, S.M.; Ferrari Bravo, D.; Sassi, R.; Impellizzeri, F.M. Validity of simple field tests as indicators of match-related physical performance in top-level professional soccer players. *Int. J. Sports Med.* **2007**, *28*, 228–235. [CrossRef] [PubMed]
3. Castagna, C.; Manzi, V.; D'Ottavio, S.; Annino, G.; Padua, E.; Bishop, D. Relation between maximal aerobic power and the ability to repeat sprints in young basketball players. *J. Strength Cond. Res.* **2007**, *21*, 1172–1176. [PubMed]

4.   Gabbett, T.; King, T.; Jenkins, D. Applied physiology of rugby league. *Sport Med.* **2008**, *38*, 119–138. [CrossRef] [PubMed]
5.   Conte, D.; Favero, T.G.; Lupo, C.; Francioni, F.M.; Capranica, L.; Tessitore, A. Time-motion analysis of Italian elite women's basketball games: Individual and team analyses. *J. Strength Cond. Res.* **2015**, *29*, 144–150. [CrossRef]
6.   Stojanović, E.; Stojiljković, N.; Scanlan, A.T.; Dalbo, V.J.; Berkelmans, D.M.; Milanović, Z. The Activity Demands and Physiological Responses Encountered During Basketball Match-Play: A Systematic Review. *Sport Med.* **2018**, *48*, 111–135. [CrossRef]
7.   Fernando, P.B.; Suarez-Arrones, L.; Rodríguez-Rosell, D.; López-Segovia, M.; Jiménez-Reyes, P.; Bachero-Mena, B.; González-Badillo, J.J. Evolution of determinant factors of repeated sprint ability. *J. Hum. Kinet.* **2016**, *54*, 115–126.
8.   Arede, J.; Vaz, R.; Gonzalo-Skok, O.; Balsalobre-Fernandéz, C.; Varela-Olalla, D.; Madruga-Parera, M.; Leite, N. Repetitions in reserve vs. maximum effort resistance training programs in youth female athletes. *J. Sports Med. Phys. Fit.* **2020**, *60*, 1231–1239. [CrossRef]
9.   Arede, J.; Gonzalo-Skok, O.; Bishop, C.; Schöllhorn, W.I.; Leite, N. Rotational flywheel training in youth female team sport athletes: Could inter-repetition movement variability be beneficial? *J. Sports Med. Phys. Fit.* **2020**, *60*, 1444–1452. [CrossRef] [PubMed]
10.   Arede, J.; Vaz, R.; Francheschi, A.; Gonzalo-Skok, O.; Leite, N. Effects of a combined strength and conditioning training program on physical abilities in adolescent male basketball players. *J. Sports Med. Phys. Fit.* **2019**, *59*, 1298–1305. [CrossRef]
11.   Beato, M.; Coratella, G.; Bianchi, M.; Costa, E.; Merlini, M. Short-Term Repeated-Sprint Training (Straight Sprint vs. Changes of Direction) in Soccer Players. *J. Hum. Kinet.* **2019**, *70*, 183–190. [CrossRef] [PubMed]
12.   Zamparo, P.; Zadro, I.; Lazzer, S.; Beato, M.; Sepulcri, L. Energetics of shuttle runs: The effects of distance and change of direction. *Int. J. Sports Physiol. Perform.* **2014**, *9*, 1033–1039. [CrossRef] [PubMed]
13.   Spencer, M.; Bishop, D.; Dawson, B.; Goodman, C. Physiological and metabolic responses of repeated-sprint activities: Specific to field-based team sports. *Sport Med.* **2005**, *35*, 1025–1044. [CrossRef]
14.   Bishop, D.; Girard, O.; Mendez-Villanueva, A. Repeated-sprint ability part II: Recommendations for training. *Sport Med.* **2011**, *41*, 741–756. [CrossRef]
15.   Schöllhorn, W. Application of system dynamic principles to technique and strength training. *Acta Acad. Olymp. Est.* **2000**, *8*, 67–85.
16.   Bauer, H.; Schöllhorn, W. Self-Organizing Maps for the Analysis of Complex Movement Patterns. *Neural Process. Lett.* **1997**, *5*, 193–199. [CrossRef]
17.   Schöllhorn, W.I.; Nigg, B.M.; Stefanyshyn, D.J.; Liu, W. Identification of individual walking patterns using time discrete and time continuous data sets. *Gait Posture* **2002**, *15*, 180–186. [CrossRef]
18.   Schöllhorn, W. Practical Consequences of Biomechanically Determined Individuality and Fluctuations on Motor Learning. In Proceedings of the International Society of Biomechanics, Calgary, AB, Canada, 8–13 August 1999.
19.   Haykin, S. *Neural Network: A Comprehensive Foundation, 1st ed*; McMaster University: Hamilton, ONT, Canada, 1994.
20.   Catalano, J.F.; Kleiner, B.M. Distant Transfer in Coincident Timing as a Function of Variability of Practice. *Percept. Mot. Ski.* **1984**, *58*, 851–856. [CrossRef]
21.   Janssen, D.; Schöllhorn, W.I.; Lubienetzki, J.; Fölling, K.; Kokenge, H.; Davids, K. Recognition of Emotions in Gait Patterns by Means of Artificial Neural Nets. *J. Nonverbal Behav.* **2008**, *32*, 79–92. [CrossRef]
22.   Janssen, D.; Schöllhorn, W.I.; Newell, K.M.; Jäger, J.M.; Rost, F.; Vehof, K. Diagnosing fatigue in gait patterns by support vector machines and self-organizing maps. *Hum. Mov. Sci.* **2011**, *30*, 966–975. [CrossRef]
23.   Burdack, J.; Horst, F.; Aragonés, D.; Eekhoff, A.; Schöllhorn, W.I. Fatigue-Related and Timescale-Dependent Changes in Individual Movement Patterns Identified Using Support Vector Machine. *Front. Psychol.* **2020**, *11*, 2273. [CrossRef]
24.   Schöllhorn, W.; Michelbrink, M.; Welminsiki, D.; Davids, K. Increasing Stochastic Perturbations Enhances Acquisition and Learning of Complex Sport Movements. In *Perspectives on Cognition and Action in Sport*; Nova Science Publishers: Hauppauge, NY, USA, 2009; pp. 59–73.
25.   Buekers, M.J.A.; Magill, R.A.; Hall, K.G. The Effect of Erroneous Knowledge of Results on Skill Acquisition when Augmented Information is Redundant. *Q. J. Exp. Psychol. Sect. A* **1992**, *44*, 105–117. [CrossRef]
26.   Henz, D.; John, A.; Merz, C.; Schöllhorn, W.I. Post-task Effects on EEG Brain Activity Differ for Various Differential Learning and Contextual Interference Protocols. *Front. Hum. Neurosci.* **2018**, *12*, 19. [CrossRef] [PubMed]
27.   Schöllhorn, W.; Horst, F. Effects of complex movements on the brain as a result of increased decision-making. *J. Complex. Health Sci.* **2020**, *2*, 40–45.
28.   Schöllhorn, W.I.; Beckmann, H.; Michelbrink, M.; Sechelmann, M.; Trockel, M.; Davids, K. Does noise provide a basis for the unification of motor learning theories? *Int. J. Sport Psychol.* **2006**, *37*, 186–206.
29.   Schöllhorn, W.; Röber, F.; Jaitner, T.; Hellstem, W.; Käubler, W. Discrete and continuous effects of traditional and differencial sprint training. In Proceedings of the 6th Annual Congress of the European College of Sport Science, Cologne, Germany, 24–28 July 2001.
30.   Feest, U. Why Replication Is Overrated. *Philos. Sci.* **2019**, *86*, 895–905. [CrossRef]
31.   Davids, K.; Glazier, P.; Araújo, D.; Bartlett, R. Movement systems as dynamical systems: The functional role of variability and its implications for sports medicine. *Sport Med.* **2003**, *33*, 245–260. [CrossRef]
32.   Gaspar, A.; Santos, S.; Coutinho, D.; Gonçalves, B.; Sampaio, J.; Leite, N. Acute effects of differential learning on football kicking performance and in countermovement jump. *PLoS ONE* **2019**, *14*, e0224280. [CrossRef]

33. Fuchs, P.X.; Fusco, A.; Bell, J.W.; von Duvillard, S.P.; Cortis, C.; Wagner, H. Effect of differential training on female volleyball spike-jump technique and performance. *Int. J. Sports Physiol. Perform.* **2020**, *15*, 1019–1025. [CrossRef]
34. Fuchs, P.X.; Fusco, A.; Cortis, C.; Wagner, H. Effects of Differential Jump Training on Balance Performance in Female Volleyball Players. *Appl. Sci.* **2020**, *10*, 5921. [CrossRef]
35. Coutinho, D.; Santos, S.; Gonçalves, B.; Travassos, B.; Wong, P.; Schöllhorn, W.; Sampaio, J. The effects of an enrichment training program for youth football attackers. *PLoS ONE* **2018**, *13*, e0199008. [CrossRef]
36. National Strength and Conditioning Association. *Essentials of Strength Training and Conditioning*, 4th ed.; Haff, G., Triplett, T., Eds.; Human Kinetics: Champaign, IL, USA, 2016.
37. Schöllhorn, W. The Nonlinear Nature of Learning-A Differential Learning Approach. *Open Sports Sci. J.* **2012**, *5*, 100–112. [CrossRef]
38. Kunugi, S.; Masunari, A.; Koumura, T.; Fujimoto, A. Altered lower limb kinematics and muscle activities in soccer players with chronic ankle instability. *Phys. Ther. Sport* **2018**, *34*, 28–35. [CrossRef] [PubMed]
39. Nimphius, S.; Callaghan, S.J.; Spiteri, T.; Lockie, R.G. Change of Direction Deficit: A More Isolated Measure of Change of Direction Performance Than Total 505 Time. *J. Strength Cond. Res.* **2016**, *30*, 3024–3032. [CrossRef] [PubMed]
40. Bishop, C.; Read, P.; Lake, J.; Chavda, S.; Turner, A. Inter-limb asymmetries: Understanding how to calculate differences from bilateral and unilateral tests. *Strength Cond. J.* **2018**, *40*, 1–6. [CrossRef]
41. Koo, T.K.; Li, M.Y. A Guideline of Selecting and Reporting Intraclass Correlation Coefficients for Reliability Research. *J. Chiropr. Med.* **2016**, *15*, 155–163. [CrossRef] [PubMed]
42. Cormack, S.; Newton, R.U.; McGuigan, M.R.; Doyle, T.L.A. Reliability of Measures Obtained During Single and Repeated Countermovement Jumps. *Int. J. Sports Physiol. Perform.* **2008**, *3*, 131–144. [CrossRef]
43. Marôco, J. *Análise Estatística com o SPSS Statistics*, 6a ed.; Pêro Pinheiro: Report Number: Análise e Gestão da Informacão; Edições Sílabo: Lisbon, Portugal, 2014.
44. Hopkins, W.; Marshall, S.W.; Batterham, A.M.; Hanin, J. Progressive Statistics for Studies in Sports Medicine and Exercise Science. *Med. Sci. Sports Exerc.* **2009**, *41*, 3–13. [CrossRef]
45. Ferguson, C.J. An Effect Size Primer: A Guide for Clinicians and Researchers. *Prof. Psychol. Res. Pract.* **2009**, *40*, 532–538. [CrossRef]
46. Lakens, D. Calculating and reporting effect sizes to facilitate cumulative science: A practical primer for t -tests and ANOVAs. *Front Psychol.* **2013**, *4*, 1–12. [CrossRef]
47. Schöllhorn, W.I. Individualität—Ein vernachlässigter Parameter? *Leistungssport* **1999**, *29*, 5–12.
48. Horst, F.; Janssen, D.; Beckmann, H.; Schöllhorn, W.I. Can Individual Movement Characteristics Across Different Throwing Disciplines Be Identified in High-Performance Decathletes? *Front. Psychol.* **2020**, *11*, 2262. [CrossRef]
49. Schöllhorn, W. Biomechanische Einzelfallanalyse im Diskuswurf. *Leistungssport* **1993**, *23*, 55–58.
50. Henz, D.; Schöllhorn, W.I. Differential Training Facilitates Early Consolidation in Motor Learning. *Front. Hum. Neurosci.* **2016**, *10*, 199. [CrossRef]
51. Pickering, C.; Kiely, J. Do Non-Responders to Exercise Exist—and If So, What Should We Do About Them? *Sport Med.* **2019**, *49*, 1–7. [CrossRef] [PubMed]
52. Kulakowski, E.; Lockie, R.G.; Johnson, Q.R.; Lindsay, K.G.; Dawes, J.J. Relationships of Lower-body Power Measures to Sprint and Change of Direction Speed among NCAA Division II Women's Lacrosse Players: An Exploratory Study. *Int. J. Exerc. Sci.* **2020**, *13*, 1667–1676. [PubMed]
53. Horst, F.; Rupprecht, C.; Schmitt, M.; Hegen, P.; Schöllhorn, W.I. Muscular Activity in Conventional and Differential Back Squat Exercise. In *Book of Abstracts 20th Annual Congress of the European College of Sport Science, Malmö, Sweden, 24–27 June 2015*; European College of Sport Science: Malmö, Sweden, 2016.
54. Shea, C.H.; Kohl, R.M. Specificity and variability of practice. *Res. Q. Exerc. Sport* **1990**, *61*, 169–177. [CrossRef]
55. Michelbrink, M.; Schöllhorn, W. Differencial learning and random walk analysis in human posture. *Gait Posture* **2005**, *21*, 148–149. [CrossRef]
56. Barbieri, F.A.; Rodrigues, S.T.; Polastri, P.F.; Barbieri, R.A.; de Paula, P.H.A.; Milioni, F.; Redkva, P.E.; Zagatto, A.M. High intensity repeated sprints impair postural control, but with no effects on free throwing accuracy, in under-19 basketball players. *Hum. Mov. Sci.* **2017**, *54*, 191–196. [CrossRef]
57. Schöllhorn, W.I. Invited commentary: Differential learning is different from contextual interference learning. *Hum. Mov. Sci.* **2016**, *47*, 240–245. [CrossRef]
58. Soria-Grila, M.; Chirosa, I.J.; Bautista, I.J.; Baena, S.; Chirosa, L.J. Effects of variable resistance training on maximal strength: A meta-analysis. *J. Strength Cond. Res.* **2015**, *29*, 3260–3270. [CrossRef]
59. Dos'Santos, T.; Thomas, C.; Jones, P.; Comfort, P. Mechanical determinants of faster change of direction speed performamce in male athletes. *J. Strength Cond. Res.* **2017**, *31*, 696–705. [CrossRef] [PubMed]
60. Spiteri, T.; Newton, R.U.; Binetti, M.; Hart, N.; Sheppard, J.M.; Nimphius, S. Mechanical determinants of faster change of direction and agility performance in female basketball athletes. *J. Strength Cond. Res.* **2015**, *29*, 2205–2214. [CrossRef] [PubMed]
61. Stevinson, C.; Plateau, C.R.; Plunkett, S.; Fitzpatrick, E.J.; Ojo, M.; Moran, M.; Clemes, S.A. Adherence and Health-Related Outcomes of Beginner Running Programs: A 10-Week Observational Study. *Res. Q. Exerc. Sport* **2020**, 1–9. [CrossRef] [PubMed]

62. Santos, S.; Coutinho, D.; Gonçalves, B.; Schöllhorn, W.; Sampaio, J.; Leite, N. Differential Learning as a Key Training Approach to Improve Creative and Tactical Behavior in Soccer. *Res. Q. Exerc. Sport* **2018**, *89*, 11–24. [CrossRef]
63. John, A.; Schöllhorn, W.I. Acute Effects of Instructed and Self-Created Variable Rope Skipping on EEG Brain Activity and Heart Rate Variability. *Front. Behav. Neurosci.* **2018**, *12*, 311. [CrossRef] [PubMed]
64. Henz, D.; Schöllhorn, W.I. Dynamic Office Environments Improve Brain Activity and Attentional Performance Mediated by Increased Motor Activity. *Front. Hum. Neurosci.* **2019**, *13*, 121. [CrossRef]
65. Neyman, J.; Pearson, E.S. On the Problem of the Most Efficient Tests of Statistical Hypotheses. In *Breakthroughs in Statistics*; Kotz, S., Johnson, N.L., Eds.; Springer: New York, NY, USA, 1992; pp. 73–108.
66. Nuzzo, R. Scientific method: Statistical errors. *Nature* **2014**, *506*, 150–152. [CrossRef]
67. Mosston, M.; Ashworth, S.; Block, B. *Teaching Physical Education*, 5th ed.; Benjamin Cummings: Boston, MA, USA, 2008.

International Journal of
Environmental Research and Public Health

MDPI

Article

Local and Remote Ischemic Preconditioning Improves Sprint Interval Exercise Performance in Team Sport Athletes

Ching-Feng Cheng [1,2,*,†], Yu-Hsuan Kuo [3,†], Wei-Chieh Hsu [2,4], Chu Chen [2,5] and Chi-Hsueh Pan [2,5]

1 Department of Athletic Performance, National Taiwan Normal University, Taipei 11677, Taiwan
2 Sports Performance Lab, National Taiwan Normal University, Taipei 11677, Taiwan; hsu.w.c1982@gmail.com (W.-C.H.); alvinc5939@gmail.com (C.C.); jasonpan8011@gmail.com (C.-H.P.)
3 Department of Physical Education, Chinese Culture University, Taipei 11114, Taiwan; gyx2@ulive.pccu.edu.tw
4 Graduate Institute of Sports Training, University of Taipei, Taipei 11153, Taiwan
5 Department of Physical Education and Sport Sciences, National Taiwan Normal University, Taipei 10610, Taiwan
* Correspondence: andescheng@ntnu.edu.tw; Tel.: +886-277496831
† C.-F.C. and Y.-H.K. contributed equally to this work.

Citation: Cheng, C.-F.; Kuo, Y.-H.; Hsu, W.-C.; Chen, C.; Pan, C.-H. Local and Remote Ischemic Preconditioning Improves Sprint Interval Exercise Performance in Team Sport Athletes. *Int. J. Environ. Res. Public Health* **2021**, *18*, 10653. https://doi.org/10.3390/ijerph182010653

Academic Editors: Bruno Gonçalves, Jorge Bravo and Hugo Folgado

Received: 20 September 2021
Accepted: 8 October 2021
Published: 12 October 2021

Publisher's Note: MDPI stays neutral with regard to jurisdictional claims in published maps and institutional affiliations.

Abstract: The aim of this study was to investigate the effects of local (LIPC) and remote (RIPC) ischemic preconditioning on sprint interval exercise (SIE) performance. Fifteen male collegiate basketball players underwent a LIPC, RIPC, sham (SHAM), or control (CON) trial before conducting six sets of a 30-s Wingate-based SIE test. The oxygen uptake and heart rate were continuously measured during SIE test. The total work in the LIPC (+2.2%) and RIPC (+2.5%) conditions was significantly higher than that in the CON condition ($p < 0.05$). The mean power output (MPO) at the third and fourth sprint in the LIPC (+4.5%) and RIPC (+4.9%) conditions was significantly higher than that in the CON condition ($p < 0.05$). The percentage decrement score for MPO in the LIPC and RIPC condition was significantly lower than that in the CON condition ($p < 0.05$). No significant interaction effects were found in pH and blood lactate concentrations. There were no significant differences in the accumulated exercise time at $\geq$80%, 90%, and 100% of maximal oxygen uptake during SIE. Overall, both LIPC and RIPC could improve metabolic efficiency and performance during SIE in athletes.

Keywords: anaerobic capacity; blood flow occlusion; fatigue resistance; high-intensity interval training

1. Introduction

In 2010, de Groot et al. [1] extended the concept of ischemic preconditioning (IPC), which involves brief cycles of ischemia and reperfusion, from the medical field to the sports science field, and discovered that IPC could help increase maximal oxygen uptake ($\dot{V}O_2$max). Since then, many studies have explored the benefits and possible mechanisms of IPC on sports performance. IPC can be divided into local (LIPC) and remote (RIPC) ischemic preconditioning according to its location of function. The difference between the two approaches is whether blood flow is occluded in the same muscle group that is subsequently used during exercise. For example, for cycling exercise, LIPC refers to application of blood flow occlusion to the legs, whereas RIPC refers to application of blood flow occlusion to the arms. At present, it is thought that both LIPC and RIPC may improve functioning of the mitochondrial ATP-dependent potassium channel, attenuate ATP depletion, promote phosphocreatine (PCr) resynthesis, enhance metabolic efficiency, and increase vasodilation, oxygen delivery and extraction, thereby enhancing subsequent endurance performance [2,3]. However, the acute effect of IPC on the performance of repeated or interval sprints, in which aerobic contribution is predominant [4], remains controversial.

Some studies have suggested that LIPC can improve subsequent repeated sprint performance, including the 12 × 6-s cycling sprint [5], 3 × (6 × [15 + 15-m]) shuttle sprints [6], and 6 × 50-m swimming sprint [7]. However, others have reported the opposite, i.e., that for repeated sprints with a short overall exercise time, including 5 × 6-s [8] and 10 × 6-s cycling sprints [9], LIPC does not enhance performance. Alternatively, application of LIPC to single short-duration sprints, such as 10–30-m sprints, revealed no obvious benefits [10,11]. Some studies have indicated that LIPC may be beneficial only in single sprints powered by glycolysis, such as 50-m swimming sprint [12] and 60-s cycling sprints [13]. In contrast, a study by Paixao et al. [14] has reported that LIPC cannot improve the power output of three sets of 30-s Wingate-based sprint with 10-min rest intervals. During a single 30-s Wingate-based sprint, PCr availability can be restored to 65% of the rest value after 1.5 min of post-exercise rest, or 86% after 6 min of post-exercise rest [15]. Therefore, LIPC cannot enhance the performance of repeated 30-s Wingate-based sprints, probably because most PCr availability can be restored through passive recovery during long rest intervals. It would be relevant to explore whether LIPC can improve the performance of the subsequent 4–6 sets of 30-s Wingate-based sprint interval exercise (SIE) with short (4-min) rest intervals, as it is an exercise intervention commonly adopted to improve aerobic capacity and performance in untrained and trained individuals [16].

For the RIPC research, some studies have suggested that RIPC does not enhance aerobic or anaerobic performance in exercises such as the 1-h cycling time trial [17], 7 × 200-m incremental swimming test [18], and 6 × 6-s and 30-s cycling sprint [19]. On the other hand, others have reported that RIPC can help to shorten the time needed for swimming 100 m [18] and can extend the time to task failure during handgrip exercise [20]. More recent studies have indicated that RIPC can increase the number of repetitions and total training volume during resistance exercise [21,22]. Kraus et al. [23] pointed out that RIPC of bilateral rather than unilateral upper limbs can increase the power output during four sets of 30-s Wingate-based SIE. They also suggested that inconsistent results may be related to the number of muscle groups involved during the application of IPC. Therefore, whether the acute effect of LIPC on the performance of SIE is superior to that of RIPC requires further clarification.

For team-sport athletes, sprint and high-intensity interval training programs have been shown to be effective approaches to enhance aerobic capacity and performance [24]. Coaches and sports scientists often find ways to further improve training quality and training adaptation. Therefore, the purpose of this study was to investigate the acute effects of LIPC and RIPC on SIE performance in athletes. We hypothesized that both LIPC and RIPC could improve SIE performance, in terms of both power output and fatigue resistance, and that LIPC would be more effective than RIPC in this respect. We also hypothesized that LIPC and RIPC could increase metabolic efficiency during SIE.

2. Materials and Methods

2.1. Participants

Fifteen male Division I collegiate basketball players (age 21 ± 2 years; height 1.87 ± 0.08 m; body mass 86 ± 13 kg; VO_2max, 51.9 ± 7.8 mL·kg^{-1}·min^{-1}; blood pressure 126/68 mmHg) were recruited to complete this repeated measured crossover study. The priori power analysis (G*Power 3.1.9.4) was used to calculate the appropriate sample size. Based on a previous meta-analysis study [3] on the overall impact of IPC on exercise performance (effect size = 0.43), the minimum sample size of $n = 9$ was required for this study ($\alpha = 0.05$, $\beta = 0.20$). All participants completed a medical history and health questionnaire, and signed informed consent forms before participating in the experiment. The participants refrained from drinking alcoholic or caffeinated beverages for 24 h before the experiments began and fasted for at least 4 h prior to visiting the laboratory, in order to reduce the interference of food in the experiment. The study was conducted according to the guidelines of the Declaration of Helsinki, and approved by the Research Ethics Committee of National Taiwan Normal University, Taipei, Taiwan (Approval code: 201612HM010).

2.2. Experimental Design and Protocols

After a familiarization trial, participants performed an incremental cycling test (GXT) and two control (CON) trials, separated by at least 3 days, on a cycling ergometer (Cyclus 2, RBM Elektronik, Automation, Leipzig, Germany). The cycling ergometer was equipped with an electromagnetically-braked system. During the subsequent visits, participants underwent LIPC, RIPC, or sham (SHAM) treatment in a randomized crossover design, separated by at least 4 days, before conducting six sets of 30-s Wingate-based SIE test (Figure 1). When they arrived at the laboratory, their body mass was measured to determine the load of the SIE test. Before treatment, participants were asked to rest supine for 10 min for baseline measurements, including their heart rate, blood lactate concentration, and pH level. The heart rate and oxygen uptake ($\dot{V}O_2$) were continuously measured during the SIE test. Blood samples for pH and lactate concentrations were drawn before (baseline) and 5-min after treatment, and 5-min after the SIE test. The participants completed all of the trials during the same time-period (± 2 h) of testing days to eliminate any effect of circadian variation.

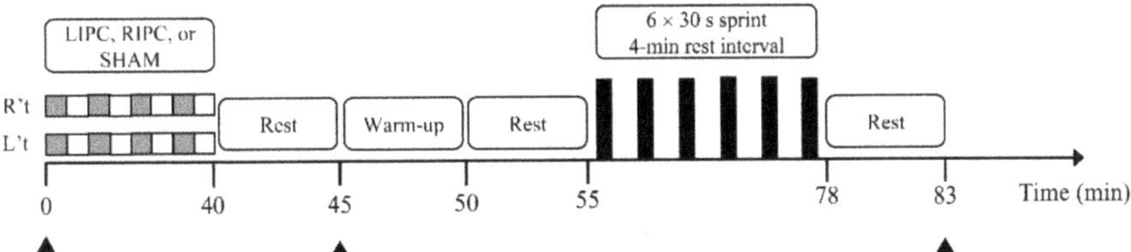

Figure 1. Protocol schematic. R't, right thigh or upper arm; L't, left thigh or upper arm; LIPC, local ischemic preconditioning; RIPC, remote ischemic preconditioning; ■, Occlusion; □, Reperfusion; ▲, Blood lactate and pH value.

2.3. Incremental Cycling Test

During each participant's first visit to the laboratory, the cycling ergometer seat and handlebars were adjusted for comfort. These same settings were restored for each consecutive exercise trial. Before the GXT, all participants were asked to cycle at different pedaling rates (70, 80, and 90 rpm) for a while (~1 min), and then chose the most comfortable rate as their self-selected cadence. Participants first performed 3 min of unloaded baseline pedaling (0 W); the load was increased by 30 W every min thereafter until volitional exhaustion. During the GXT, the participants maintained a self-selected cadence for as long as possible. Strong verbal encouragement was provided throughout the trial. Exhaustion was defined as a pedaling rate of 10 rpm lower than the self-selected cadence, lasting for 10 s or more.

Pulmonary gas exchanges were measured breath-by-breath throughout the GXT by having the participants wear a face mask (7400 Vmask series, Hans Rudolph, Kansas City, MO, USA) attached to a portable gas analysis system (Cortex Metamax 3B; Cortex Biophysik, Leipzig, Germany). Before the test, the system was calibrated according to the manufacturer's guidelines against known concentrations of cylinder gases (15% oxygen, 5% carbon dioxide) and a 3-L calibration syringe (5530 series, Hans Rudolph, Kansas City, MO, USA). Heart rates were monitored using a telemetry system with a wireless chest strap (Polar S810i; Polar Electro, Inc., Oy, Kempele, Finland) and were continuously measured through a link to the Cortex gas analysis system during the exercise test. The greatest $\dot{V}O_2$ value (averaged every 10 s) measured during the GXT was recorded as the $\dot{V}O_2$max value.

2.4. IPC Protocols

The LIPC and SHAM treatments were performed in supine position using 14-cm-wide blood pressure cuffs placed on the most proximal portions of the upper thighs. Bilateral occlusion was performed simultaneously on the right and left thighs. The cuff was rapidly inflated to 220 mmHg for the LIPC and 20 mmHg for the SHAM conditions, respectively, by means of a cuff inflator (CK-113P, Spirit Corp, Taipei, Taiwan). Absolute cuff pressure was used in SHAM to allow direct comparison with relevant studies [5,12–14]. This occlusion procedure was repeated 4 times, each separated by 5 min of reperfusion, as this protocol has been successfully applied in previous studies investigating the ergogenic effects of LIPC on exercise performance [5,6,13]. In RIPC treatment, the same blood pressure cuffs were placed bilaterally on the proximal parts of upper arms. Both cuffs were simultaneously inflated to 30 mmHg above systolic arterial pressure for 5 min, followed by 5 min deflation. The average cuff pressure in RIPC was 156 ± 7 mmHg. The cuffs were inflated and deflated 4 times in total, according with a previous study about the effect of RIPC on four sets of 30-s Wingate-based SIE performance [23]. In CON treatment, participants performed a 5-min standardized warm-up followed by a 5-min passive rest without any blood occlusion before SIE test.

2.5. Wingate-Based 6 × 30-s Sprint Interval Exercise

Prior to each trial, all participants performed a standardized warm-up comprising 5 min of submaximal cycling at 50 W (60–75 rpm), in which three bouts of maximal accelerations (approximately 5 s) at the end of the second, third and fourth minutes were performed. After the warm-up, participants rested for 5 min on the cycle ergometer. The SIE protocol involved participants to complete six sets (S1–S6) of 30-s Wingate-based SIE with a 4-min rest interval against the given load ([0.7 × body mass]/0.173) as fast as possible [25]. Thirty seconds before starting each sprint, participants were informed to ride at a moderate pedal cadence (50–60 rpm), from a seated position. Ten seconds before the initiation of each sprint, participants were instructed to increase pedaling rate to 80–85 rpm until given the signal to start pedaling maximally. During the last 2 s of the recovery period, participants were required to increase the pedaling rate to over 100 rpm (0 N) for the convenience of the subsequent 30-s cycling sprint test. At the end of the sprint, the participants remained seated on the bike for recovering between sets. Participants were given strong verbal encouragement by the same researchers throughout the entire trial.

The total work, peak (PPO) and mean (MPO) power outputs, and percentage decrement score (100 − [{total sprint power outputs/ideal sprint power outputs} × 100]) were calculated at each cycling sprint. The total sprint power was calculated as the sum of the peak/mean power outputs from all sprints. The ideal sprint power was defined as the number of sprints × highest peak/mean power output. The accumulated exercise time at ≥80%, 90%, and 100% $\dot{V}O_2$max were also calculated during SIE.

2.6. Blood Sampling and Analysis

Capillary blood samples were taken by ear lobe puncture to evaluate lactate concentrations. The first blood samples were discarded, and the second blood samples (ca. 0.3 µL) were used to analyze the lactate concentrations using a lactate chemistry analyzer (Lactate Pro2, Arkray, Inc., Kyoto, Japan). To measure blood pH values, whole blood samples (ca. 1 mL) were drawn from an antecubital venous catheter and assessed by a blood gas analyzer (OPTI CCA-TS; OPTI Medical System, Inc., Roswell, GA, USA).

2.7. Statistical Analysis

The Shapiro–Wilk normality test was performed to determine the homogeneity of the sample. The performance data (total work, PPO, MPO, percentage decrement score, and accumulated exercise time), blood lactate and pH values were assessed using repeated-measures ANOVA. The LSD post-hoc test was applied if a significant difference was found. The effect size (Cohen's *d*) was calculated by dividing the difference between the mean values of the conditions by the pooled SD. Cohen's *d* of <0.5, 0.5–0.79, and ≥0.8 were considered as small, moderate, and large effects, respectively. The intraclass correlation coefficient (ICC) was used to assess the test–retest reliability of the SIE test in CON conditions. Statistical significance was set at $p < 0.05$ and all procedures were conducted using SPSS for Windows (Version 17, IBM SPSS Inc., Chicago, IL USA).

3. Results

The PPO (ICC = 0.93–0.98, $p < 0.05$) and MPO (ICC = 0.77–0.91, $p < 0.05$) at each sprint, and total work (ICC = 0.92, $p < 0.05$) demonstrated good to excellent test–retest reliabilities. The total work done during the SIE test differed significantly among treatments ($F = 3.307$, $p < 0.05$). The total work in the LIPC (+2.2%, $d = 0.98$) and RIPC (+2.5%, $d = 1.10$) conditions was significantly higher than that in the CON (Table 1). In Figure 2A, the MPO of the third and fourth sprint in LIPC (S3, +4.9%, $d = 1.28$; S4, +4.0%, $d = 1.10$) and RIPC (S3, +4.7%, $d = 1.26$; S4, +5.1%, $d = 1.18$) were significantly higher than those in the CON condition ($p < 0.05$). Table 1 also indicates that the percentage decrement scores of MPO in the LIPC ($d = 1.08$) and RIPC ($d = 1.13$) conditions were significantly lower than that in the CON condition ($F = 3.534$, $p < 0.05$). No significant interaction effect of PPO was found ($F = 0.998$, $p > 0.05$; Figure 2B); however, the percentage decrement score of PPO in the LIPC condition was significantly lower than that in the CON condition ($d = 1.27$, $p < 0.05$; Table 1).

There were no significant interaction effects in pH ($F = 1.405$, $p > 0.05$) and lactate ($F = 1.035$, $p > 0.05$) concentrations (Table 1). No significant differences in peak $\dot{V}O_2$ were found during SIE among treatment conditions (LIPC vs. RIPC vs. SHAM vs. CON, 55.4 ± 7.7 vs. 55.6 ± 7.7 vs. 54.3 ± 10.2 vs. 55.7 ± 7.2 mL·kg^{-1}·min^{-1}, $F = 0.296$, $p > 0.05$). Moreover, there were no significant differences in the accumulated exercise time at $\geq 80\%$ ($F = 0.679$, $p > 0.05$), 90% ($F = 0.422$, $p > 0.05$), and 100% ($F = 0.741$, $p > 0.05$) $\dot{V}O_2$max during SIE test (Table 1).

Table 1. Effects of local and remote ischemic preconditioning on exercise performance and physiological responses during sprint interval exercise test.

	LIPC ($n = 15$)	RIPC ($n = 15$)	SHAM ($n = 15$)	CON ($n = 15$)
Total work (kJ)	108.3 ± 8.9 *	108.4 ± 6.9 *	107.1 ± 8.6	106.0 ± 8.6
Percentage decrement score (%)				
PPO	5.4 ± 2.3 *	6.2 ± 3.1	6.5 ± 2.9	7.6 ± 3.8
MPO	11.9 ± 4.7 *	11.9 ± 4.6 *	13.1 ± 4.1	15.2 ± 5.3
Accumulated exercise time (s)				
$\geq 80\% \dot{V}O_2$max	95.5 ± 52.7	99.9 ± 53.9	81.2 ± 62.0	95.2 ± 59.5
$\geq 90\% \dot{V}O_2$max	34.5 ± 28.0	36.7 ± 33.9	28.9 ± 32.9	37.5 ± 40.1
$\geq 100\% \dot{V}O_2$max	9.1 ± 11.7	10.5 ± 13.1	7.5 ± 11.5	11.6 ± 17.3
Lactate (mmol/L)				
Baseline	0.96 ± 0.21	1.01 ± 0.23	0.95 ± 0.27	0.95 ± 0.24
5 min after treatment	1.02 ± 0.24	1.04 ± 0.17	0.95 ± 0.15	-
5 min after SIE	11.09 ± 1.73	10.83 ± 2.02	10.62 ± 2.89	11.59 ± 2.21
pH				
Baseline	7.37 ± 0.02	7.39 ± 0.03	7.37 ± 0.02	7.38 ± 0.02
5 min after treatment	7.39 ± 0.02	7.40 ± 0.02	7.39 ± 0.02	-
5 min after SIE	7.20 ± 0.05	7.20 ± 0.05	7.20 ± 0.05	7.19 ± 0.05

LIPC, local ischemic preconditioning; RIPC, remote ischemic preconditioning; SHAM, sham; CON, control; PPO, peak power output; MPO, mean power output; $\dot{V}O_2$max, maximal oxygen uptake; SIE, sprint interval exercise; * $p < 0.05$, compared with CON.

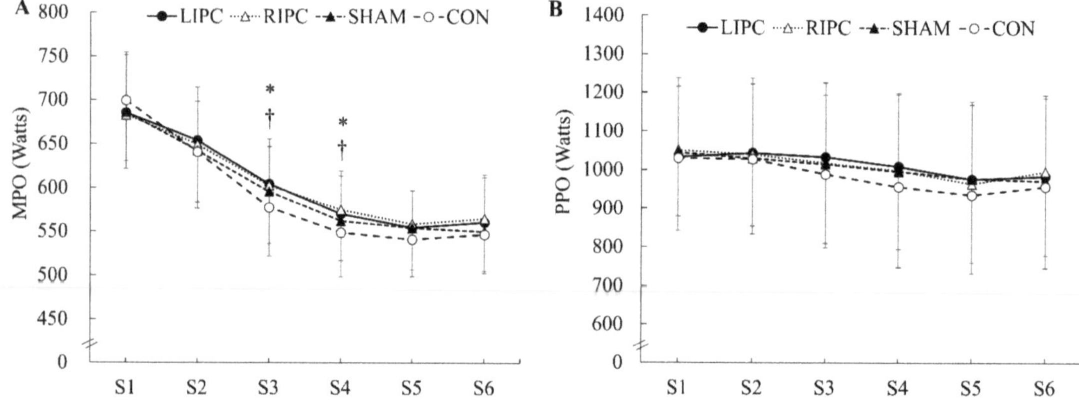

Figure 2. Effects of local (LIPC) and remote (RIPC) ischemic preconditioning on (**A**) mean power outputs (MPO) and (**B**) peak power outputs (PPO) during a sprint interval exercise (SIE) test. * $p < 0.05$, significant differences between LIPC and control (CON); † $p < 0.05$, significant differences between RIPC and CON; S1–S6, sprint 1–6.

4. Discussion

The purpose of this study was to examine the acute effects of IPC on the SIE performance in athletes. We hypothesized that both LIPC and RIPC could improve metabolic efficiency, and increase power output and fatigue resistance during SIE. To the best of our knowledge, no previous study has investigated the acute effects of LIPC and RIPC on the performance of six sets of SIE. The main finding of this study was that, for trained athletes, both LIPC and RIPC were effective in improving the total work of high-intensity interval sprints (ca. 2%) as well as the MPO during exercise (ca. 4%). In addition, compared to CON treatment, LIPC and RIPC treatment could also slow down the decline in power output. The reason for the escalation in SIE performance is likely that IPC increased metabolic efficiency during the high-intensity exercise.

Previous studies have reported that LIPC can improve the performance of six sets of 50-m swimming sprints with 3-min rest intervals [7], while RIPC can improve the PPO and MPO of four sets of Wingate-based sprints with 2-min rest intervals [23]. Our results were consistent with these findings, in that both LIPC and RIPC could improve the total work and the MPO of six sets of Wingate-based sprints with 4-min rest intervals. Previous studies have found that IPC can promote vasodilation as well as oxygen delivery and extraction by increasing the amount of vascular nitric oxide that can simulate vascular endothelial cells [5,26,27]. In addition, animal [28] and human trials [29] have discovered that IPC can facilitate the rate of PCr resynthesis. Reduced PCr availability has been considered as a limiting factor for power output recovery during multiple sprint exercises [30]. Although Paixao et al. [14] have reported that LIPC cannot improve the power output of three sets of Wingate-based sprint (with a rest interval of 10 min), it is more likely due to a long rest interval than to the effect of IPC on PCr resynthesis, as 6 min of passive rest after high-intensity exercise can restore PCr availability to 90% of the rest value [15]. Therefore, both LIPC and RIPC can improve the power output during SIE by promoting oxygen delivery and extraction as well as by facilitating PCr resynthesis during rest intervals.

Previous studies have suggested that LIPC cannot boost the percentage decrement score of 5–12 × 6-s cycling sprints [5,8,9]. In contrast, our study found that, compared to the CON condition, both LIPC and RIPC can increase the percentage decrement score. This inconsistent result may be attributed to different exercise patterns. First, among studies that have introduced multiple sets of repeated sprint exercises, Griffin et al. [6] reported that LIPC can escalate the percentage decrement score of three sets of exercises consisting of 6 × (15 + 15-m) shuttle sprints. In addition, a meta-analysis study by Salvador et al. [3] has indicated that, while IPC may have a >99% and a 57.9% chance of benefiting aerobic (>90 s) and anaerobic (10–90 s) exercise performance, respectively, its effect on the performance of sprints with a duration of less than 10 s is negligible. Similarly, other studies have pointed out that LIPC cannot improve the performance of 10–30 m sprints [10,11], while RIPC cannot enhance the power output and the fatigue index of a single 30-s Wingate sprint [19]. However, LIPC has been shown to improve the power output of both 60 s [13] and 3 min all-out [31] cycling sprint. Therefore, it appears that IPC can only facilitate repeated sprint exercises, with an extended single exercise time or total exercise time. Second, some studies have found that, for SIE, neither LIPC [14] nor RIPC [23] can consistently improve the fatigue index of every single 30-s Wingate sprint, in contradiction to our results. This is possibly due to different calculation methods, as these studies [14,23] only calculated the individual fatigue index of each sprint. As mentioned earlier, LIPC and RIPC can increase the total work as well as the percentage decrement score, thereby improving the fatigue resistance of SIE. Although this is probably associated with the promotion of muscle blood flow and PCr resynthesis during the rest interval, further research is required to explore the physiological and biochemical effects of LIPC and RIPC on the rest interval of SIE.

For further improvement in cardiopulmonary endurance, the exercise protocols must allow athletes to work for as much time as possible at an intensity close to or very close to $\dot{V}O_2$max to expand the stimulating effect on the cardiopulmonary system [32]. The present study revealed that, compared to the CON condition, LIPC and RIPC did not change the accumulated time spent at high-intensity during SIE, and did not alter the blood lactate concentration or the H^+ concentration post-SIE. This is consistent with the results of previous studies, which reported that the blood lactate concentration remains the same after high-intensity exercise if LIPC or RIPC is adopted [5–8,14,23,31]. Thus, LIPC and RIPC can improve the total work and the fatigue resistance of SIE without contributing to the acidic environment in the body, which indicates that LIPC and RIPC can improve the metabolic efficiency of SIE. Similarly, Incognito et al. [2] reported that, by increasing metabolic efficiency, IPC can improve the performance of subsequent exercises. Since LIPC and RIPC can escalate metabolic efficiency and total work, they can strengthen the training quality of a sprint interval training (SIT) program. This finding has partly supported the result of a study conducted by Paradis-Deschênes et al. [33], who found that the inclusion of LIPC in a 4-week SIT program could improve the aerobic and anaerobic performance of endurance athletes. However, at present, there has been no research on the chronic effect of RIPC interventions. Based on the results of this study, the inclusion of RIPC in the long-term SIT program in future studies is expected to provide additional adaptation benefits.

One limitation of this study was that participants were not completely blinded during the study. The presence of the placebo effect in IPC experiments remains controversial, with some studies finding [34,35] and others not findings evidence for this [7,36]. The use of placebo is a standard control component of most clinical trials, and it attempts to clarify the potential effect of treatment. It should be noted that a placebo effect relies on the assumption that participants believe (through conscious or unconscious cues) an intervention will change outcomes [37]. In general, previous IPC studies used low-pressure control conditions (e.g., 20 mmHg) as placebo conditions [5,12–14]. Although we did not inform participants which intervention we thought could improve their exercise performance, the participants could still distinguish the pressure difference between IPC and SHAM. The study of Marocolo et al. [35], who informed participants that both IPC and SHAM (20 mmHg cuff pressure) can increase exercise performance before starting the experiment, found that compared with CON, both IPC and SHAM could significantly increase the number of repetitions during the leg extension exercise. However, the study by Ferreira et al. [7], who told participants that both IPC and SHAM can enhance performance prior to experiment, using 1-min blood flow occlusions (200 mmHg cuff pressure) as SHAM to simulate the potential nocebo effect of IPC. The results indicated that SHAM induced similar levels of discomfort to IPC but comparable exercise performance outcomes to the CON [7]. In the study of Cheung et al. [36], they informed participants that the effects of IPC and SHAM on exercise performance were unclear before starting the study, and used the therapeutic ultrasound procedure as SHAM condition. They reported that 69% of participants believed that IPC would hinder exercise performance, whereas SHAM ultrasound could improve performance [36]. Nevertheless, the results indicated that only IPC could significantly increase time to exhaustion during incremental cycling test [36]. Therefore, the placebo/nocebo effects still seem to be a limitation of studies focusing on IPC and exercise performance.

Another limitation of this study was that the external validity and applicability of the results of this study are limited, as they can be influenced by the particular exercise mode as well as the training status, sex, and age of the participant. For example, participants in this study received 6 weeks of structured training using SIT 6 months before the experiment. Therefore, it needs to be verified whether the same promotion effect is observed when LIPC and RIPC are applied to untrained and obese individuals or diabetes patients, who are common research participants in SIT programs.

5. Conclusions

This study found that, for trained athletes, LIPC and RIPC were both ergogenic aids that could improve the performance and fatigue resistance of SIE. In addition, they did not alter the stimulus that SIE demonstrated on the cardiopulmonary system, which indicated that IPC could increase metabolic efficiency during high-intensity interval exercise. Further research is required to specify the possible mechanism by which IPC improves power output and delays fatigue during SIE.

Author Contributions: C.-F.C. and Y.-H.K. conceived and designed the study as well as drafted the manuscript. C.-F.C., Y.-H.K. and W.-C.H. performed the data analysis and interpretation. Y.-H.K., W.-C.H. and C.C. conducted the experiments. C.C. and C.-H.P. contributed methodological advice and data interpretation. C.-F.C. is the main corresponding author. C.-F.C. and Y.-H.K. contributed equally to this work. All authors have read and agreed to the published version of the manuscript.

Funding: This study was supported by a grant from the Ministry of Science and Technology, Taiwan (MOST 106-2628-H-003-008-MY2).

Institutional Review Board Statement: The study was conducted according to the guidelines of the Declaration of Helsinki, and approved by the Research Ethics Committee of National Taiwan Normal University, Taipei, Taiwan (Approval code: 201612HM010).

Informed Consent Statement: Informed consent was obtained from all subjects involved in the study.

Data Availability Statement: Data sharing not applicable.

Acknowledgments: The authors would like to thank the participants who contributed their time and effort to undertake this study, as well as Yu-Hsiang Chiang and Yu-Sheng Lin for their assistance in data collection.

Conflicts of Interest: The authors declare that they have no conflict of interest.

References

1. De Groot, P.C.; Thijssen, D.H.; Sanchez, M.; Ellenkamp, R.; Hopman, M.T. Ischemic preconditioning improves maximal performance in humans. *Eur. J. Appl. Physiol.* **2010**, *108*, 141–146. [CrossRef] [PubMed]
2. Incognito, A.V.; Burr, J.F.; Millar, P.J. The effects of ischemic preconditioning on human exercise performance. *Sports Med.* **2016**, *46*, 531–544. [CrossRef] [PubMed]
3. Salvador, A.F.; De Aguiar, R.A.; Lisboa, F.D.; Pereira, K.L.; Cruz, R.S.; Caputo, F. Ischemic preconditioning and exercise performance: A systematic review and meta-analysis. *Int. J. Sports Physiol. Perform.* **2016**, *11*, 4–14. [CrossRef] [PubMed]
4. Bishop, D.; Spencer, M. Determinants of repeated-sprint ability in well-trained team-sport athletes and endurance-trained athletes. *J. Sports Med. Phys. Fitness* **2004**, *44*, 1–7. [PubMed]
5. Patterson, S.D.; Bezodis, N.E.; Glaister, M.; Pattison, J.R. The effect of ischemic preconditioning on repeated sprint cycling performance. *Med. Sci. Sports Exerc.* **2015**, *47*, 1652–1658. [CrossRef] [PubMed]
6. Griffin, P.J.; Hughes, L.; Gissane, C.; Patterson, S.D. Effects of local versus remote ischemic preconditioning on repeated sprint running performance. *J. Sports Med. Phys. Fitness* **2019**, *59*, 187–194. [CrossRef] [PubMed]

7.   Ferreira, T.N.; Sabino-Carvalho, J.L.; Lopes, T.R.; Ribeiro, I.C.; Succi, J.E.; AC, D.A.S.; Silva, B.M. Ischemic preconditioning and repeated sprint swimming: A placebo and nocebo study. *Med. Sci. Sports Exerc.* **2016**, *48*, 1967–1975. [CrossRef] [PubMed]
8.   Gibson, N.; Mahony, B.; Tracey, C.; Fawkner, S.; Murray, A. Effect of ischemic preconditioning on repeated sprint ability in team sport athletes. *J. Sports Sci.* **2015**, *33*, 1182–1188. [CrossRef] [PubMed]
9.   Cocking, S.; Ihsan, M.; Jones, H.; Hansen, C.; Timothy Cable, N.; Thijssen, D.; Wilson, M.G. Repeated sprint cycling performance is not enhanced by ischaemic preconditioning or muscle heating strategies. *Eur. J. Sport Sci.* **2021**, *21*, 166–175. [CrossRef]
10.   Gibson, N.; White, J.; Neish, M.; Murray, A. Effect of ischemic preconditioning on land-based sprinting in team-sport athletes. *Int. J. Sports Physiol. Perform.* **2013**, *8*, 671–676. [CrossRef]
11.   Thompson, K.M.A.; Whinton, A.K.; Ferth, S.; Spriet, L.L.; Burr, J.F. Ischemic preconditioning: No influence on maximal sprint acceleration performance. *Int. J. Sports Physiol. Perform.* **2018**, *13*, 986–990. [CrossRef]
12.   Lisbôa, F.D.; Turnes, T.; Cruz, R.S.; Raimundo, J.A.; Pereira, G.S.; Caputo, F. The time dependence of the effect of ischemic preconditioning on successive sprint swimming performance. *J. Sci. Med. Sport* **2017**, *20*, 507–511. [CrossRef]
13.   Cruz, R.S.; de Aguiar, R.A.; Turnes, T.; Salvador, A.F.; Caputo, F. Effects of ischemic preconditioning on short-duration cycling performance. *Appl. Physiol. Nutr. Metab.* **2016**, *41*, 825–831. [CrossRef]
14.   Paixao, R.C.; da Mota, G.R.; Marocolo, M. Acute effect of ischemic preconditioning is detrimental to anaerobic performance in cyclists. *Int. J. Sports Med.* **2014**, *35*, 912–915. [CrossRef]
15.   Bogdanis, G.C.; Nevill, M.E.; Boobis, L.H.; Lakomy, H.K.; Nevill, A.M. Recovery of power output and muscle metabolites following 30 s of maximal sprint cycling in man. *J. Physiol.* **1995**, *482*, 467–480. [CrossRef]
16.   Gibala, M.J.; McGee, S.L. Metabolic adaptations to short-term high-intensity interval training: A little pain for a lot of gain? *Exerc. Sport Sci. Rev.* **2008**, *36*, 58–63. [CrossRef]
17.   Cocking, S.; Landman, T.; Benson, M.; Lord, R.; Jones, H.; Gaze, D.; Thijssen, D.; George, K. The impact of remote ischemic preconditioning on cardiac biomarker and functional response to endurance exercise. *Scand. J. Med. Sci. Sports* **2017**, *27*, 1061–1069. [CrossRef]
18.   Jean-St-Michel, E.; Manlhiot, C.; Li, J.; Tropak, M.; Michelsen, M.M.; Schmidt, M.R.; McCrindle, B.W.; Wells, G.D.; Redington, A.N. Remote preconditioning improves maximal performance in highly trained athletes. *Med. Sci. Sports Exerc.* **2011**, *43*, 1280–1286. [CrossRef]
19.   Lalonde, F.; Currier, D.Y. Can anaerobic performance be improved by remote ischemic preconditioning? *J. Strength Cond. Res.* **2015**, *29*, 80–85. [CrossRef]
20.   Barbosa, T.C.; Machado, A.C.; Braz, I.D.; Fernandes, I.A.; Vianna, L.C.; Nobrega, A.C.; Silva, B.M. Remote ischemic preconditioning delays fatigue development during handgrip exercise. *Scand. J. Med. Sci. Sports* **2015**, *25*, 356–364. [CrossRef]
21.   Da Silva Novaes, J.; da Silva Telles, L.G.; Monteiro, E.R.; da Silva Araujo, G.; Vingren, J.L.; Silva Panza, P.; Reis, V.M.; Laterza, M.C.; Vianna, J.M. Ischemic preconditioning improves resistance training session performance. *J. Strength Cond. Res.* **2020**. online ahead of print. [CrossRef]
22.   Da Silva Telles, L.G.; Carelli, L.C.; Bráz, I.D.; Junqueira, C.; Monteiro, E.R.; Reis, V.M.; Vianna, J.M.; da Silva Novaes, J. Effects of ischemic preconditioning as a warm-up on leg press and bench press performance. *J. Hum. Kinet.* **2020**, *75*, 267–277. [CrossRef]
23.   Kraus, A.S.; Pasha, E.P.; Machin, D.R.; Alkatan, M.; Kloner, R.A.; Tanaka, H. Bilateral upper limb remote ischemic preconditioning improves anaerobic power. *Open Sports Med. J.* **2015**, *9*, 1–6. [CrossRef]
24.   Buchheit, M.; Laursen, P.B. High-intensity interval training, solutions to the programming puzzle: Part I: Cardiopulmonary emphasis. *Sports Med.* **2013**, *43*, 313–338. [CrossRef]
25.   Cheng, C.F.; Hsu, W.C.; Kuo, Y.H.; Chen, T.W.; Kuo, Y.C. Acute effect of inspiratory resistive loading on sprint interval exercise performance in team-sport athletes. *Respir. Physiol. Neurobiol.* **2020**, *282*, 103531. [CrossRef]
26.   Costa, F.; Christensen, N.J.; Farley, G.; Biaggioni, I. NO modulates norepinephrine release in human skeletal muscle: Implications for neural preconditioning. *Am. J. Physiol. Regul. Integr. Comp. Physiol.* **2001**, *280*, R1494–R1498. [CrossRef]
27.   Rassaf, T.; Totzeck, M.; Hendgen-Cotta, U.B.; Shiva, S.; Heusch, G.; Kelm, M. Circulating nitrite contributes to cardioprotection by remote ischemic preconditioning. *Circ. Res.* **2014**, *114*, 1601–1610. [CrossRef]
28.   Kida, M.; Fujiwara, H.; Ishida, M.; Kawai, C.; Ohura, M.; Miura, I.; Yabuuchi, Y. Ischemic preconditioning preserves creatine phosphate and intracellular pH. *Circulation* **1991**, *84*, 2495–2503. [CrossRef] [PubMed]
29.   Andreas, M.; Schmid, A.I.; Keilani, M.; Doberer, D.; Bartko, J.; Crevenna, R.; Moser, E.; Wolzt, M. Effect of ischemic preconditioning in skeletal muscle measured by functional magnetic resonance imaging and spectroscopy: A randomized crossover trial. *J. Cardiovasc. Magn. Reson.* **2011**, *13*, 32. [CrossRef]
30.   Bogdanis, G.C.; Nevill, M.E.; Boobis, L.H.; Lakomy, H.K. Contribution of phosphocreatine and aerobic metabolism to energy supply during repeated sprint exercise. *J. Appl. Physiol.* **1996**, *80*, 876–884. [CrossRef] [PubMed]
31.   Griffin, P.J.; Ferguson, R.A.; Gissane, C.; Bailey, S.J.; Patterson, S.D. Ischemic preconditioning enhances critical power during a 3 min all-out cycling test. *J. Sports Sci.* **2018**, *36*, 1038–1043. [CrossRef] [PubMed]
32.   Midgley, A.W.; McNaughton, L.R.; Wilkinson, M. Is there an optimal training intensity for enhancing the maximal oxygen uptake of distance runners?: Empirical research findings, current opinions, physiological rationale and practical recommendations. *Sports Med.* **2006**, *36*, 117–132. [CrossRef] [PubMed]

33. Paradis-Deschênes, P.; Joanisse, D.R.; Mauriège, P.; Billaut, F. Ischemic preconditioning enhances aerobic adaptations to sprint-interval training in athletes without altering systemic hypoxic signaling and immune function. *Front. Sports Act. Living* **2020**, *2*, 41. [CrossRef] [PubMed]
34. Marocolo, M.; da Mota, G.R.; Pelegrini, V.; Appell Coriolano, H.J. Are the beneficial effects of ischemic preconditioning on performance partly a placebo effect? *Int. J. Sports Med.* **2015**, *36*, 822–825. [CrossRef]
35. Marocolo, M.; Willardson, J.M.; Marocolo, I.C.; da Mota, G.R.; Simão, R.; Maior, A.S. Ischemic preconditioning and placebo intervention improves resistance exercise performance. *J. Strength Cond. Res.* **2016**, *30*, 1462–1469. [CrossRef]
36. Cheung, C.P.; Slysz, J.T.; Burr, J.F. Ischemic preconditioning: Improved cycling performance despite nocebo expectation. *Int. J. Sports Physiol. Perform.* **2019**, *15*, 354–360. [CrossRef]
37. Jensen, K.B.; Kaptchuk, T.J.; Kirsch, I.; Raicek, J.; Lindstrom, K.M.; Berna, C.; Gollub, R.L.; Ingvar, M.; Kong, J. Nonconscious activation of placebo and nocebo pain responses. *Proc. Natl. Acad. Sci. USA* **2012**, *109*, 15959–15964. [CrossRef]

International Journal of
**Environmental Research
and Public Health**

MDPI

Article

Repetition without Repetition or Differential Learning of Multiple Techniques in Volleyball?

Julius B. Apidogo [1,2], Johannes Burdack [1,*] and Wolfgang I. Schöllhorn [1]

[1] Department of Training and Movement Science, Institute of Sport Science, Johannes Gutenberg-University Mainz, 55099 Mainz, Germany; japidogo@uni-mainz.de (J.B.A.); wolfgang.schoellhorn@uni-mainz.de (W.I.S.)

[2] Akanten Appiah-Menka University of Skills Training and Entrepreneurial Development, Kumasi AK-039, Ghana

* Correspondence: burdack@uni-mainz.de

Abstract: A variety of approaches have been proposed for teaching several volleyball techniques to beginners, ranging from general ball familiarization to model-oriented repetition to highly variable learning. This study compared the effects of acquiring three volleyball techniques in parallel with three approaches. Female secondary school students (N = 42; 15.6 ± 0.54 years) participated in a pretest for three different volleyball techniques (underhand pass, overhand pass, and overhead serve) with an emphasis on accuracy. Based on their results, they were parallelized into three practice protocols, a repetitive learning group (RG), a differential learning group (DG), and a control group (CG). After a period of six weeks with 12 intervention sessions, all participants attended a posttest. An additional retention test after two weeks revealed a statistically significant difference between DG, RG, and CG for all single techniques as well as the combined multiple technique. In each technique—the overhand pass, the underhand pass, the overhand service, and the combination of the three techniques—DG performed best (each $p < 0.001$).

Keywords: motor learning; differential learning; volleyball; overhand service; overhand pass; underhand pass; multiple techniques; skill acquisition

Citation: Apidogo, J.B.; Burdack, J.; Schöllhorn, W.I. Repetition without Repetition or Differential Learning of Multiple Techniques in Volleyball? *Int. J. Environ. Res. Public Health* **2021**, *18*, 10499. https://doi.org/10.3390/ijerph181910499

Academic Editor: Ricardo J. Fernandes

Received: 8 September 2021
Accepted: 3 October 2021
Published: 6 October 2021

Publisher's Note: MDPI stays neutral with regard to jurisdictional claims in published maps and institutional affiliations.

1. Introduction

Coaches and physical education teachers are always faced with the challenge of teaching multiple techniques and fostering athlete performance in the most time efficient manner. To achieve this goal, coaches and athletic trainers are always looking for the most effective and efficient learning approaches. The four most popular and widely used approaches to teaching and improving performance and learning, which include innovative elements that had not been previously considered, are listed in a historical order:

(a) The repetition method approach. This method was first mentioned by Plato (450 B.C.) in the context of learning by contrast and was later investigated in more detail by Gentile [1]. The repetitive method approach is based on the assumption that there is an ideal type of movement that can be perfected by several repetitions of the target movement during the learning process. This method is still considered the method of choice by many physical education teachers and coaches.

(b) The original purpose of the methodical series of exercise approach [2] is to learn more complex target movements through a streamlined (blocked) sequence of preliminary exercises increasingly similar to the target movement. In this process, each preparatory exercise follows the logic of the RM. This method is still chosen the most for learning singular complex movements.

(c) The variability of practice approach [3] is based on Schmidt's schema theory [4], which coarsely indicates that invariant elements such as relative timing or relative forces of an already automatized movement become more stable when trained in combination with variable parameters such as absolute forces or absolute durations. Nevertheless, each exercise is trained repetitively (=blocked), oriented on subgoals as prototype. Although the area of application was limited to movements without the influence of gravitational forces [5], this approach inspired teachers and coaches to make the training of a single technique more variable once it has been learned.

(d) The contextual interference was originally operationalized by Battig [6,7] in the context of verbal learning and later applied to fine motor learning by Shea and Morgan [8]. From its origin, contextual interference approach is a learning approach in which one skill is practiced in the context of other skills. The approach is typically associated with two phenomena, namely impaired acquisition on the one hand, and enhanced learning on the other [9,10]. Three models from cognitive psychology have been proposed to explain these phenomena: the elaboration [11], the reconstruction [12], and the retroactive inhibition hypothesis [13]. All three models assume that a higher cognitive effort is required for the random schedule compared to a blocked schedule, which is typically associated with immediately poorer performance due to working memory overload but leads to better retention.

Originally focused only on the learning of a single (text) motion interspersed with additional motions (=context), the contextual interference approach is now primarily investigated for the parallel learning of multiple movements. The approach is still struggling with its application in sports practice, since, among other things, systematic effects have only been found for movements with a small number of degrees of freedom [10,14] despite isolated evidence of positive effects in movements with more degrees of freedom [15].

(e) The differential learning approach [16,17] assumes that improving the performance or learning of a movement depends largely on an individual's characteristics and experiences, which are assumed to be embodied in individual neuro-(muscular) structures that need to be stimulated individually in varying contexts in order to achieve an effective restructuring for changes in behavior. The reciprocal matching of the exercises provided by the trainer to the nature or extent of the learner's individual variations is described by the principle of stochastic resonance [18–20]. Because the differential learning approach is the most recent approach proposed to increase technical performance and since it is the primary subject of the study, it is discussed in some detail below.

The parallel observation of analogies related to fluctuations in three research areas served as the inspiration for the differential learning approach. First, within the research on the identification of individual movement patterns, constant fluctuations of biomechanical parameters were observed [21–23]; second, fluctuations in the field of dissipative dynamic systems were assigned an essential role especially in phase transitions [24]; third, in the field of research on artificial neural networks, it was known that they perform better when added with noisy information during the training phase [25–27]. It is postulated that the learner's behavior should be the focus of interest rather than the idea of a collective movement ideal. Supposedly destructive deviations from the movement ideal became reinterpreted as constructive fluctuations that should make the learning system unstable and enable self-organized learning [17,19]. Whereas differential learning was initially applied and studied only in sports for learning and improving individual techniques [18,28–30], there are now also confirming studies on its effectiveness in fine motor [31–33] or everyday movements, as well as in the field of tactics [34,35], strength [36,37], and endurance training. Isolated studies on the parallel acquisition of two techniques [38,39] suggest its application in the learning of multiple techniques as well.

The only approach that so far tries to explain the learning of multiple techniques is the contextual interference approach. However, the studies on the simultaneous acquisition of multiple techniques in volleyball using this approach have led to ambiguous results. The acquisition of two volleyball techniques showed partial or no support for benefits of contextual interference in the form of random compared to blocked order [40]. Fialho, Benda, and Ugrinovich [41] could not find significant differences between blocked and interleaved training groups for either the post or the retention test when training two service techniques. Similar results were provided by a study on training two techniques (overhand and underhand service) [42] under the contextual interference approach. Several other studies [43–45] also failed to find significant effects of training condition in acquisition or retention performance when training the three basic volleyball techniques. In contrast, when training the same three techniques, Bortoli et al. [46] reported better transfer for the random and serial practice groups than for the blocked group.

Interestingly, all of these studies were conducted on adolescent participants with an average age between 12.4 years [45] and 16.3 years [41]. None of these studies could fully substantiate the two contextual interference related phenomena; only one study [46] partially verified the advantageous learning effect. In contrast, a study of three volleyball techniques with adult students with an average age of 21.5 years found verification of the full contextual interference effect with impaired acquisition and increased retention [47]. Taken together, all these studies on volleyball suggest that the contextual interference approach should be restricted to adults [48]. Whereas most contextual interference studies on athletic movements investigated the parallel training of similar techniques within a sport and found largely consistent changes for this, the parallel acquisition of a running, a jumping, and a throwing movement showed discipline-specific trajectories during the learning process [46].

Apparently, the contextual interference approach does not provide a model that can explain the different results comprehensively. As suggested above, the differential learning approach may provide a more general and appropriate framework for understanding movement learning, at least in movements with more degrees of freedom. In order to increase external validity by further approximating practical, realistic learning, this study aims to investigate the effect of differential learning training on the parallel acquisition and learning of the three volleyball techniques mainly used and taught by beginners. The expectation, based on previous studies, is that students taught using the differential learning approach will increase their performance on the posttest and retention test more than students using the repetitive training method or the general ball familiarization. To what extent an extension of the applications of the differential learning approach for novices can be recommended and to what extent the performance developments of the three approaches differ are the questions that will be investigated.

2. Material and Methods

2.1. Participants

A total of 42 female volleyball novices (15.6 ± 0.54 years) from several Ghanaian state high schools in Kumasi voluntarily participated in this study. After being informed of the content and purpose of the study, the participants' parents provided written informed consent. All procedures were conducted according to the guidelines of the Declaration of Helsinki and approved by the Institutional Review Board of Akanten Appiah-Menka University of Skills Training and Entrepreneurial Development (AAMUSTED/K/RO/L.1/219, 31 August 2021).

A pretest in all three techniques was conducted with them in blocked sequence, and the individual scores for each technique were summed up to form individual total scores for each of the participants. Based on the individual results, they were parallelized into three groups of 14 participants each: a repetitive learning group (RG), a differential learning group (DG), and a control group (CG). The individual scores were summed up to represent the group score (group means).

2.2. Design

A pre-posttest design with additional retention test (see Section 2.2.2) was chosen for this investigation. The pretest was followed by an intervention period of six weeks with a subsequent posttest and a retention test after another two weeks without intervention. A standardized warm-up was performed before all tests. During the intervention phase, the participants trained twice a week (always on Mondays and Thursdays), where each training session lasted one hour.

2.2.1. Intervention

The RG trained according to the Federation International de volleyball Coaches manual [49]. Each training session was preceded by a five-minute warm-up activity, which consisted of minor games, such as "three-on-one", "chase and catch", and "seven-on-one". After completing the warm-up activities, the participants proceeded directly to practice in their respective groups. After the group training, the session was finished. The RG training was characterized by taking one of the techniques per session repeating it 15 times in the blocked form, from overhand service (S) to overhand pass (O), to underhand pass (U) (SSS ... , OOO ... , UUU ...) following that order repeatedly with corrective feedback per session.

The DG training corresponded to the training sequence of the RG in block; however, their training was characterized first by no repetitions by adding stochastic perturbations to the three techniques to be learned and second by no corrections. Appendix A Table A1 contains a list of all given tasks from which a number was randomly selected to instruct the group. Each participant in both intervention groups had 15 trials per training session for each technique. In total, each participant had 180 relevant ball contacts over the entire period.

The control group (CG) engaged in ball familiarization games that were not directly related to volleyball, such as ball throwing and catching games.

2.2.2. Test Design

The test as presented in Figure 1 comprised of three subtests, each corresponding to one of the techniques to be learned: underhand pass, overhand pass, overhand service. All subtests were carried out according to the AAHPERD volleyball skill test manual [50] on a regular outdoor volleyball court.

Subtest underhand pass (Figure 1A): To test the underhand pass accuracy, the student stood in a 2 m² square on the right-hand side of the volleyball court (zone Z5) and received a ball thrown from zone 2 of the other court and passed the ball over a rope (height 2.24 m) into a 3 m × 2 m target area in zone Z2 of the participant's court for which 4 points are awarded if ball lands in the target area and 2 points if it lands on the lines of the target area.

Subtest overhand pass (Figure 1B): To test the accuracy of the overhand pass, the participant stood in zone Z2, received the ball from zone Z6, and passed the ball over a 2.24 m high rope into two 1 m × 2 m target areas, with the one farther from the participant scoring 4 points, the one closer scoring 2 points, and the line in between scoring 3 points.

Subtest overhand service (Figure 1C): The participants stood at the end of the field in a central 2 m wide area and served the ball over their head to the other field over the 2.24 m high net into the 2 m × 2 m rectangular target areas, with points awarded for each area. The further back and sideways the target area that was hit, the more points a serve resulted in, ranging from 1 to 4 points. In between the zones, the two zone points were added and divided by two and the points given.

For all subtests, a score of zero was given if the ball did not land within the target zones or did not touch any of the lines of the marked target areas. The participants performed 10 trials in each subtest. The maximum score for each subtest was 40 points and a minimum of 0. The test was performed in the order from underhand pass to overhand pass to overhand service and on the same day under comparable conditions.

Six research assistants were trained to assist in the process of training and conducting the test. The execution of an attempt was counted only if the ball thrown by the research assistant was receivable by the participant within the marked area. Otherwise, the attempt was repeated.

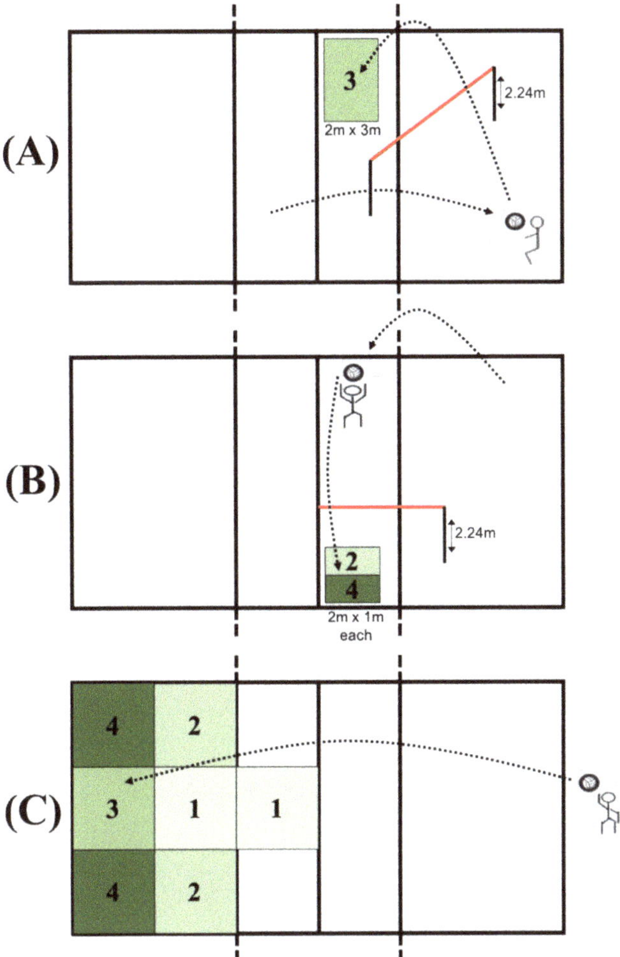

Figure 1. Test designs for the three volleyball techniques including scores. Each subtest corresponds to one technique: (**A**) underhand pass, (**B**) overhead pass, and (**C**) overhead service.

2.3. Data Analysis

The groups were compared statistically based on their results in each technique and in combined multiple techniques. To determine the combined multiple techniques, the mean values of the three individual techniques (overhand pass, underhand pass, and overhand service) were adjusted using z-standardization. To check the internal consistency of the tests for the respective techniques, 10 participants each performed the respective test at intervals of one week. Cronbach's alpha was determined based on the values from weeks 1 and 2.

Analyses of the data using Shapiro–Wilk tests revealed that some variables violated the assumption of normal distribution. Consequently, the development of the groups across the measurement time points and the comparison of the groups at the respective measurement time points were performed using non-parametric statistical tests.

For the analysis of the development within the groups in the respective techniques at pre-, post-, and retention-test, the results of the tests were statistically compared using

Friedman ANOVA. In case of significant results, pairwise Bonferroni-corrected post hoc Dunn–Bonferroni tests were performed.

In order to compare the different groups at the respective pre-, post-, and retention test, the test results of the specific techniques were compared statistically using Kruskal–Wallis tests. The comparison at the time of the pretest here also represents the basis of the test for homogeneity. Significant results were further statistically compared using pairwise Bonferroni-corrected post hoc Dunn–Bonferroni tests.

In addition, the effect size r was calculated for the pairwise post-hoc tests of the Friedman and Kruskal–Wallis tests, respectively. Thereby, $0.1 \leq r < 0.3$ corresponds to a weak effect, $0.3 \leq r < 0.5$ to a medium effect, and $r \geq 0.5$ to a strong effect [51].

The p-value at which it is considered worthwhile to continue research [52] was set at $p = 0.05$, with decreasing p increasing the probability that the null hypothesis does not explain all the facts.

3. Results

The proof for internal consistency of the tests showed acceptable or good results for the overhand pass ($\alpha = 0.774$), underhand pass ($\alpha = 0.812$), and for combined multiple techniques ($\alpha = 0.889$) tests. Only the overhand service test was just below the threshold in the questionable interval ($\alpha = 0.678$). The test results of each technique and the combined z-standardized values of each test are shown in Figure 2A–D. The results of the statistical analyses are presented in Table 1.

Table 1. Statistical comparisons at the three measurement time points within and between groups.

Comparison	Friedman-Test or Kruskal-Wallis-Test (Rank Scores)	Post Hoc Dunn-Bonferroni-Tests
	Overhand Pass	
RG: Pre—Post—Ret	$\chi^2(2) = 3.720, p = 0.156$ (Pre: 1.61; Post: 2.25; Ret: 2.14)	–
DG: Pre—Post—Ret	$\chi^2(2) = 25.529, p < 0.001$ *** (Pre: 1.04; Post: 1.96; Ret: 3.00)	Pre vs. Ret: $p < 0.001$ ***, r = 0.544 +++ Post vs. Ret: $p = 0.024$ *, r = 0.288 +
CG: Pre—Post—Ret	$\chi^2(2) = 11.306, p = 0.004$ ** (Pre: 1.57; Post: 2.68; Ret: 1.75)	Pre vs. Post: $p = 0.010$ *, r = 0.296 + Post vs. Ret: $p = 0.042$ *, r = 0.248 +
Pre: RG—DG—CG	$\chi^2(2) = 1.709, p = 0.426$ (RG: 24.14; DG: 22.11; CG: 18.25)	–
Post: RG—DG—CG	$\chi^2(2) = 7.758, p = 0.021$ * (RG: 18.04; DG: 28.89; CG: 17.57)	CG vs. DG: $p = 0.042$ *, r = 0.465 ++
Ret: RG—DG—CG	$\chi^2(2) = 15.508, p < 0.001$ *** (RG: 18.36; DG: 31.42; CG: 13.96)	RG vs. DG: $p = 0.013$ *, r = 0.550 +++ DG vs. CG: $p < 0.001$ ***, r = 0.732 +++
	Underhand Pass	
RG: Pre—Post—Ret	$\chi^2(2) = 21.714, p < 0.001$ *** (Pre: 1.07; Post: 2.29; Ret: 2.64)	Pre vs. Post: $p = 0.004$ **, r = 0.324 ++ Pre vs. Ret: $p < 0.001$ ***, r = 0.420 ++
DG: Pre—Post—Ret	$\chi^2(2) = 23.306, p < 0.001$ *** (Pre: 1.00; Post: 2.19; Ret: 2.81)	Pre vs. Post: $p = 0.007$ **, r = 0.319 ++ Pre vs. Ret: $p < 0.001$ ***, r = 0.483 ++
CG: Pre—Post—Ret	$\chi^2(2) = 14.000, p < 0.001$ *** (Pre: 1.82; Post: 2.64; Ret: 1.54)	Post vs. Ret: $p = 0.010$ *, r = 0.296 +
Pre: RG—DG—CG	$\chi^2(2) = 0.392, p = 0.822$ (RG: 22.79; DG: 20.86; CG: 20.86)	
Post: RG—DG—CG	$\chi^2(2) = 31.014, p < 0.001$ *** (RG: 20.36; DG: 34.50; CG: 9.64)	RG vs. DG: $p = 0.005$ **, r = 0.597 +++ RG vs. CG: $p = 0.05$ *, r = 0.452 ++ DG vs. CG: $p < 0.001$ ***, r = 1.049 +++
Ret: RG—DG—CG	$\chi^2(2) = 36.687, p < 0.001$ *** (RG: 21.43; DG: 35.00; CG: 7.57)	RG vs. DG: $p < 0.008$ **, r = 0.577 +++ RG vs. CG: $p < 0.005$ **, r = 0.589 +++ CG vs. DG: $p < 0.001$ ***, r = 1.165 +++
	Overhand Service	
RG: Pre—Post—Ret	$\chi^2(2) = 8.000, p = 0.018$ * (Pre: 1.86; Post: 2.43; Ret: 1.71)	–
DG: Pre—Post—Ret	$\chi^2(2) = 21.347, p < 0.001$ *** (Pre: 1.00; Post: 2.35; Ret: 2.65)	Pre vs. Post: $p = 0.002$ **, r = 0.360 ++ Pre vs. Ret: $p < 0.001$ ***, r = 0.442 ++
CG: Pre—Post—Ret	$\chi^2(2) = 9.172, p = 0.010$ * (Pre: 1.96; Post: 1.61; Ret: 2.43)	–

Table 1. *Cont.*

Comparison	Friedman-Test or Kruskal-Wallis-Test (Rank Scores)	Post Hoc Dunn-Bonferroni-Tests
Pre: RG—DG—CG	$\chi^2(2) = 14.235, p < 0.001$ *** (RG: 16.39; DG: 17.86; CG: 30.25)	RG vs. CG: $p = 0.002$ **, r = 0.649 +++ DG vs. CG: $p = 0.006$ **, r = 0.580 +++
Post: RG—DG—CG	$\chi^2(2) = 29.276, p < 0.001$ *** (RG: 14.39; DG: 35.43; CG: 14.68)	RG vs. DG: $p < 0.001$ ***, r = 0.892 +++ DG vs. CG: $p < 0.001$ ***, r = 0.879 +++
Ret: RG—DG—CG	$\chi^2(2) = 35.229, p < 0.001$ *** (RG: 8.21; DG: 34.85; CG: 20.93)	RG vs. DG: $p < 0.001$ ***, r = 1.142 +++ RG vs. CG: $p = 0.012$ *, r = 0.546 +++ DG vs. CG: $p = 0.006$ **, r = 0.597 +++
Combined multiple techniques		
RG: Pre—Post—Ret	$\chi^2(2) = 18.582, p < 0.001$ *** (Pre: 1.07; Post: 2.54; Ret: 2.39)	Pre vs. Post: $p < 0.001$ ***, r = 0.391 ++ Pre vs. Ret: $p = 0.001$ **, r = 0.353 ++
DG: Pre—Post—Ret	$\chi^2(2) = 24.571, p < 0.001$ *** (Pre: 1.00; Post: 2.14; Ret: 2.86)	Pre vs. Post: $p = 0.007$ **, r = 0.305 ++ Pre vs. Ret: $p < 0.001$ ***, r = 0.496 ++
CG: Pre—Post—Ret	$\chi^2(2) = 11.259, p = 0.004$ * (Pre: 1.57; Post: 2.71; Ret: 1.71)	Pre vs. Post: $p = 0.007$ **, r = 0.305 ++ Post vs. Ret: $p = 0.024$ *, r = 0.267 +
Pre: RG—DG—CG	$\chi^2(2) = 0.288, p = 0.866$ (RG: 16.39; DG: 17.86; CG: 30.25)	–
Post: RG—DG—CG	$\chi^2(2) = 28.127, p < 0.001$ *** (RG: 14.39; DG: 35.43; CG: 14.68)	RG vs. DG: $p < 0.001$ ***, r = 0.775 +++ DG vs. CG: $p < 0.001$ ***, r = 0.938 +++
Ret: RG—DG—CG	$\chi^2(2) = 30.205, p < 0.001$ *** (RG: 8.21; DG: 34.85; CG: 20.93)	RG vs. DG: $p = 0.001$ ***, r = 0.700 +++ DG vs. CG: $p < 0.001$ ***, r = 1.015 +++

Note. All *p*-values of the post hoc tests are Bonferroni-corrected. RG = repetitive learning group; DG = differential learning group; CG = control group; Pre = pretest; Post = posttest; Ret = retention test. * $p \leq 0.05$. ** $p \leq 0.01$. *** $p \leq 0.001$. + $0.1 \leq r < 0.3$. ++ $0.3 \leq r < 0.5$. +++ $r \geq 0.5$.

3.1. Development within Groups over Measurement Time Points

The DG improved statistically significantly over the course of the study in all three techniques and also in the combined multiple techniques ($p < 0.001$) and the effect size was at least medium each time (r > 0.442). In the subtests for the underhand pass, the overhand service, and the combined multiple techniques, there was a statistically significant improvement with a medium effect size in each case in the acquisition phase ($p < 0.007$, r > 0.305) and a further, however, not significant, improvement in the retention phase. Solely in the case of the overhand pass there was only a statistical trend in the acquisition phase ($p = 0.056$, r = 0.256), although there was a significant increase in the retention phase ($p = 0.024$, r = 0.288).

In the overall course, the performance level of RG tended to develop similarly to the DG in the techniques of the underhand pass and in the combined multiple technique. The results showed a significant improvement with a medium effect size ($p < 0.001$, r > 0.391). In both techniques, significant improvement was also observed in the acquisition phase ($p < 0.004$, r > 0.324). However, no significant improvement was shown in the overhand service and the overhand pass.

The performance level of CG never changed significantly, neither positively nor negatively, over the course of the study. Nonetheless, significant improvements from pretest to posttest were seen in the underhand pass and the overhand pass, each followed by significant decreases to the retention test. In the case of the overhand service, the exact opposite development was observed.

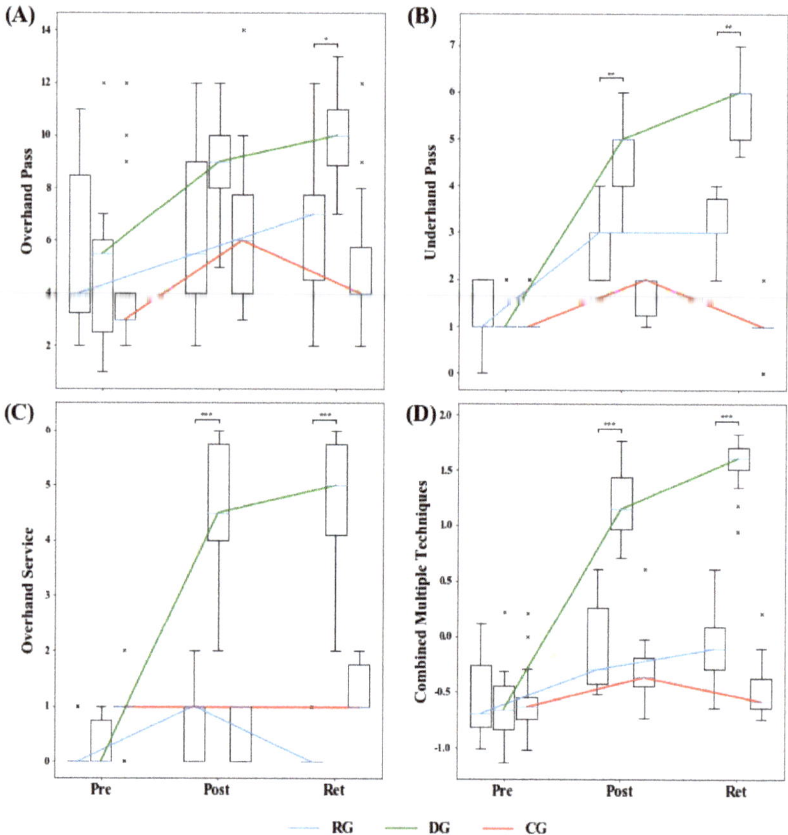

Figure 2. Development of the groups in the test on the respective techniques over the duration of the examination. Values are considered as outliers if they are outside the interval [Q1 − 1.5 * (Q3 − Q1), Q3 + 1.5 * (Q3 − Q1)]. × = outlier (each × stands for one outlier). Brackets show significant differences between RG and DG only. (* $p \leq 0.05$; ** $p \leq 0.01$; *** $p \leq 0.001$). Shown are the boxplots of the overhand pass (**A**), underhand pass (**B**), overhand service (**C**), and combined multiple techniques (**D**). For the clarity of the development of the groups, the median curves are also shown by line plots. RG = repetitive learning group; DG = differential learning group; CG = control group; Pre = pretest; Post = posttest; Ret = retention test.

3.2. Comparison between Groups across Measurement Time Points

An examination of the homogeneity of the three groups on the pre-test using the Kruskal–Wallis test revealed no statistically significant differences ($p \geq 0.42$) for the overhand and underhand pass as well as for the combined techniques. Only for the overhand service the groups differed significantly ($p < 0.001$), and pairwise Bonferroni-corrected post hoc comparisons revealed significant significantly larger values of the control group to the RG ($p = 0.002$, r = 0.649) and DG ($p = 0.006$, r = 0.580); there were no differences between RG and DG ($p = 1.000$).

Statistically significant global differences were found between groups for the overhand pass, underhand pass, overhand service, and combined multiple techniques in both the posttest and retention test ($p < 0.021$).

For the overhand pass technique, the posttest showed a significant difference with a medium effect size between the DG and the CG ($p = 0.042$, r = 0.465), whereas for the retention test, the DG performed significantly better than the RG ($p = 0.013$, r = 0.550) and the CG ($p < 0.001$, r = 0.732) with each a strong effect size.

Pairwise post hoc comparisons in the underhand pass technique showed that in the posttest and retention test, the DG performed significantly better and with strong effect size than the RG ($p = 0.005$, r = 0.597) and the CG ($p < 0.001$, r = 1.165), with the RG also performing significantly with a medium effect size better than the CG ($p = 0.050$, r = 0.452).

The post hoc tests for the overhand service, although the CG was still significantly better than the RG and DG at pretest, showed that the DG performed significantly better than the RG ($p < 0.001$, r > 0.892) and CG ($p < 0.006$, r > 0.597) in both the post and retention tests with a strong effect size. The RG also scored significantly better with a strong effect size than the CG at the retention test ($p = 0.012$, r = 0.546).

The same picture of DG outperforming RG ($p < 0.001$, r > 0.700) and CG ($p < 0.001$, r > 0.938) in both the post and retention test with strong effect sizes can also be seen in the combined multiple techniques; there was no statistical difference between RG and CG.

4. Discussion

The purpose of this study was to compare the effects of the repetition-oriented learning (RG) approach with the differential learning (DG) approach of teaching three volleyball techniques (underhand pass, overhand pass, and overhand service) to adolescent female novices in parallel, compared to general ball familiarization exercises (CG). All three groups started from the same performance level but developed differently depending on the learning approach. The changes of the CG, whose activities can be understood as having no direct relation to volleyball techniques, are statistically within the chance level and represent a fair reference to the other two interventions. The performance of the RG and DG each improved from pre to posttest, with the DG performing better than the RG on average for all comparisons. From the post to the retention test, the DG improved a further time in each technique, although only statistically significantly in the overhand pass. At the time of the retention test, the DG outperformed the RG in every single technique as well as in the multiple technique in a statistically significant manner.

First, we consider the results in the acquisition phase from pretest to posttest. With respect to each of the techniques individually, these findings are in accordance with earlier studies on the comparison of differential learning with repetitive-corrective learning [18,53–56]. Looking at the results from pre- to posttest as a whole, however, it is somewhat surprising how clearly the DG outperformed the RG in the individual techniques as well as in the combined multiple technique in the posttest and later in the retention test. This suggests that the parallel learning of multiple techniques in one sport with the differential learning approach might have had a particularly positive effect on learning the respective single techniques.

In order to understand the increased learning outcomes of the combination of individual techniques in a theoretical framework, it seems reasonable to use the contextual interference approach, which is currently the most widely used explanatory approach. Interestingly, however, the results contradict explanatory models within the contextual interference theory. These models predicted posttest advantages for the repetition-oriented groups, because the working memories were so adversely overloaded in the DG that it was even more likely to perform worse. The effect had to be larger, because the DG was exposed to such a large number of different movements, which has never been observed in studies on the contextual interference effect. Surprisingly, the absence of posttest benefits on the side of blocked-training groups can also be observed in several contextual interference studies on athletic movements [43,46]. Within the contextual interference framework, the favorable results of the DG group at posttest are even more surprising from another point of view. The theory postulates at least a certain number of correct repetitions of the prototypical movement as a mandatory prerequisite for successful acquisition and learning [11].

In contrast, however, the results of the combined techniques can also be explained by means of the theory of differential learning. According to this approach, training is associated with successful learning even without having performed the "ideal" prototypical movement once, just by practicing mainly movements surrounding the theoretically presumed target movement. According to differential learning theory, a neural network becomes more robust to perturbations, which can be considered as the deviations from a given ideal, when trained with variable input. According to the knowledge of the behavior of artificial neural networks, which derive their original principles from the properties of the neurons [25,26,57,58], differential learning theory expects the system to be trained with additional noise so that the individual not only finds a more global solution [19], but also prepares the system for more and larger deviations from a mentally constructed prototype that are highly likely to occur in the future [59]. Moreover, this noise could be amplified by learning multiple techniques in parallel to the correct degree, which in turn could provide a rationale for more efficient learning. By training an athlete's neural network for a wider solution space, the system uses the ability to interpolate, which appears to be superior to extrapolation [17]. In addition to what is now generally known about this property in artificial neural networks, Catalano and Kleiner [60] provided evidence for its validity in human behavior. They showed that in speed-based tasks variable-trained participants performed better than block-trained ones not only within the trained range, but also outside of it. Analogously, differential learning theory recommends increasing the range of experience to have a higher probability of using the interpolation in the next movement, which is sure to contain something new. In this context, it is important to mention that the space of solution is an abstract image that is highly dimensional and spanned by all joint angles, angular velocities, angular accelerations, spatial limb orientation, muscles activations, and many other influencing variables. Due to the high dimensionality of the solution space, it seems very unlikely to find variables that do not change individually and situationally. A very first computational approach was suggested with the Uncontrolled Manifold Theory [61], which could so far only be successfully applied to fine motor movements due to the limited number of influencing variables.

Nevertheless, the contextual interference theory has been developed for explaining the interference phenomenon after acquisition that can be systematically observed in fine-motor movements in the laboratory, such as keypress or barrier knock-down tasks. Since differential learning theory has failed to explain this so far, it is strongly recommended to develop differentiated explanatory models for fine and for gross motor movements and abstain from attempts to generalize the theory beyond the original model purpose [62,63]. One starting point could be the differentiated inclusion of various relative demands on different sensory systems during the specific movement task. Whereas the majority of contextual interference experiments for small motor movements contain dominant visual-spatial components (sequence of buttons, sequence of wooden blocks to be thrown) that are primarily processed sequentially, the majority of experiments for large motor sports movements contain a dominant proprioceptive and kinesthetic component that is primarily processed in parallel. When the importance of the visual component increases in gross-motor movements, such as in baseball [64], where the batter's decision is highly dependent on the ability to read approaching balls with different speeds, spin, and directions, or in basketball [65] and pistol shooting [66], with the problem of estimating the distance to the basket or the launch angle, the likelihood of finding support for the contextual interference paradigm also seems to increase. Whether the relative importance of visual and kinesthetic influence shifts over the course of a long-term learning process or is also age-dependent in volleyball needs to be clarified by future studies. A suggestion to explore the interdependencies of the visual and motor systems is given by Gestalt psychologists with their analysis of the "motor outline of optical shapes" [67].

The relative retention test performances of the RG and the DG can be explicated by both theories, the differential learning and the contextual interference theory. The group with more variation during acquisition performs significantly better than the other.

The contextual interference theory, however, only expects better results in the relative comparison, which would occur if both groups decreased in performance, the RG more than the DG. In contrast, based on the majority of experiments to date, differential learning is expected to show a further increase in performance on the retention test compared to the posttest.

If, in addition to the results of the two experimental groups RG and DG, we take into account the results of the control group CG for analysis, we can notice a relation to both the "specificity–versus–generalization" problem [68,69] and the resonance problem [18,19,70], which, strictly speaking, can be considered as the same problem on different scales. Whereas the specificity–versus–generalization problem is binary coded (specific/general), the stochastic resonance with its optimum function is at first sight ternary coded (too much/exact/too little). Because of its continuous frequency and learning rate scale, the SR model can actually be viewed as a decimal scale that provides information on how much noise needs to be added or reduced to achieve optimal resonance. Looking at the higher improvement rates of the two volleyball-specific approaches (RG and DG) after acquisition, it is obvious that the interventions of the CG for volleyball were too general and most likely outside the specific solution space. The interpretation based on the stochastic resonance phenomenon takes it one step further. The previous distinction becomes more sophisticated and, analogous to the other two, identifies too much noise for the CG, but too little noise for the RG and in between the optimal noise in the interventions for the DG.

From a more application-oriented point of view, this study presents a picture that alleged errors in movement during the acquisition process should not be regarded as detrimental to learning, but rather as an advantage for learning [16], indicative of the fact that the learning base has been broadened. This is because in the differential learning paradigm, error identification and correction during acquisition were completely absent, which allowed for self-organization. In the same context, a few issues in motor learning need more attention in terms of classical learning theories:

(1) Traditional motor learning theories point out errors and seek to correct them (motor learning), implying that there is a specific way to perform a movement, technique, or skill [1,11]. These error corrections aim to improve or perfect performance within a feedback loop [71]. However, from the results of the study, it appears that these corrections limit the potential of learners by making them stick to certain movement patterns that are considered the best movements in the techniques to be learned. This in a way controls the actual potential of the learner and leads to role model learning that does not allow the learner's originality and innovation to shine.

(2) Learning theories also attempt to repeat the movement being learned multiple times in order to perfect the movement. Considering judgment errors only as fluctuations has the potential to destabilize the system to allow true self-organization [16,17,72]. An intermediate step between teacher-oriented and self-organized learning is already offered by reform pedagogical approaches according to Basedow, Pestalozzi, or Dewey, where discovery-based learning is allowed within given limits. Either way, coaches and trainers may have to reconsider what errors are and to rethink the idea of role modeling (ideal movement), because learners are not given the freedom and opportunity to learn and develop naturally as individuals. Again, the physical education teacher's task to teach gross-motor movements in sports to beginners may involve revising their approach towards their lessons, as well as the methods they employ.

The individual technique performance by the three groups shows that the DG had more participants who improved their performance throughout the experiment than the RG and CG. This is at least an indication that constant repetition—even without "repetition" in Bernstein's sense [73,74]—is not the only approach to motor skill acquisition, does not really improve performance quickly, can limit the learner, and hardly allows for system adaptation.

This study also had some attendant challenges; (1) the age of the participants, who were female, indicated that they were undergoing physical maturation and development, so they were reluctant to be active during the exercises and tests at a time when some of them had physical changes. (2) Facilities such as playing courts and volleyballs were not enough; at least three playing courts would be better. (3) The single-hour physical education class might have been too short to practice adequately.

5. Conclusions

This study compared the effects of the general (CG), repetitive (RG), and differential motor learning (DG) approaches in parallel acquisition of three volleyball techniques (underhand pass, overhand pass, and overhand service). The results indicate the advantages of the differential learning approach in comparison to the general ball familiarization and the repetitive prescriptive approaches not only during the acquisition, but also during the learning phase. Through differential learning, where no movement was repeated, the adolescent girls who were absolute beginners in volleyball seemed to have experienced a broader spectrum of movements compared to before, which allowed their neuro-motor system to adapt more efficiently to the demands of the three techniques.

Previous models of variable [4] or interleaved practice [8] fail to explain this phenomenon, since they assume either memorization of the to-be learned movements or at least parts (invariants) of the movements through correct repetitions. Noisy training is associated with the theory and behavior of artificial neural networks in connection with the principles of system dynamics. In the first case, noise leads to more stability in the subsequent application, and in the second case, the increase of noise is a condition for a self-organized change of states. Although more research will be necessary to understand in detail the form and extent of the variations (noise) on the situations (e.g., learning phases) and individuals or groups, the results in conjunction with the findings of other studies indicate an essential role of increased noise in learning processes of multiple movements. Learning multiple movements in parallel appears to further positively influence this "differential learning effect". In addition, the highly variable differential learning exercise protocol could expand the space of experience and thus better prepare athletes to solve future problems more adequately, as indicated by the retention tests.

Although the statistics applied do not allow for generalization, the numerous significant results with corresponding effect sizes may, on the one hand, encourage researchers (according to Fisher's original interpretation of his statistics) to continue the study of differential learning and, on the other hand, provide coaches and physical education teachers with an effective and time-saving method to support motor learning.

Author Contributions: Conceptualization, J.B.A. and W.I.S.; methodology, J.B.A. and W.I.S.; validation, J.B.A., W.I.S. and J.B.; formal analysis, J.B.A. and J.B.; investigation, J.B.A.; resources, J.B.A. and W.I.S.; data curation, J.B.A.; writing—original draft preparation, J.B.A., J.B. and W.I.S.; writing—review and editing, J.B.A., W.I.S. and J.B.; visualization, J.B.A., J.B. and W.I.S.; supervision, W.I.S.; project administration, W.I.S.; funding acquisition, None. All authors have read and agreed to the published version of the manuscript.

Funding: This research received no external funding.

Institutional Review Board Statement: The study was conducted according to the guidelines of the Declaration of Helsinki and approved by the Institutional Review Board of Akanten Appiah-Menka University of Skills Training and Entrepreneurial Development (AAMUSTED/K/RO/L.1/219, 31 August 2021).

Informed Consent Statement: Informed consent was obtained from all subjects involved in the study.

Data Availability Statement: The data that support the findings of this study are available from the corresponding author, J.B., upon reasonable request.

Conflicts of Interest: The authors declare no conflict of interest.

Appendix A

Table A1. Training exercises for the DG for the techniques underhand pass, overhand pass, and overhand service divided into different categories.

	Standing Position
1	Stand with one leg forward, change leg position while performing.
2	Stand on one leg, change leg while executing.
3	Both legs parallel.
4	Both knees bent, but parallel.
5	Bend knees while performing.
6	Both knees bent, but one leg in front, change leg position while executing.
7	Legs slightly apart.
8	Spread legs slightly while performing.
9	Extend one leg forward.
10	Leg raised, knee bent to chest height, switch with other leg.
11	As in Task 10, but change angle of legs.
12	Stand on the balls of the feet.
13	Stand on the heels of the feet.
	Trunk Movement
14	Forward movement during the execution.
15	Sideways movement during execution.
16	Backward movement during execution.
17	Rounding in during execution.
18	Straighten during the execution.
	Head Movement
19	Look up.
20	Look down.
21	Head circling.
22	One eye closed.
23	Both eyes blinking.
	Hand/Arm Movement
24	Arms higher.
25	Arms (more) forward.
26	Arms sideways.
27	Elbow position slightly back to the side.
28	Elbow position slightly forward and inward.
29	Elbows bent.
30	Elbows extended.
31	Arms crossed.
32	Hands on top of each other.
33	Hands parallel
34	Hands supinated.
35	Hands pronated.

Table A1. *Cont.*

	Standing Position
36	Hands as fists.
37	Hands wide open.
	Mass Movement (Velocity Change)
38	One step forward during the execution.
39	Two steps forward during the execution.
40	One step backward during execution.
41	Two steps backward during the execution.
42	Side steps to the left and right during execution.
43	Moving one leg forward, backward, left, right during the execution.
44	During execution one quick step forward.
45	During execution one quick step backward.
46	During the execution two quick steps to the front.
47	During execution two quick steps backward.
48	During the execution quick lateral steps to the left and right.
49	Quick one-legged movement forward, backward, left, right during execution.
	Jumps/Hops
50	During the execution single-leg hops to the front, back, left, right.
51	While performing two-legged hop forward, backward, left, right.
52	Jump with execution of the technique before landing.
53	Jump with execution of the technique exactly at the landing.
	Running
54	Fast runs to the execution position.
55	Fast and slow runs during the performance.
	Twist and Turns
56	Half turn left/right on command immediately before execution.
57	Execution with hip position 60°, 120°, 180°, 240°, 300° to the stroke direction.
	Position Changes
58	Execution in seated position with legs crossed, legs forward, hurdle seat position, legs wide apart, one leg bent.
59	Execution in a seated position with legs crossed, legs forward, hurdle seat position, legs wide apart, one leg bent.
60	Execution lying bent, stretched, on the side.
61	60, but faster.
62	60, but slower.
63	Execution backwards.
	Combine at Least two of all Above
64	Jump and turn left while performing.
65	27 and 46 while execution
66	21 and 32 while execution
67	Different type of balls.
68	Different terrains (e.g., on sand, grass, soft floor mat, …)

References

1. Gentile, A.M. A working model of skill acquisition with application to teaching. *Quest* **1972**, *17*, 3–23. [CrossRef]
2. Gaulhofer, K.; Streicher, M. *Grundzüge des Österreichischen Schulturnens*; Deutscher Verlag für Jugend und Volk: Wien, Austria, 1924.
3. Moxley, S.E. Schema: The Variability of Practice Hypothesis. *J. Mot. Behav.* **1979**, *11*, 65–70. [CrossRef]
4. Schmidt, R.A. A schema theory of discrete motor skill learning. *Psychol. Rev.* **1975**, *82*, 225–260. [CrossRef]
5. Schmidt, R.A. Motor Schema Theory after 27 Years: Reflections and Implications for a New Theory. *Res. Q. Exerc. Sport* **2003**, *74*, 366–375. [CrossRef] [PubMed]
6. Battig, W.F. Facilitation and interference. In *Acquisition of Skill*; Bilodeau, A., Ed.; Academic Press: New York, NY, USA, 1966; pp. 215–224.
7. Battig, W.F. The flexibility of human memory. In *Levels of Processing in Human Memory*; Cermak, L.S., Craik, F.I., Eds.; Erlbaum: Hillsdale, NJ, USA, 1979; pp. 23–44.
8. Shea, J.B.; Morgan, R.L. Contextual interference effects on the acquisition, retention, and transfer of a motor skill. *J. Exp. Psychol. Hum. Learn. Mem.* **1979**, *5*, 179–187. [CrossRef]
9. Magill, R.A.; Hall, K.G. A review of the contextual interference effect in motor skill acquisition. *Hum. Mov. Sci.* **1990**, *9*, 241–289. [CrossRef]
10. Brady, F. Contextual Interference: A Meta-Analytic Study. *Percept. Mot. Skills* **2004**, *99*, 116–126. [CrossRef]
11. Shea, J.B.; Zimny, S.T. Context effects in memory and learning movement information. In *Memory and Control of Action*; Magill, R.A., Ed.; North-Holland: New York, NY, USA, 1983; pp. 345–366.
12. Lee, T.D.; Magill, R.A. The locus of contextual interference in motor-skill acquisition. *J. Exp. Psychol. Learn. Mem. Cogn.* **1983**, *9*, 730–746. [CrossRef]
13. Shewokis, P.A.; Rey, P.D.; Simpson, K.J. A Test of Retroactive Inhibition as an Explanation of Contextual Interference. *Res. Q. Exerc. Sport* **1998**, *69*, 70–74. [CrossRef]
14. Barreiros, J.; Figueiredo, T.; Godinho, M. The contextual interference effect in applied settings. *Eur. Phys. Educ. Rev.* **2007**, *13*, 195–208. [CrossRef]
15. Goode, S.; Magill, R.A. Contextual Interference Effects in Learning Three Badminton Serves. *Res. Q. Exerc. Sport* **1986**, *57*, 308–314. [CrossRef]
16. Schöllhorn, W.I. Individualitaet—Ein vernachlässigter Parameter. *Leistungssport* **1999**, 7–12.
17. Schöllhorn, W.I. Applications of systems dynamic principles to technique and strenght training. *Acta Acad. Olympiquae Est.* **2000**, 67–85.
18. Schöllhorn, W.I.; Beckmann, H.; Michelbrink, M.; Sechelmann, M.; Trockel, M.; Davids, K. Does noise provide a basis for the unification of motor learning theories? *Int. J. Sport Psychol.* **2006**, *37*, 186–206.
19. Schöllhorn, W.I.; Mayer-Kress, G.; Newell, K.M.; Michelbrink, M. Time scales of adaptive behavior and motor learning in the presence of stochastic perturbations. *Hum. Mov. Sci.* **2009**, *28*, 319–333. [CrossRef]
20. Beckmann, H.; Schöllhorn, W.I. Increasing noise improves signal-noise-ratio in motor learning. In *2nd European Workshop on Movement Sciences*; Peham, C., Schöllhorn, W.I., Verwey, W., Eds.; Sport & Buch Strauß: Köln, Germany, 2005; pp. 139–140.
21. Schöllhorn, W.I.; Bauer, H.U. Identifying individual movement styles in high performance sports by means of self-organizing Kohonen maps. In Proceedings of the XVI Annual Conference of the International Society for Biomechanics in Sport, Konstanz, Germany, 21–25 July 1998; 1998; pp. 574–577.
22. Schöllhorn, W.I.; Nigg, B.M.; Stefanyshyn, D.J.; Liu, W. Identification of individual walking patterns using time discrete and time continuous data sets. *Gait Posture* **2002**, *15*, 180–186. [CrossRef]
23. Horst, F.; Eekhoff, A.; Newell, K.M.; Schöllhorn, W.I. Intra-individual gait patterns across different time-scales as revealed by means of a supervised learning model using kernel-based discriminant regression. *PLoS ONE* **2017**, *12*, e0179738. [CrossRef] [PubMed]
24. Schöner, G.; Haken, H.; Kelso, J.A.S. A stochastic theory of phase transitions in human hand movement. *Biol. Cybern.* **1986**, *53*, 247–257. [CrossRef]
25. Haykin, S. *Neural Networks: A Comprehensive Foundation*, 1st ed.; McMaster University: Toronto, OT, Canada, 1994.
26. Miglino, O.; Lund, H.H.; Nolfi, S. Evolving mobile robots in simulated and real environments. *Artif. Life* **1995**, *2*, 417–434. [CrossRef] [PubMed]
27. Schöllhorn, W.I.; Stefanysbyn, D.J.; Nigg, B.M.; Liu, W. Recognition of individual walking patterns by means of artificial neural nets. *Gait Posture* **1999**, *10*, 85–86. [CrossRef]
28. Römer, J.; Schöllhorn, W.I.; Jaitner, T.; Preiss, R. Differenzielles Lernen bei der Aufschlagannahme im Volleyball. In *Messplätze, Messplatztraining, Motorisches Lernen*; Krug, J., Müller, T., Eds.; Academia Verlag: Sankt Augustin, Germany, 2003; pp. 129–133.
29. Beckmann, H.; Schöllhorn, W.I. Differencial learning in shot put. In *1st European Workshop on Movement*; Schöllhorn, W.I., Bohn, C., Jäger, J.M., Schaper, H., Alichmann, M., Eds.; Sport & Buch Strauß: Köln, Germany, 2003; p. 68.
30. Savelsbergh, G.J.P.; Kamper, W.J.; Rabius, J.; De Koning, J.J.; Schöllhorn, W. New methods to learn to start in speed skating: A differencial learning approach. *Int. J. Sport Psychol.* **2010**, *41*, 415–427.
31. Pabel, S.-O.; Pabel, A.-K.; Schmickler, J.; Schulz, X.; Wiegand, A. Impact of a Differential Learning Approach on Practical Exam Performance: A Controlled Study in a Preclinical Dental Course. *J. Dent. Educ.* **2017**, *81*, 1108–1113. [CrossRef] [PubMed]

32. Pabel, S.-O.; Freitag, F.; Hrasky, V.; Zapf, A.; Wiegand, A. Randomised controlled trial on differential learning of toothbrushing in 6- to 9-year-old children. *Clin. Oral Investig.* **2018**, *22*, 2219–2228. [CrossRef]

33. Serrien, B.; Tassignon, B.; Verschueren, J.; Meeusen, R.; Baeyens, J.-P. Short-term effects of differential learning and contextual interference in a goalkeeper-like task: Visuomotor response time and motor control. *Eur. J. Sport Sci.* **2020**, *20*, 1061–1071. [CrossRef] [PubMed]

34. Santos, S.; Coutinho, D.; Gonçalves, B.; Schöllhorn, W.I.; Sampaio, J.; Leite, N. Differential Learning as a Key Training Approach to Improve Creative and Tactical Behavior in Soccer. *Res. Q. Exerc. Sport* **2018**, *89*, 11–24. [CrossRef] [PubMed]

35. Poureghbali, S.; Arede, J.; Rehfeld, K.; Schöllhorn, W.I.; Leite, N. Want to Impact Physical, Technical, and Tactical Performance during Basketball Small-Sided Games in Youth Athletes? Try Differential Learning Beforehand. *Int. J. Environ. Res. Public Health* **2020**, *17*, 9279. [CrossRef] [PubMed]

36. Hegen, P.; Polywka, G.; Schöllhorn, W.I. The Differential Learning Approach in Strength Training. In Proceedings of the Book of Abstract of 20th annual Congress of the European College of Sport Science, Malmö, Sweden, 24–27 June 2015; Radmann, A., Hedenborg, S., Tsolakidis, E., Eds.; University of Malmö: Malmö, Sweden, 2015; p. 590.

37. Horst, F.; Rupprecht, C.; Schmitt, M.; Hegen, P.; Schöllhorn, W.I. Muscular Activity in Conventional and Differential Back Squat Exercise. In Proceedings of the 21st Annual Congress of the European College of Sport Science, Vienna, Austria, 6–9 July 2016; p. 516.

38. Schöllhorn, W.I.; Paschke, M.; Beckmann, H. Differenzielles Training im Volleyball beim Erlernen von zwei Techniken. In *Volleyball 2005—Beach-WM*; Langolf, K., Roth, R., Eds.; Czwalina: Hamburg, Germany, 2006; pp. 97–105.

39. Hegen, P.; Schöllhorn, W.I. Gleichzeitig in verschiedenen Bereichen besser werden, ohne zu wiederholen? Paralleles differenzielles Training von zwei Techniken im Fußball. *Leistungssport* **2012**, *42*, 17–23.

40. French, K.; Rink, J.; Rikard, L.; Mays, A.; Lynn, S.; Werner, P. The Effects of Practice Progressions on Learning Two Volleyball Skills. *J. Teach. Phys. Educ.* **1991**, *10*, 261–274. [CrossRef]

41. Fialho, J.V.A.P.; Benda, R.N.; Ugrinowitsch, H. The contextual interference effect in a serve skill acquisition with experienced volleyball players. *J. Hum. Mov. Stud.* **2006**, *50*, 65–77.

42. Meira, J.R.; Tani, G. Contextual interference effects assessed by extended transfer trials in the acquisition of the volleyball serve. *J. Hum. Mov. Stud.* **2003**, *45*, 449–468.

43. French, K.E.; Rink, J.E.; Werner, P.H. Effects of Contextual Interference on Retention of Three Volleyball Skills. *Percept. Mot. Ski.* **1990**, *71*, 179–186. [CrossRef]

44. Jones, L.; French, K. Effects of contextual interference on acquisition and retention of three volleyball skills. *Percept. Mot. Skills* **2007**, *105*, 883–890. [CrossRef] [PubMed]

45. Zetou, E.; Michalopoulou, M.; Giazitzi, K.; Kioumourtzoglou, E. Contextual Interference Effects in Learning Volleyball Skills. *Percept. Mot. Skills* **2007**, *104*, 995–1004. [CrossRef]

46. Bortoli, L.; Robazza, C.; Durigon, V.; Carra, C. Effects of Contextual Interference on Learning Technical Sports Skills. *Percept. Mot. Skills* **1992**, *75*, 555–562. [CrossRef] [PubMed]

47. Kalkhoran, J.F.; Shariati, A. The Effects of Contextual Interference on Learning Volleyball Motor Skills. *J. Sport Sci.* **2015**, *6*, 12–24.

48. Pollock, B.J.; Lee, T.D. Dissociated contextual interference effects in children and adults. *Percept. Mot. Skills* **1997**, *84*, 851–858. [CrossRef]

49. FIVB Coaches Manual. Available online: https://www.fivb.com/en/development/education/toolsandresourcecentre (accessed on 4 October 2021).

50. American Association for Health, Physical Education, and Recreation (AAHPERD). *AAHPER Skills Test Manual: Volleyball*; American Association for Health, Physical Education, and Recreation: Washington, DC, USA, 1969.

51. Cohen, J. *The Concepts of Power Analysis BT—Statistical Power Analysis for the Behavioral Sciences (Revised Edition)*; Hillsdale, N.J., Ed.; Lawrence Erlbaum Associates Inc.: Mahwah, NJ, USA, 1988; ISBN 9780805802832.

52. Fisher, R.A. On the Mathematical Foundations of Theoretical Statistics. *Philos. Trans. R. Soc. A Math. Phys. Eng. Sci.* **1922**, *222*, 309–368. [CrossRef]

53. Reynoso, S.R.; Sabido Solana, R.; Reina Vaíllo, R.; Moreno Hernández, F.J. Differential learning applied to volleyball serves in novice athletes. *Apunt. Educ. Fis. Esports* **2013**, *114*, 45–52.

54. Savelsbergh, G.J.P.; Cañal-Bruland, R.; van der Kamp, J. Error Reduction during Practice: A Novel Method for Learning to Kick Free-Kicks in Soccer. *Int. J. Sports Sci. Coach.* **2012**, *7*, 47–56. [CrossRef]

55. Wagner, H.; Müller, E. The effects of differential and variable training on the quality parameters of a handball throw. *Sport. Biomech.* **2008**, *7*, 54–71. [CrossRef]

56. Fuchs, P.; Fusco, A.; Bell, J.; Von Duvillard, S.; Cortis, C.; Wagner, H. Effect of Differential Training on Female Volleyball Spike-Jump Technique and Performance. *Int. J. Sports Physiol. Perform.* **2020**, *15*, 1019–1025. [CrossRef] [PubMed]

57. Horak, M. The utility of connectionism for motor learning: A reinterpretation of contextual interference in movement schemas. *J. Mot. Behav.* **1992**, *24*, 58–66. [CrossRef]

58. Shea, J.B.; Graf, R.C. A Model for Contextual Interference Effects in Motor Learning. *Cogn. Assess.* **1994**, 73–87. [CrossRef]

59. Schöllhorn, W.I. Practical consequences of biomechanically determined individuality and fluctuations on motor learning. In Proceedings of the XVII Conference of the International Society of Biomechanics, Calgary, AB, Canada, 8–13 August 1999.

60. Catalano, J.F.; Kleiner, B.M. Distant Transfer in Coincident Timing as a Function of Variability of Practice. *Percept. Mot. Skills* **1984**, *58*, 851–856. [CrossRef]
61. Scholz, J.P.; Schöner, G. The uncontrolled manifold concept: Identifying control variables for a functional task. *Exp. Brain Res.* **1999**, *126*, 289–306. [CrossRef]
62. Newell, K.M.; Kugler, P.N.; Van Emmerik, R.E.A.; McDonald, P.V. Search strategies and the acquisition of coordination. *Adv. Psychol.* **1989**, *61*, 85–122. [CrossRef]
63. Wulf, G.; Shea, C.H. Principles derived from the study of simple skills do not generalize to complex skill learning. *Psychon. Bull. Rev.* **2002**, *9*, 185–211. [CrossRef]
64. Hall, K.G.; Domingues, D.A.; Cavazos, R. Contextual Interference Effects with Skilled Baseball Players. *Percept. Mot. Skills* **1994**, *78*, 835–841. [CrossRef]
65. Porter, C.; Greenwood, D.; Panchuk, D.; Pepping, G.-J. Learner-adapted practice promotes skill transfer in unskilled adults learning the basketball set shot. *Eur. J. Sport. Sci.* **2019**, *20*, 61–71. [CrossRef]
66. Boyce, B.A.; Del Rey, P. Designing applied research in a naturalistic setting using a contextual interference paradigm. *J. Hum. Mov. Stud.* **1990**, *18*, 189–200.
67. Kern, G.; Krüger, F. Motorische Umreißung optischer Gestalten. *Neue Psychol. Stud.* **1933**, *9*, 65–104.
68. Shea, C.H.; Kohl, R.; Indermill, C. Contextual interference: Contributions of practice. *Acta Psychol.* **1990**, *73*, 145–157. [CrossRef]
69. Willey, C.R.; Liu, Z. Limited generalization with varied, as compared to specific, practice in short-term motor learning. *Acta Psychol.* **2018**, *182*, 39–45. [CrossRef]
70. McDonnell, M.D.; Stocks, N.G.; Pearce, C.E.M.; Abbott, D. *Stochastic Resonance: From Suprathreshold Stochastic Resonance to Stochastic Signal Quantization*; University Press: Cambridge, UK, 2008; ISBN 978-0-521-88262-0.
71. Adams, J.A. Historical review and appraisal of research on the learning, retention, and transfer of human motor skills. *Psychol. Bull.* **1987**, *101*, 41–74. [CrossRef]
72. Kelso, J.A.S.; Zanone, P.G. Coordination dynamics of learning and transfer across different effector systems. *J. Exp. Psychol. Hum. Percept. Perform.* **2002**, *28*, 776–797. [CrossRef] [PubMed]
73. Bernstein, N.A. *The Co-Ordination and Regulation of Movements*; Pergamon Press: Oxford, UK, 1967.
74. Ranganathan, R.; Lee, M.-H.; Newell, K.M. Repetition Without Repetition: Challenges in Understanding Behavioral Flexibility in Motor Skill. *Front. Psychol.* **2020**, *11*, 2018. [CrossRef] [PubMed]

International Journal of
Environmental Research and Public Health

Article

The Effect of Eight-Week Sprint Interval Training on Aerobic Performance of Elite Badminton Players

Haochong Liu [1], Bo Leng [2], Qian Li [2], Ye Liu [1], Dapeng Bao [1,*] and Yixiong Cui [3,*]

[1]   China Institute of Sport and Health Science, Beijing Sport University, Beijing 100084, China; liuhaochong1011@gmail.com (H.L.); olliejanessatdu62@gmail.com (Y.L.)
[2]   Sports Coaching College, Beijing Sport University, Beijing 100084, China; lengbo@bsu.edu.cn (B.L.); 2020110023@bsu.edu.cn (Q.L.)
[3]   AI Sports Engineering Lab, School of Sports Engineering, Beijing Sport University, Beijing 100084, China
*   Correspondence: baodp@bsu.edu.cn (D.B.); cuiyixiong@bsu.edu.cn (Y.C.)

Abstract: This study was aimed to: (1) investigate the effects of physiological functions of sprint interval training (SIT) on the aerobic capacity of elite badminton players; and (2) explore the potential mechanisms of oxygen uptake, transport and recovery within the process. Thirty-two elite badminton players volunteered to participate and were randomly divided into experimental (Male-SIT and Female-SIT group) and control groups (Male-CON and Female-CON) within each gender. During a total of eight weeks, SIT group performed three times of SIT training per week, including two power bike trainings and one multi-ball training, while the CON group undertook two Fartlek runs and one regular multi-ball training. The distance of YO-YO IR2 test (which evaluates player's ability to recover between high intensity intermittent exercises) for Male-SIT and Female-SIT groups increased from 1083.0 ± 205.8 m to 1217.5 ± 190.5 m, and from 725 ± 132.9 m to 840 ± 126.5 m ($p < 0.05$), respectively, which were significantly higher than both CON groups ($p < 0.05$). For the Male-SIT group, the ventilatory anaerobic threshold and ventilatory anaerobic threshold in percentage of VO_2max significantly increased from 3088.4 ± 450.9 mL/min to 3665.3 ± 263.5 mL/min ($p < 0.05$),and from $74 \pm 10\%$ to $85 \pm 3\%$ ($p < 0.05$) after the intervention, and the increases were significantly higher than the Male-CON group ($p < 0.05$); for the Female-SIT group, the ventilatory anaerobic threshold and ventilatory anaerobic threshold in percentage of VO_2max were significantly elevated from 1940.1 ± 112.8 mL/min to 2176.9 ± 78.6 mL/min, and from $75 \pm 4\%$ to $82 \pm 4\%$ ($p < 0.05$) after the intervention, which also were significantly higher than those of the Female-CON group ($p < 0.05$). Finally, the lactate clearance rate was raised from $13 \pm 3\%$ to $21 \pm 4\%$ ($p < 0.05$) and from $21 \pm 5\%$ to $27 \pm 4\%$ for both Male-SIT and Female-SIT groups when compared to the pre-test, and this increase was significantly higher than the control groups ($p < 0.05$). As a training method, SIT could substantially improve maximum aerobic capacity and aerobic recovery ability by improving the oxygen uptake and delivery, thus enhancing their rapid repeated sprinting ability.

Keywords: interval training; badminton; aerobic; repeated sprint; testing

Citation: Liu, H.; Leng, B.; Li, Q.; Liu, Y.; Bao, D.; Cui, Y. The Effect of Eight-Week Sprint Interval Training on Aerobic Performance of Elite Badminton Players. *Int. J. Environ. Res. Public Health* **2021**, *18*, 638. https://doi.org/10.3390/ijerph 18020638

Received: 27 November 2020
Accepted: 8 January 2021
Published: 13 January 2021

Publisher's Note: MDPI stays neutral with regard to jurisdictional claims in published maps and institutional affiliations.

1. Introduction

Badminton is a fast and dynamic sport, which has high requirements for the player's rapid reaction, fast action and high-speed hitting ability. Studies have shown that there are on average 5–9 strokes in badminton games [1]. Due to its fastball speed, high swing frequency and short interval time, badminton requires players to mainly compete with fast running, sudden acceleration, abrupt stop, change of direction and continuously high intensity of multiple rallies, which requires a player's well-developed aerobic endurance [2,3]. Due to the influence of players' strength level, badminton competition is easy to form a multi-round competition, which requires higher aerobic working capacity. Particularly, a relevant study has indicated that badminton players usually reach an average heart rate of over 90% of their HRmax during competitive games, which is demanding to both aerobic and anaerobic systems: 60–70% on the aerobic system and 30% on the anaerobic system, with a greater demand on alactic metabolism [3].

Previous research [3,4] has shown that various physiological parameters had a strong correlation with badminton performance. Particularly, aerobic capacity and intermittent exercise performance are positively correlated, involving VO_2max, lactate/anaerobic threshold and running efficiency [5]. However, practically, due to the limited training time, the traditional long-period aerobic endurance training would not be the most suitable modality for the actual needs of current competition. Therefore, it is essential to explore time-efficient and badminton-specific fitness training programs.

As one of the advocated alternatives to traditional continuous aerobic training, Sprint Interval Training (SIT) is a training approach that asks athletes to complete the required actions at maximum effort in a short period of time, and takes active rest with limited recovery time between two sets of training. Meanwhile, as the active rest often lasts only 3–5 min between multiple short and full sprint training, SIT can effectively improve the performance of athletes in intermittent sports with substantially lower overall training volume [6].

In recent years, there have been many studies on the training effect of SIT on sports performance in other intermittent sports such as soccer, basketball, volleyball and field hockey [7]. Among them, Buchan [8] and Bayati et al. [9] conducted a 6-week 30-s full-speed sprint running and rowing training experiment. Each training was carried out in 4–6 rounds, with 4 min of low-intensity activities serving as an active rest interval between groups. The results showed that the maximum oxygen uptake, peak power, average power and aerobic capacity significantly improved compared with those of the control group. Meanwhile, the study by Jone et al. [10] proved that field hockey players' muscle oxygenation kinetics and performance during the 30–15 intermittent fitness test (30-s shuttle runs with 15-s passive recovery) were significantly improved after 6-week of Sprint Interval Cycling. Additionally, Burgomaster [11] and Gibala et al. [12] also conducted similar comparative experiments between traditional endurance training and SIT on young healthy individuals. Their results showed that the SIT group shortened the training time by about 80%, and the participants' aerobic capacity was significantly improved.

Nonetheless, currently, there are few attempts to investigate the application of SIT to badminton training. This study was, therefore aimed to explore the effect of SIT on improving players' aerobic capacity, as well as the mechanism of oxygen uptake and transport, by testing the changes of badminton players' rapid and repeated sprinting ability and related aerobic capacity parameters before and after 8 weeks of SIT. It was hypothesized that such training would induce greater improvement in before-mentioned parameters compared to traditional continuous aerobic training.

2. Materials and Methods

2.1. Participants

Thirty-two elite players from who had played in or beyond the semi-finals of Badminton Championship at National level volunteered to participate in the study. There were sixteen male and female players, respectively, and they were randomly divided into male Sprint Interval Training (SIT) group ($n = 8$) and control (CON) group ($n = 8$), and female SIT group ($n = 8$) and CON group ($n = 8$). Detailed information about different groups can be found in Table 1. All subjects were in good health and had no severe injuries during the last six months before the study. Prior to the formal experiment and test, the nature and possible risks were explained to the participants, and they provided their written informed consent to participate. The tests were conducted at least 48 h after competitive match or heavy training session. The subjects participated in all the training sessions as well as pre- and post- training tests. All procedures were approved by Research Ethics Committee of Beijing Sport University (Approval number: 2020008H). All procedures were conducted in accordance with the Declaration of Helsinki.

Table 1. Personal information of participating players.

Group	N	Age (year)	Height (cm)	Weight (kg)	Training Age (year)	HRmax (bpm)
Male-SIT	8	20.0 ± 1.3	179.6 ± 3.6	73.8 ± 6.9	12.1 ± 2.2	190.7 ± 8.8
Male-CON	8	21.5 ± 2.2	177.1 ± 7.1	72.4 ± 6.7	13.2 ± 3.2	191.8 ± 6.2
Female-SIT	8	20.5 ± 1.4	168.5 ± 4.2	62.6 ± 4.2	9.5 ± 1.2	181.9 ± 8.9
Female-CON	8	19.4 ± 1.5	168.2 ± 4.8	61.3 ± 4.2	9.8 ± 1.5	180.4 ± 8.5

Note: SIT = Sprint Interval Training; CON = Control; HRmax = maximum heart rate.

2.2. Procedures

For eight weeks, the CON group followed previous training routines of two Fartlek running sessions and one regular multi-ball feeding training per week, which was a traditionally employed aerobic training protocol for these badminton players. Meanwhile, the SIT group carried out sprint interval training three times a week, including two power bicycle training sessions with a Monark 894E exercise bike (Monark Exercise AB, Vansbro, Sweden), which has high reliability of weight loading for anaerobic testing or training [13], and one SIT-specific multi-ball training session. Pedaling is a closed-chain exercise, and is relatively easier for the players to acquire correct technique and to achieve expected training effect from an injury-prevention perspective. Moreover, it is practically applicable to indoor badminton courts training during winter or in bad weather condition. The training intervention was designed and modified based on the previous literature [10,14,15]. The detailed training plan and description are shown in Table 2.

The Polar Team2 System (Polar Electro Oy, Kemple, Finland) was used to monitor the heart rate of each player throughout each training session, with data later extracted from custom-specific software (Polar Team2, Electro Oy, Kemple, Finland), in order to obtain maximum heart rate (HRmax), time spent in each HRmax% zone and Training impulse (TRIMP). TRIMP takes into account the training duration and intensity at the same time, and reflects the comprehensive effect of training on the internal and external load of the athlete's body, as well as the load of medium and high intensity training. The method to determine the athlete's TRIMP in the current study is based on the formula proposed by Edwards [16], where the time in each HRmax% zone is multiplied by the corresponding weighting factor for that zone and the results summated (see Table 3 for detailed description of the zone and factors). The HRmax of each player was established using the peak value recorded by the monitoring system during the training.

Table 2. Weekly training plan for two groups during the study.

Group	Monday	Wednesday	Friday
SIT		*SIT Cycling Training* 1–2 min 50 W cycling, prepare to 30 s cycling with full force, the load is 0.075/kg of weight individualized to each player's body weight [17,18], between-group rest: 5 min 5 groups in total	*SIT-specific Multiple Balls Training* 30 s × 8 groups × 2 rounds of multi-ball training, intensity: > 90% HRmax between-group rest: 5 min between-round rest: 8 min
CON		*Traditional Training:* 40 min of Fartlek Run (Intensity: 65–79% HRmax)	*Traditional Multiple Ball Training:* 1 min × 4 groups × 2 rounds of continuous multi-ball training between-group rest: 5 min between-round rest: 8 min

Table 3. HRmax% zones and corresponding weighting factors.

Zone	Weighting Factor	HRmax%
I	1	50–60%
II	2	60–70%
III	3	70–80%
IV	4	80–90%
V	5	90–100%

2.3. Test Program

Before and after 8 weeks of training, four groups all participated in a set of testing, which included YO-YO IR2 intermittent recovery test, analysis of the increasing load gas metabolism and lactate clearance rate test.

2.3.1. YO-YO Intermittent Recovery Test Level 2 (YO-YO IR2 Test)

Speed endurance level is generally reflected by short bursts of repetitive sprints (RS), which requires subjects to try their best to accomplish the fastest speed in each repetitive sprint, and this ability is generally evaluated via the YO-YO IR2 test during field-test [6]. The test is based on increasing and intermittent load protocol, and evaluates player's ability to recover between high intensity intermittent exercises. Moreover, it has been proven to validly monitor training effects [19].

After dynamic warm-up, players perform a combination of running to and fro on a 20 m course and a 10-s interval of active rest after 40 m, and players quit the test when the subjective exhaustion occurs or when they drop behind the required pace or make one of the errors listed below for a second time:

(i)   does not come to a complete stop before starting the next 40 m run;
(ii)   starts the run before the audio signal;
(iii)   does not reach/either line before the audio signal;
(iv)   turns at the 20 m mark without touching or crossing the line (therefore running short).

The starting speed starts at 13 km/h, and increases to 15 km/h, 16 km/h, and then increases by 0.5 km/h thereafter. The final running distance is then recorded. The speed of each bout is controlled by an audio recorder. All subjects were familiarized with the test within a one-minute trial.

2.3.2. Analysis of Increasing Load of Gas Metabolism and Test of Lactate Clearance Rate

An incremental load test was performed using an incremental load treadmill (H/P Cosmos, Germany). Warm-up exercises should be performed for 5–10 min before each test. At the beginning of the test, the starting speed of the treadmill was set at 6 km/h, increasing by 1 km/h per minute, until 16 km/h, when the speed was stopped and the slope increased by 1.5% per minute, until the subject was exhausted. Relevant ventilation indicators such as maximum oxygen uptake (VO_2max), ventilatory anaerobic threshold (VT-VO_2) and ventilatory anaerobic threshold in percentage of VO_2max (VT/VO_2max) were measured using a gas metabolism analyzer (Max I, Physio-Dyne Instrument Corp., New York, USA). Among them, VT is determined according to the following criteria:

In the incremental load test, the VT value is determined (i) when the ratio of ventilation (VE) to carbon dioxide production (VCO2) shows a non-linear increase in the inflection point, and (ii) when the load intensity reaches a certain level, and the ratio of VE to oxygen consumption (VO_2) increases sharply [20]. VT is determined by two independent investigators. When they are not coherent, and if the difference between the two selected results is remarkable, the value of VT needs to be determined again, while, if the difference could be overlooked, the average value is taken.

Next, in order to analyze the aerobic recovery speed of athletes after increasing load and to evaluate their recovery ability after aerobic exercise, blood samples were collected for rest (before testing with players being seated) and 0, 1, 3, 5, 7 and 10 min immediately after the increasing load test via a volume of 20 microliters of fingertip blood. The EKF Biosen s-line automatic blood lactate analyzer (EKF-diagnostic GmbH, Barleben, Germany) was used to measure blood lactate, with the results being later recorded with the lactate clearance rate being calculated using the following formula [21]:

$$LA_{10}\% = \frac{LA_{max} - LA_{10}}{LA_{max} - LA_{rest}} \times 100\%$$

where $LA_{10}\%$ means the lactate clearance rate at 10 min after testing, LA_{max} represents the peak lactic value after testing, LA_{10} is the lactate value at 10 min after testing and LA_{rest} the value of lactate before testing.

2.4. Statistical Analysis

Experimental data were processed by SPSS statistical software package (version 23.0, Chicago, IL, USA); all test results before and after training were presented using the average $\pm$ standard deviation ($\bar{x} \pm s$). The normality of the tests results was checked before the subsequent analysis. A repetitive measure analysis of variance was then used to compare the within and between group difference in test outcomes for both genders, with the statistical significance level defined as $p < 0.05$. Pairwise differences and post hoc comparisons were tested with the Bonferroni post hoc test. Besides, the effect size (ES) was calculated using Cohen's d to quantify the amount of change before and after each group of training and to reflect the comparison of training effects between SIT and CON groups based on the following scales: <0.2 trivial, 0.2–0.6 small, 0.6–1.2 moderate, 1.2–2.0 large and >2.0 very large [22].

3. Results

3.1. Training Intensity and Time Used During Training

Table 4 shows the descriptive statistics of heartrate and time within the 8-week training, and the results show that the average heart rate and maximum heart rate of both male and female SIT groups during training were significantly higher than those of the CON groups ($p < 0.05$). Moreover, the effective training time of the former was significantly less than that of the latter ($p < 0.05$).

Table 4. Intensity monitoring during training.

Group	N	Avg HR (bpm)	HRmax (bpm)	Total Training Time (min)	Effective Training Time (min)
Male-SIT	8	132.7 ± 7.3 *	190.7 ± 8.8 *	52.7 ± 4.1	19.8 ± 3.0 *
Male-CON	8	126.0 ± 10.2	169.8 ± 6.2	78.5 ± 4.5	40.2 ± 1.8
Female-SIT	8	134.1 ± 6.0 *	181.9 ± 8.9 *	52.7 ± 4.1	19.8 ± 3.0 *
Female-CON	8	115.4 ± 8.4	169.4 ± 8.5	78.5 ± 4.5	40.2 ± 1.8

Note: Values are expressed as means ± SD. * indicates significant difference between SIT and CON group, $p < 0.05$.

During the 8-week training, the mean weekly effective training time (time spent within 50–100% HRmax zone) and TRIMP in the 80–100% HRmax intensity range of the Male-SIT group were significantly higher than those in the Male-CON group ($p < 0.05$), while the total weekly effective training time and TRIMP for the former were significantly lower than the latter ($p < 0.05$). As for female players, the average weekly effective training time and TRIMP in the 90–100%HRmax intensity range for the Female-SIT group were significantly higher than those in the Female-CON group ($p < 0.05$). However, in the intensity range of 80–90% HRmax, no differences were found between the F-SIT and F-CON groups. The overall effective training time and TRIMP for Female-SIT were significantly lower than those in Female-CON as well ($p < 0.05$), as is shown in Figure 1.

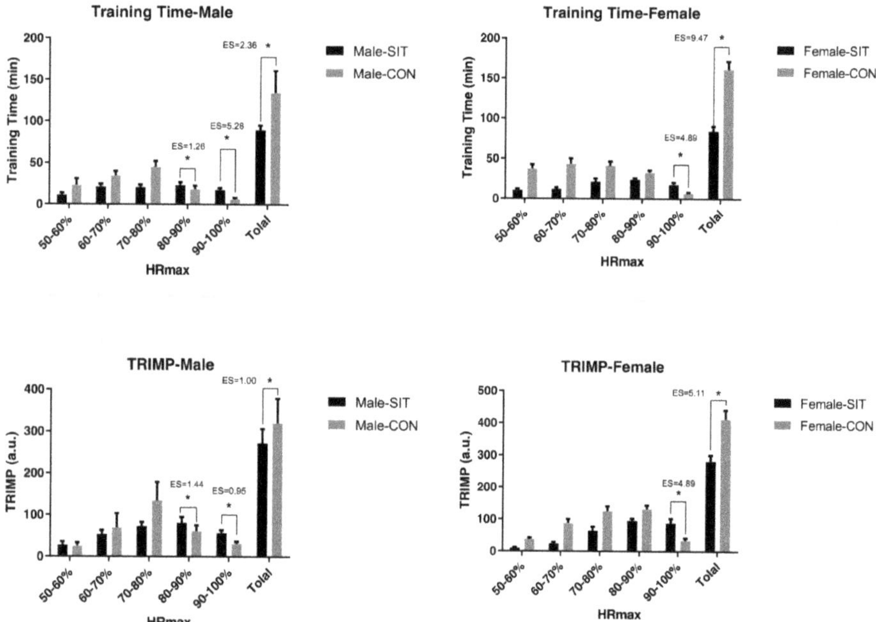

Figure 1. Comparisons of weekly effective training time and training impulse (TRIMP) between SIT and CON groups. Note: * Indicates a significant difference between SIT and CON group, $p < 0.05$.

3.2. Comparisons of Testing Results Before and After Training Intervention

After training, the running distance of the YO-YO IR2 and the lactate clearance rate at 10 min after testing ($LA_{10}\%$) significantly increased in both the Male-SIT and the Female-SIT group ($p < 0.05$), and such improvement was significantly higher than that of the CON groups ($p < 0.05$), as is shown in Figure 2.

Meanwhile, as Table 5 demonstrates, VO_2max, $VT\text{-}VO_2$ and VT/VO_2max for the SIT group significantly improved after the intervention ($p < 0.05$), and the improvement was significantly higher than that in the Male-CON group and Female-CON group ($p < 0.05$), as is shown in Table 4.

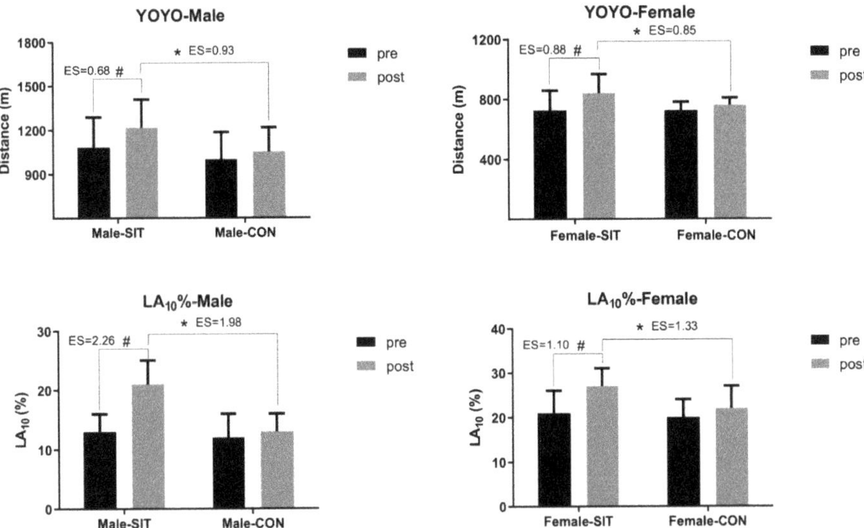

Figure 2. Within- and between-group differences in YO-YO Intermittent Recovery Test Level 2 (IR2) distance and lactate clearance rate for SIT and CON groups before and after intervention. Note: * Indicates significant difference between SIT and CON group, $p < 0.05$; # indicates significant within-group difference before and after intervention, $p < 0.05$; ES: effect size.

Table 5. Within- and between-group differences in gas metabolism analysis for SIT and CON groups before and after intervention.

Group		VO_2max (mL/kg/min)	VT-VO_2 (mL/min)	VT/VO_2max (%)
Male-SIT	pre	56.8 ± 7.0	3088.4 ± 450.9	73.8 ± 9.7
	post	63.6 ± 4.7 [#,*]	3665.3 ± 263.5 [#,*]	84.8 ± 3.37 [#,*]
	ES	1.14	1.56	1.51
Male-CON	pre	55.8 ± 8.0	2962.5 ± 743.4	80.2 ± 1.5
	post	57.7 ± 6.7	3004.1 ± 738.1	80.8 ± 2.3
	ES	0.26	0.06	0.31
Female-SIT	pre	42.5 ± 2.9	1940.1 ± 112.9	0.75 ± 0.04
	post	46.2 ± 3.0 [*,#]	2176.9 ± 78.6 [*,#]	0.82 ± 0.04 [*,#]
	ES	1.28	1.11	1.75
Female-CON	pre	42.9 ± 1.6	1930.7 ± 151.8	0.75 ± 0.06
	post	43.3 ± 2.1	2055.3 ± 160.7	0.78 ± 0.08
	ES	0.21	0.79	0.42

Note: * Indicates significant difference between SIT and CON group, $p < 0.05$; # indicates significant within-group difference before and after intervention, $p < 0.05$; ES: effect size.

4. Discussion

This study was aimed to explore the effect of 8-weeks of SIT on the aerobic capacity of badminton players. The results showed that their performance in the YO-YO IR2 test, the lactate clearance rate, VO_2max, VT-VO_2 and VT/VO_2max were significantly enhanced in a time-efficient manner, compared to the control group, which confirms the hypothesis of this research.

The badminton match is highly demanding to a player's aerobic capacity due to the differences in individual physical fitness and the appearance of the new scoring model [4,5]. Under such situation, the competition rhythm is obviously accelerated and the proportion of multiple rallies is gradually increased, which forces players to endure longer periods of rapid and repeated accelerations and decelerations [23]. In the Male-SIT and Female-SIT groups, the time spent and TRIMP values in the 80–100% HRmax intensity interval accounted for the highest proportion, suggesting that SIT enables the body to complete multiple short-time and high-intensity outputs under the continuous incomplete recovery state, which is more in line with the current badminton competition characteristics and demands. In contrast to the previous training programs where all subjects routinely undertook Fartlek running, the 30-s SIT is a training mode closer to the maximum physiological load intensity of players, and the features of its time structure are also closer to the actual combat of badminton competition. It is an effective training program to improve a player's aerobic capacity. Previously, it was reported that SIT could induce skeletal muscle metabolism, increase capillaries and mitochondrial proliferation, enhance oxidase activity and improve peripheral vascular function and peripheral fitness of skeletal muscle [24]. When training intensity exceeded 90%VO$_2$max, SIT could simultaneously improve oxygen uptake and transport ability of the cardiopulmonary system and skeletal muscle [25]. Ermanno et al., found that intermittent exercise could activate the energy supply of the aerobic system in advance and reduce the proportion of the energy supply of the anaerobic system, thus delaying the generation of fatigue [25]. These changes in the body were physiological feedback for SIT. While the training improved the player's ability to maintain high-intensity exercise for a long time in competition and training, their ability to recover from fast running could be improved, consequently achieving the goal of improving aerobic capacity.

Moreover, this study found that after 8 weeks of SIT, players' VO$_2$max, VT-VO$_2$ and VT/VO$_2$max increased significantly, implying that the proportion of exercise intensity lower than the anaerobic threshold for the body was increased under the same testing protocol. The time players take to enter the anaerobic glycolytic process would be postponed, thus reducing the consumption of glycogen. At the same time, the movement of the body would become more efficient, and eventually the maximum aerobic capacity of the players would be improved [26]. Previous studies showed that by inducing skeletal muscle metabolism, SIT could increase capillary proliferation, mitochondrial proliferation, enhance the activity and oxidation of glycolytic oxidase and improve peripheral vascular function and skeletal muscle peripheral adaptability [14,27]. Studies with similar schemes applied to the general population showed that the oxygen uptake and transport capacity were improved via a series of changes, such as increased capillary density and blood volume, decreased heart rate and increased stroke output, when the same exercise intensity was completed. At this time, the body showed certain adaptability. In practice, with the improvement of the body's oxygen uptake ability, elite badminton players could prolong the time of oxygen supply and enter the hypoxemia state later in the competition, which could effectively improve their match performance during the competition. Besides, although some research showed certain discrepancies in results, we found that after the SIT, no significant increase in VO$_2$max was indicated. It would be inferred that the effect of such modality would be conditioned by factors such as the level of training, whether the subjects undertake regular training and the body weight.

Aerobic recovery ability has a direct impact on players' on-court performance. High-intensity and high-load activity during competition would produce physiological fatigue and large amount of lactate accumulation in the skeletal muscle. Changes in the internal responses of the body may cause players' physical dysfunction and decline in athletic performance [28,29]. Therefore, rapid recovery ability is the key prerequisite for decent physical and technical performance during the competition. This study analyzed the changes in skeletal muscle's oxygen recovery ability from both physiological and biochemical perspectives.

Blood lactate is one of the most commonly used biochemical indicators to detect the body fatigue recovery status [30], and the accumulation of lactate may indirectly lead to reduced performance, because the conversion of lactic acid to lactate releases H$^+$ that leads to a metabolic acidosis with subsequent inhibition of glycolytic rate-limiting enzymes, lipolysis and contractility of the skeletal muscles [31]. From the results, it was shown that SIT performed at a higher level of intensity could positively influence the clearance of lactate after exercise, increasing intra-cellular alkali reserve and slowing the pH reduction in muscles, and delaying the onset of fatigue [21]. Consequently, players' ability to recover from intermittent activities was enhanced and they would be better prepared for the next point and game [32]. In particular, at the final game and the last points of each game, each point would be ended with prolonged multiple-strokes and high-intensity movements, which might even last couple of minutes. Under such circumstances, possessing rapid aerobic recovery would become a key factor determining elite player's aerobic endurance and technical-tactical performance in the next point [33]. In the study conducted by Jones et al. [10], near-infrared spectroscopy was used to measure muscle oxygenation of the vastus lateralis of elite women hockey players for SIT groups and endurance training groups. Their results showed that there were significant increases in tissue deoxyhaemoglobin + deoxymyoglobin (HHb + HMb) and tissue oxygenation (TSI%), and a significant decrease in tissue oxyhaemoglobin + oxymyoglobin (HbO2 + MbO2), which indicated 'positive peripheral muscle oxygen adaptations' occurring in response to SIT training. Moreover, existing literature also stated that the higher exercise intensity provided during SIT would increase the probability of favorable adaptations in both type one and type two fibers as opposed to the generally lower intensity of endurance training [34]. Although as a limitation, the current study was unable to measure blood saturation, it could be implied that the SIT protocol might promote the skeletal muscle oxidative capacity of badminton players after the training. Nonetheless, future studies should look into the changes of EPOC (excess post-exercise oxygen consumption), body temperature and ventilation to comprehensively verify the improvement in recovery after such intervention.

5. Conclusions

Eight-week SIT effectively improved the aerobic exercise capacity of elite badminton players, particularly considering oxygen uptake and recovery ability, and the adaptability of skeletal muscle to exercising load. Eventually, the rapidly repeated sprint ability and physical performance of players were enhanced. The study has provided evidence-based findings that as a time-efficient training alternative, SIT could be suitable to be included in the training routine for badminton players.

However, it is acknowledged that this study also has certain limitations. The technical and tactical performance was not considered, which might be another representative indicator of improved aerobic capacity. Moreover, anaerobic endurance training, strength training and functional training are also of vital importance for badminton players and their joint effect on aerobic training was not investigated within the current program. Future research is suggested to look into these aspects to better inform sport-specific training prescription.

Author Contributions: Conceptualization, H.L., Q.L. and D.B.; methodology, H.L., B.L., Q.L. and D.B.; software, Q.L. and Y.C.; validation, H.L., Q.L., D.B. and Y.C.; formal analysis, H.L., Q.L. and Y.C.; investigation, H.L., Q.L. and Y.L.; resources, B.L. and D.B.; data curation, H.L. and Q.L.; writing—original draft preparation, H.L., Q.L., D.B. and Y.C.; writing—review and editing, H.L., Q.L. and Y.C.; visualization, Q.L. and Y.C.; supervision, D.B. and Y.C.; project administration, D.B. and Y.C.; funding acquisition, D.B. and Y.C. All authors have read and agreed to the published version of the manuscript.

Funding: This work was supported in part by the National Key Research and Development Program of China under grants 2020AAA0103404 and 2018YFC2000600, and by National Natural Science Foundation of China under grant 72071018. The corresponding author (Y.C.) was supported by the China Postdoctoral Science Foundation (2020T130067).

Institutional Review Board Statement: The study was conducted according to the guidelines of the Declaration of Helsinki, and approved by the Ethics Committee of Beijing Sport University (2020008H, 17/01/2020).

Informed Consent Statement: Informed consent was obtained from all subjects involved in the study.

Conflicts of Interest: The authors declare no conflict of interest.

References

1. Jan, C.; Petr, S. Serve and Return in Badminton: Gender Differences of Elite Badminton Players. *Int. J. Phys. Educ. Fit. Sports* **2020**, *9*, 44–48. [CrossRef]
2. Nhan, D.T.; Klyce, W.; Lee, R.J. Epidemiological Patterns of Alternative Racquet-Sport Injuries in the United States, 1997–2016. *Orthop. J. Sports Med.* **2018**, *6*, 2325967118786237. [CrossRef] [PubMed]
3. Phomsoupha, M.; Laffaye, G. The science of badminton: Game characteristics, anthropometry, physiology, visual fitness and biomechanics. *Sports Med.* **2015**, *45*, 473–495. [CrossRef] [PubMed]
4. Laffaye, G.; Phomsoupha, M.; Dor, F. Changes in the Game Characteristics of a Badminton Match: A Longitudinal Study through the Olympic Game Finals Analysis in Men's Singles. *J. Sports Sci. Med.* **2015**, *14*, 584–590. [PubMed]
5. Faude, O.; Meyer, T.; Rosenberger, F.; Fries, M.; Huber, G.; Kindermann, W. Physiological characteristics of badminton match play. *Eur. J. Appl. Physiol.* **2007**, *100*, 479–485. [CrossRef]
6. Bangsbo, J.; Iaia, F.M.; Krustrup, P. The Yo-Yo intermittent recovery test: A useful tool for evaluation of physical performance in intermittent sports. *Sports Med.* **2008**, *38*, 37–51. [CrossRef]
7. Kelly, D.T.; Tobin, C.; Egan, B.; McCarren, A.; O'Connor, P.L.; McCaffrey, N.; Moyna, N.M. Comparison of Sprint Interval and Endurance Training in Team Sport Athletes. *J. Strength Cond. Res.* **2018**, *32*, 3051–3058. [CrossRef]
8. Buchan, D.S.; Ollis, S.; Young, J.D.; Thomas, N.E.; Cooper, S.M.; Tong, T.K.; Nie, J.; Malina, R.M.; Baker, J.S. The effects of time and intensity of exercise on novel and established markers of CVD in adolescent youth. *Am. J. Hum. Biol.* **2011**, *23*, 517–526. [CrossRef]
9. Bayati, M.; Farzad, B.; Gharakhanlou, R.; Agha-Alinejad, H. A practical model of low-volume high-intensity interval training induces performance and metabolic adaptations that resemble 'all-out' sprint interval training. *J. Sports Sci. Med.* **2011**, *10*, 571–576.
10. Jones, B.; Hamilton, D.K.; Cooper, C.E. Muscle Oxygen Changes following Sprint Interval Cycling Training in Elite Field Hockey Players. *PLoS ONE* **2015**, *10*, e0120338. [CrossRef]
11. Burgomaster, K.A.; Heigenhauser, G.J.; Gibala, M.J. Effect of short-term sprint interval training on human skeletal muscle carbohydrate metabolism during exercise and time-trial performance. *J. Appl. Physiol.* **2006**, *100*, 2041–2047. [CrossRef] [PubMed]
12. Gibala, M.J.; Little, J.P.; van Essen, M.; Wilkin, G.P.; Burgomaster, K.A.; Safdar, A.; Raha, S.; Tarnopolsky, M.A. Short-term sprint interval versus traditional endurance training: Similar initial adaptations in human skeletal muscle and exercise performance. *J. Physiol.* **2006**, *575*, 901–911. [CrossRef] [PubMed]
13. Lunn, W.R.; Axtell, R.S. Validity and Reliability of the Lode Excalibur Sport Cycle Ergometer for the Wingate Anaerobic Test. *J. Strength Cond. Res.* **2019**. [CrossRef] [PubMed]
14. Gist, N.H.; Fedewa, M.V.; Dishman, R.K.; Cureton, K.J. Sprint interval training effects on aerobic capacity: A systematic review and meta-analysis. *Sports Med.* **2014**, *44*, 269–279. [CrossRef]
15. Hostrup, M.; Gunnarsson, T.P.; Fiorenza, M.; Mørch, K.; Onslev, J.; Pedersen, K.M.; Bangsbo, J. In-season adaptations to intense intermittent training and sprint interval training in sub-elite football players. *Scand. J. Med. Sci. Sports* **2019**, *29*, 669–677. [CrossRef]
16. Edwards, S. The Heart Rate Monitor Book. *Med. Sci. Sports Exerc.* **1994**, *26*, 647. [CrossRef]
17. Bar-Or, O.; Dotan, R.; Inbar, O. A 30 second all-out ergometric test-its reliability and validity for anaerobic capacity. *Isr. J. Med. Sci.* **1977**, *13*, 126.
18. Laurent, C.M.; Meyers, M.C.; Robinson, C.A.; Green, J.M. Cross-validation of the 20-versus 30-s Wingate anaerobic test. *Eur. J. Appl. Physiol.* **2007**, *100*, 645–651. [CrossRef]
19. Schmitz, B.; Pfeifer, C.; Kreitz, K.; Borowski, M.; Faldum, A.; Brand, S.M. The Yo-Yo Intermittent Tests: A Systematic Review and Structured Compendium of Test Results. *Front. Physiol.* **2018**, *9*, 870. [CrossRef]
20. Binder, R.K.; Wonisch, M.; Corra, U.; Cohen-Solal, A.; Vanhees, L.; Saner, H.; Schmid, J.-P. Methodological approach to the first and second lactate threshold in incremental cardiopulmonary exercise testing. *Eur. J. Cardiovasc. Prev. Rehabil.* **2008**, *15*, 726–734. [CrossRef]
21. Wang, J.; Qiu, J.; Yi, L.; Hou, Z.; Benardot, D.; Cao, W. Effect of sodium bicarbonate ingestion during 6 weeks of HIIT on anaerobic performance of college students. *J. Int. Soc. Sports Nutr.* **2019**, *16*, 1–10. [CrossRef]
22. Hopkins, W.; Marshall, S.; Batterham, A.; Hanin, J. Progressive statistics for studies in sports medicine and exercise science. *Med. Sci. Sports Exerc.* **2009**, *41*, 3–12. [CrossRef] [PubMed]
23. Gomez, M.A.; Rivas, F.; Connor, J.D.; Leicht, A.S. Performance Differences of Temporal Parameters and Point Outcome between Elite Men's and Women's Badminton Players According to Match-Related Contexts. *Int. J. Environ. Res. Public Health* **2019**, *16*. [CrossRef]
24. Buchheit, M.; Abbiss, C.R.; Peiffer, J.J.; Laursen, P.B. Performance and physiological responses during a sprint interval training session: Relationships with muscle oxygenation and pulmonary oxygen uptake kinetics. *Eur. J. Appl. Physiol.* **2012**, *112*, 767–779. [CrossRef] [PubMed]
25. Koral, J.; Oranchuk, D.J.; Herrera, R.; Millet, G.Y. Six Sessions of Sprint Interval Training Improves Running Performance in Trained Athletes. *J. Strength Cond. Res.* **2018**, *32*, 617–623. [CrossRef] [PubMed]

26. Connolly, D.A. The anaerobic threshold: Over-valued or under-utilized? A novel concept to enhance lipid optimization! *Curr Opin. Clin. Nutr. Metab. Care* **2012**, *15*, 430–435. [CrossRef] [PubMed]
27. Phillips, S.M.; Sproule, J.; Turner, A.P. Carbohydrate ingestion during team games exercise: Current knowledge and areas for future investigation. *Sports Med.* **2011**, *41*, 559–585. [CrossRef] [PubMed]
28. Meeusen, R.; Watson, P.; Hasegawa, H.; Roelands, B.; Piacentini, M.F. Central fatigue: The serotonin hypothesis and beyond. *Sports Med.* **2006**, *36*, 881–909.
29. Welsh, R.S.; Davis, J.M.; Burke, J.R.; Williams, H.G. Carbohydrates and physical/mental performance during intermittent exercise to fatigue. *Med. Sci. Sports Exerc.* **2002**, *34*, 723–731. [CrossRef]
30. Robergs, R.A.; McNulty, C.R.; Minett, G.M.; Holland, J.; Trajano, G. Lactate, not lactic acid, is produced by cellular cytosolic energy catabolism. *Physiology* **2018**, *33*, 10–12.
31. Menzies, P.; Menzies, C.; McIntyre, L.; Paterson, P.; Wilson, J.; Kemi, O.J. Blood lactate clearance during active recovery after an intense running bout depends on the intensity of the active recovery. *J. Sports Sci.* **2010**, *28*, 975–982. [CrossRef] [PubMed]
32. Fukuoka, Y.; Iihoshi, M.; Nazunin, J.T.; Abe, D.; Fukuba, Y. Dynamic Characteristics of Ventilatory and Gas Exchange during Sinusoidal Walking in Humans. *PLoS ONE* **2017**, *12*, e0168517. [CrossRef] [PubMed]
33. Andersen, L.L.; Larsson, B.; Overgaard, H.; Aagaard, P. Torque–velocity characteristics and contractile rate of force development in elite badminton players. *Eur. J. Sport Sci.* **2007**, *7*, 127–134. [CrossRef]
34. Hood, D.A. Invited Review: Contractile activity-induced mitochondrial biogenesis in skeletal muscle. *J. Appl. Physiol.* **2001**, *90*, 1137–1157. [CrossRef]

International Journal of
Environmental Research and Public Health

Article

Using Anthropometric Data and Physical Fitness Scores to Predict Selection in a National U19 Rugby Union Team [†]

Luis Vaz [1,2,*], Wilbur Kraak [3], Marco Batista [4], Samuel Honório [4] and Hélder Miguel Fernandes [2,5]

1   Department of Sport Science, Exercise and Health, School of Life and Environmental Sciences, University of Trás-os-Montes and Alto Douro, 5000-801 Vila Real, Portugal
2   Research Centre in Sports Sciences, Health Sciences and Human Development (CIDESD), 5000-801 Vila Real, Portugal; hmfernandes@gmail.com
3   Department of Sport Science, Faculty of Medicine and Health Sciences, Stellenbosch University, Stellenbosch 7600, South Africa; kjw@sun.ac.za
4   Department of Sports and Well-Being, Polytechnic Institute of Castelo Branco, SHERU (Sport, Health and Exercise Research Unit), 6000-084 Castelo Branco, Portugal; marco.batista@ipcb.pt (M.B.); samuelhonorio@ipcb.pt (S.H.)
5   Research in Education and Community Intervention, RECI, 3500 Viseu, Portugal
*   Correspondence: lvaz@utad.pt; Tel.: +351-965079370
†   A preliminary version of this paper was presented as oral communication in the International Seminar of Physical Education, Leisure and Health, 17–19 June 2019. Castelo Branco, Portugal.

Abstract: The purpose of this study was to compare measures of anthropometry characteristics and physical fitness performance between rugby union players (17.9 ± 0.5 years old) recruited (n = 39) and non-recruited (n = 145) to the Portuguese under-19 (U19) national team, controlling for their playing position (forwards or backs). Standardized anthropometric, physical, and performance assessment tests included players' body mass and height, push up and pull-up test, squat test, sit-and-reach test, 20 m shuttle run test, flexed arm hang test, Sargent test, handgrip strength test, Illinois agility test, and 20-m and 50-m sprint test. Results showed that recruited forwards players had better agility scores ($p = 0.02$, ES = −0.55) than the non-recruited forwards, whereas recruited backs players had higher right ($p < 0.01$, ES = 0.84) and left ($p = 0.01$, ES = 0.74) handgrip strength scores than their counterparts. Logistic regression showed that better agility (for the forwards) and right handgrip strength scores (for the backs) were the only variables significantly associated with an increased likelihood of being recruited to the national team. In sum, these findings suggest that certain well-developed physical qualities, namely, agility for the forwards players and upper-body strength for the back players, partially explain the selection of U19 rugby players to their national team.

Keywords: talent identification; rugby union; prediction of performance outcomes; selective factors.

check for
updates

Citation: Vaz, L.; Kraak, W.; Batista, M.; Honório, S.; Miguel Fernandes, H. Using Anthropometric Data and Physical Fitness Scores to Predict Selection in a National U19 Rugby Union Team. *Int. J. Environ. Res. Public Health* **2021**, *18*, 1499. https://doi.org/10.3390/ijerph18041499

Academic Editor: Mathieu Gayda
Received: 21 December 2020
Accepted: 26 January 2021
Published: 5 February 2021

Publisher's Note: MDPI stays neutral with regard to jurisdictional claims in published maps and institutional affiliations.

1. Introduction

The expanded professionalism in rugby union has elicited fast changes in the athletic profile of rugby elite players [1]. Effective talent identification is an important issue in rugby union, as success at the international level conveys significant commercial benefits. Usually, the recruitment and development of rugby union players have focused on training interventions and game-play performance, with less attention on the anthropometric and physical performance characteristics [2]. However, the understanding of the rugby union movement patterns during competition, and the related physical demands, is critical to developing effective rugby training programs aiming to improve both athletes' and teams' performance [3]. Previous investigations have shown differences in fitness and morphological aspects between teams [4,5], levels of competition [6,7], and players' positions [8,9].

According to this reasoning, some studies [10] reported that elite rugby players (backs and forwards) have specific skills, motor and physical abilities, and anthropometrical characteristics that distinguish them from the non-elite players. Under this scope, many studies have stressed the importance of anthropometric characteristics on team success in rugby union [11,12]. Indeed, previous studies demonstrated that teams with stronger players who possess heavier forwards and faster backs frequently achieve high success in elite international and club competitions; some physical fitness and player performance measures have also been found to be associated with game-related statistics [13]. Indeed, anthropometric characteristics, such as sprinting, tackling, muscular strength, and aerobic and anaerobic power, were reported as important factors to play at the elite level and illustrate the heterogeneous nature among rugby players.

The efficacy of training in youth and adolescent rugby players has been investigated [14], showing that substantial developments can be achieved with different approaches of effective rugby training programs to improve performance. However, the specific demands of competition differ markedly between forwards and backs [15], making it necessary to better understand these (potential) differences across the different playing positions.

Considering the important implications for rugby union, it is understandable that a degree of reluctance occurs concerning the explicit requirements in the prediction of performance outcomes in young talented rugby union players. Therefore, it is important to further investigate how rugby players' anthropometry characteristics and physical and performance assessment tests can help to recruit, select, and develop rugby union elite players progression. The assessment of elite rugby players performance is frequently performed as part of their routine monitoring procedures, both to improve high-performance models for the elite player and monitor the player success according to training regimens. Considering the abovementioned information, the purpose of the current study is to compare measures of anthropometry characteristics and physical fitness performance between recruited and non-recruited rugby union players of under-19 (U19) national team, controlling for their playing position (forwards or backs).

2. Materials and Methods

2.1. Subjects

A total of 184 male junior rugby players from 2017/2018 Portuguese rugby academies were evaluated (age 17.9 ± 0.5 years old, height 1.79 ± 0.7 m, and body mass 84.2 ± 13.5 kg). Subjects were separated into two groups, based on recruited (n = 39) and non-recruited (n = 145), to represent the 2018 Portuguese rugby U19 national team, as well as according to their playing position (forwards or backs).

To be recruited, eligibility criteria to represent U19 national team included skills, selection policy, and principles established from "checklists" of technical and medical staff selectors (e.g., general knowledge and practical knowledge of the game, eligibility, time and availability, purpose of selection, personal and team profile ideas and systems, attributes, rugby experience, physical conditioning relative to rugby and specific positions, attitude, development, and game plan). All subjects were in the four-week pre-season period of their training plan, which included developing physical conditioning, strength and explosive power, acceleration, and power endurance and stamina. Throughout the preparation for the national competition, subjects performed 5 to 6 rugby training sessions with their clubs, usually 10 to 12 h weekly.

During the training camp week, all recruited players trained with the U19 national team twice a day (training volume of 4 to 6 h per day) and completed specific strength training sessions focused on the development of player´s physical conditioning.

After being fully informed about the experimental procedures, subjects agreed to participate in this study, and written guardian or parental consent was obtained before subjects were permitted to participate. Additionally, the subjects were informed that they were free to withdraw at any time without penalty.

None of the subjects reported any physical, psychological, or orthopedic limitation or injury that could limit their full participation. The study protocol was performed following the guidelines stated in the Declaration of Helsinki and was approved by the institutional University Ethical Committee (CE.UTAD 27/2016).

2.2. Anthropometric, Physical, and Performance Assessment Tests

Fitness testing results from the rugby elite development national academies representative squads were collected for each participant during the pre-season phase in the national camp each year. The total sample executed a set of anthropometric measures and physical tests on the initial day of the training camp week, as planned by the Portugal rugby U19 technical staff. All the tests were accomplished in a single day and under similar environmental conditions; the local temperature in October was $14.0 \pm 2\,°C$, compared to $16.1 \pm 2\,°C$ in March. While the anthropometrics assessments started in the morning ($\approx$9 a.m.) in a kinanthropometry laboratory, the physical performance assessment tests were held in the afternoon and were performed in outdoor natural turf rugby pitch and on running track (finished at $\approx$5 p.m.).

All assessments were carried out by medical and technical supervision of the staff of Portuguese rugby U19 team. Testing was led by competent technical staff who were adequately trained to use the devices and had the ability to perform the testing protocols.

A full dynamic warm-up drill with a gradual progression through a series of rugby movements was performed, including running at a moderate intensity, squats for 5-min, followed by 5-min of lunge twist active stretching and detailed exercises (e.g., submaximal jumps and accelerations before Sargent test, agility test, and acceleration and speed tests). After the warm-up ($\pm$3-min), players were required to perform the tests. Moreover, stretching exercises and a standardized recovery protocol were completed by all players after the final test sessions of the day. As part of these players' training routines in rugby academies, all subjects were highly familiarized with the anthropometric and physical performance tests.

Players were allowed to ingest water during recovery periods (3 ± 2 min.). Anthropometrics measurements only included body height (m) and body mass (kg), whereas fitness testing battery included pull-up test, push up test, and free-standing squat [16]. The weight lifted and the number of repetitions done was used to calculate the 1RM [17]. The outdoor assessment tests included: Sargent test, flexed arm hang test, sit-and-reach test, and maximal aerobic power (20 m shuttle-run test) [18]. The VO_2 max was estimated using regression equations described by [19]. Handgrip strength test (left and right hand) was performed as described by [20]. Acceleration and speed (10, 20, and 50 m sprint) using electronic timing gates and Illinois agility test (10×5 m) [21] were performed. A more detailed description of all tests may also be found in Table S1 (Supplementary Material). Test-retest reliability, typical error, and validity of measurement of all assessment tests used for this study have been previously confirmed in literature and their descriptions are documented elsewhere [22–24].

2.3. Procedures and Data Processing

Initially, data were inspected for normality using visual inspection (histograms and Q-Q plots of residuals) and analysis of skewness and kurtosis. Both procedures showed appropriate distributions and values.

Independent t-tests were used to identify significant differences between groups. Cohen's d effect size statistics were calculated (ES), and respective 95% confidence intervals were also computed. Magnitudes of observed effects, and thresholds, were 0–0.20, trivial; 0.20–0.60, small; 0.60–1.20, moderate; 1.20–2.00, large; and >2.00, very large [25].

Multivariable forward logistic regression analyses [26] were performed separately for each playing position in order to identify potential significant predictors of being selected to the national team.

Statistical significance was set at $p < 0.05$, and calculations were carried out using SPSS software, 23.0, Version (IBM SPSS Statistics for Windows, Armonk, NY, USA: IBM Corp.).

3. Results

The descriptive results of the anthropometry characteristics and physical fitness measures are shown in Table 1 for the total sample and for playing positions (forwards and backs).

Table 1. Descriptive statistics of the anthropometry characteristics and physical fitness performance measures for the total sample and for playing positions.

Variables	Total Sample (n = 184) M ± SD	Forwards (n = 106) M ± SD	Backs (n = 78) M ± SD
Body mass (kg)	84.28 ± 13.53	87.58 ± 13.38	79.81 ± 12.48
Body height (m)	1.79 ± 0.07	1.81 ± 0.07	1.70 ± 0.05
Push up test (reps)	43.25 ± 11.10	44.68 ± 12.00	41.30 ± 9.46
Pull up test (reps)	11.83 ± 4.29	10.96 ± 4.67	13.00 ± 3.40
Flexed arm hang test (s)	40.42 ± 13.63	39.75 ± 13.99	41.33 ± 13.15
Handgrip strength—right (kg)	48.00 ± 4.88	47.84 ± 4.16	48.21 ± 5.75
Handgrip strength—left (kg)	46.03 ± 4.65	45.28 ± 4.00	47.06 ± 5.27
Free standing squat (1RM)	140.77 ± 31.09	143.80 ± 31.89	136.67 ± 29.68
Sargent test (cm)	44.95 ± 5.06	45.07 ± 5.49	44.80 ± 4.45
20 m sprint (s)	2.99 ± 0.18	3.08 ± 0.13	2.89 ± 0.18
50 m sprint (s)	6.85 ± 0.27	6.91 ± 0.27	6.78 ± 0.26
Illinois agility test (s)	15.15 ± 0.96	15.37 ± 1.15	14.86 ± 0.49
Sit-and-reach test (cm)	32.93 ± 4.88	31.85 ± 5.26	34.38 ± 3.90
VO_2max (mL kg^{-1} min^{-1})	49.69 ± 3.96	49.43 ± 3.93	50.03 ± 3.99

Forward players had greater body mass ($p < 0.001$, ES = 0.60), body height ($p < 0.01$, ES = 0.48), and push up performance ($p = 0.04$, ES = 0.31), whereas back players showed greater pull up ($p < 0.01$, ES = -0.49), left handgrip strength ($p = 0.01$, ES = -0.39), 20 m sprint ($p < 0.001$, ES = 1.26), 50 m sprint ($p < 0.001$, ES = 0.51), agility ($p < 0.001$, ES = 0.56), and flexibility performance ($p < 0.001$, ES = -0.53). No between-group significant differences were observed in the flexed arm hang (ES = -0.12), right handgrip strength (ES = -0.08), free standing squat (ES = 0.23), Sargent (ES = 0.05), and VO_2max tests (ES = -0.15).

Table 2 presents the comparison results of the anthropometry characteristics and physical fitness scores between the recruited (n = 23) and non-recruited forwards (n = 83).

Table 2. Descriptive and comparative statistics of the anthropometry characteristics and physical fitness performance between recruited and non-recruited forwards.

Variables	Recruited (n = 23) M ± SD	Non-Recruited (n = 83) M ± SD	p	ES (95%CI)
Body mass (kg)	89.97 ± 14.36	86.91 ± 13.11	0.33	0.23 (-0.24, 0.69)
Body height (m)	1.82 ± 0.11	1.81 ± 0.07	0.31	0.24 (-0.22, 0.71)
Push up test (reps)	43.83 ± 14.56	44.92 ± 11.29	0.70	-0.09 (-0.55, 0.37)
Pull up test (reps)	11.78 ± 5.13	10.74 ± 4.54	0.34	0.22 (-0.24, 0.69)
Flexed arm hang test (s)	36.60 ± 10.44	40.62 ± 14.76	0.22	-0.29 (-0.75, 0.18)
Handgrip strength—right (kg)	48.53 ± 5.39	47.65 ± 3.76	0.38	0.21 (-0.25, 0.67)
Handgrip strength—left (kg)	46.35 ± 5.04	44.98 ± 3.64	0.15	0.34 (-0.12, 0.81)
Free standing squat (1RM)	140.57 ± 34.36	144.70 ± 31.33	0.59	-0.13 (-0.59, 0.33)
Sargent test (cm)	44.96 ± 4.71	45.10 ± 5.71	0.91	-0.03 (-0.49, 0.44)
20 m sprint (s)	3.05 ± 0.14	3.09 ± 0.13	0.16	-0.33 (-0.80, 0.13)
50 m sprint (s)	6.88 ± 0.31	6.92 ± 0.26	0.55	-0.14 (-0.60, 0.32)
Illinois agility test (s)	14.89 ± 1.40	15.50 ± 1.04	0.02	-0.55 (-1.01, -0.08)
Sit-and-reach test (cm)	32.09 ± 5.36	31.80 ± 5.26	0.81	0.06 (-0.41, 0.52)
VO_2max (mL kg^{-1} min^{-1})	49.34 ± 4.77	49.46 ± 3.69	0.90	-0.03 (-0.49, 0.43)

Descriptive and comparative results showed that the recruited forwards players had significantly better agility scores than their non-recruited counterparts, with a small to moderate size effect ($p = 0.02$, ES = -0.55). No other significant differences were observed between these groups.

Table 3 presents the comparison results of the anthropometry characteristics and physical fitness scores between the recruited (n = 16) and non-recruited backs (n = 62).

Table 3. Descriptive and comparative statistics of the anthropometry characteristics and physical fitness performance between recruited and non-recruited backs.

Variables	Recruited (n = 16) M ± SD	Non-Recruited (n = 62) M ± SD	p	ES (95%CI)
Body mass (kg)	82.05 ± 8.38	79.23 ± 13.33	0.42	0.23 (−0.33, 0.78)
Body height (m)	1.77 ± 0.06	1.77 ± 0.05	0.69	−0.11 (−0.66, 0.44)
Push up test (reps)	40.94 ± 8.87	41.39 ± 9.67	0.87	−0.05 (−0.60, 0.50)
Pull up test (reps)	13.19 ± 3.64	12.95 ± 3.36	0.81	0.07 (−0.48, 0.62)
Flexed arm hang test (s)	42.00 ± 9.64	41.15 ± 13.97	0.82	0.06 (−0.49, 0.61)
Handgrip strength—right (kg)	51.85 ± 7.66	47.27 ± 4.78	<0.01	0.84 (0.27, 1.40)
Handgrip strength—left (kg)	50.03 ± 6.70	46.29 ± 4.59	0.01	0.74 (0.17, 1.30)
Free standing squat (1RM)	144.81 ± 26.95	134.57 ± 30.19	0.22	0.35 (−0.21, 0.90)
Sargent test (cm)	44.63 ± 6.00	44.84 ± 4.02	0.87	−0.05 (−0.60, 0.50)
20 m sprint (s)	2.86 ± 0.10	2.90 ± 0.19	0.47	−0.20 (−0.75, 0.35)
50 m sprint (s)	6.79 ± 0.15	6.77 ± 0.28	0.85	0.05 (−0.50, 0.60)
Illinois agility test (s)	14.78 ± 0.33	14.88 ± 0.53	0.49	−0.20 (−0.74, 0.36)
Sit-and-reach test (cm)	34.44 ± 3.25	34.37 ± 4.10	0.95	0.02 (−0.53, 0.57)
VO$_2$max (mL kg^{-1} min^{-1})	50.01 ± 3.65	50.03 ± 4.11	0.99	−0.01 (−0.56, 0.54)

Results indicated that the recruited back players had significantly higher handgrip strength scores than their non-recruited counterparts. Both right ($p < 0.01$, ES = 0.84) and left ($p = 0.01$, ES = 0.74) values showed moderate size effects.

In order to better understand the potential effects of the anthropometry data and the physical fitness scores on the likelihood of rugby players being recruited for the U19 national team, separate logistic regressions were performed for each playing position.

With respect to the forward players, the logistic regression model was statistically significant ($\chi_{(2)} = 8.88$, $p = 0.01$), explained 12.4% of the variance in the national team selection, and correctly classified 80.2% of cases. Better (lower) scores on the agility test were associated with an increased likelihood of being recruited to the national team ($\beta = -0.56$, $p = 0.02$).

When analyzing the back players, the logistic regression model was also statistically significant ($\chi_{(1)} = 7.48$, $p < 0.01$), explained 14.3% of the variance in the national team selection, and correctly classified 82.1% of cases. Only better (higher) scores on the right handgrip strength test were associated with an increased likelihood of being recruited to the national team ($\beta = 0.13$, $p < 0.01$).

4. Discussion

This study aimed to quantify and explore how anthropometry and physical performance outcomes, according to playing position (forwards or backs), can help to recruit elite junior rugby union players. The results highlight an important association between handgrip strength and agility ability, and team recruitment in junior elite rugby union, and suggest that certain well-developed physical qualities, namely, agility for the forwards and upper-body strength for the backs, can help to predict the selection of U19 rugby players to their national team. Our findings are consistent with previous studies [27] describing forwards as having greater body mass and body height and exhibiting greater push-up performance when compared with faster, leaner, and smaller backs, who display greater values in 20 m and 50 m sprint, agility, and flexibility performance. No other significant differences were observed between these groups. The lack of clear differences illustrates the importance of supervision if the rugby development program aims to improve physical attributes. Therefore, physical characteristics should be developed from an early age to ensure the rugby player is physically ready for elite professional rugby union. Body composition, greater levels of strength and power, agility, and sprint ability appear to be important physical characteristics in rugby union players due to superior performances at higher playing levels and their relationship with game behaviors. Furthermore, the logistic regression results showed better agility scores for forwards and better right handgrip strength scores for backs. These variables were significantly associated with an increased likelihood of being recruited for the U19 Portugal team. The results indicate that hand strength can be considered an important capability and selective factor in elite junior rugby union players, especially considering that no previous study has examined the importance of the handgrip strength test in rugby players. In general, the evaluation of the importance of the handgrip strength used in rugby is not explored; however, hand dynamometry is simple, not expensive, and is a well-established method for assessing the strength of hand muscles.

In fact, the reliable evaluation of handgrip strength in rugby players appears to be an essential and somehow neglected component in strength monitoring, planning of strength training programs, as well as injury prevention and recovery. This test is suitable to assess players as it is a simple, inexpensive, and important diagnostic criterion. Frequently measured as a proxy for global muscle strength, this test is a well-established and recognized protocol and method for assessing hand strength [28].

In order to better understand the potential effects of the anthropometry data and the physical fitness scores on the likelihood of rugby players being recruited for the U19 national team, separate logistic regressions were performed for each playing position. Overall, the results were generally consistent with those of other rugby union studies [29,30] which demonstrated that different characteristics between forwards and backs players can be caused by specific player-position demands. Backs need to be leaner, faster, and aerobically fitter to defend in more open spaces, cover more distance during a rugby union match, perform more accelerations, and try to score more opportunities [31]. Conversely, forwards [32] need to be larger and stronger for scrumming with more force and to assist teammates winning line outs. Moreover, they need to gain and retain possession of the ball and receive more collisions (i.e., tackles, ball carries, and collisions).

Although we limited our investigation to the anthropometric characteristics and physical performances that discriminated between recruited and non-recruited rugby players according to playing position (forwards or backs) to U19 national team, it was previously reported that technical, tactical, and psychological skills were positively associated with game-specific position and performance measures; therefore, they can help to discriminate rugby elite players [33]. Indeed, the ability to make the appropriate decisions in game-play and during interactions between players is also a significant aspect of performance [34].

The non-use of motivation, learning effect, motor coordination, individual training status, and biomechanical factors (e.g., hand size) is another limitation of the current investigation. It is a possible claim that concentric or eccentric measures should have been evaluated rather than isometric strength. However, it should be noted that assessment of isometric strength can be easily obtained in practical contexts. To date, no studies have tested the possibility of using these study measures to understand this conflict's effects to help improve the task of effective selection in elite junior rugby union players. Therefore, further research is required to understand the contribution of handgrip strength with other variables (e.g., physical training programs, player progress, strength control practices, and game behaviors). Lastly, a small sample size of subjects within each positional group (forwards and backs) would be useful for establishing a reliable testing protocol and/or the normative values for rugby players and to help selectors in the evaluation and recruitment of rugby union players.

5. Conclusions

In sum, the results of this study demonstrate the importance of handgrip strength and agility ability for the rugby national team selection process. Furthermore, objective measures can be useful for quantifying and evaluating player anthropometric characteristics and physical fitness performance progress. Thus, by combining further objective measures, rugby player recruitment and progress will be better monitored.

Indeed, the variables used allowed us to distinguish between recruited and non-recruited rugby union players for the Portugal U19 team. In addition to normal age-group criteria, subjective evaluation, or training and game perception aspects of performance, coaches can use this information to help in player recruitment approaches. The main findings have several implications for the effective selection process design, particularly by helping to identify and to improve the accuracy of elite junior rugby talent identification programs through a physical fitness assessment.

Supplementary Materials: The following are available online at https://www.mdpi.com/1660-4601/18/4/1499/s1, Table S1: Anthropometric, physiological and performance assessment test measurements/procedures.

Author Contributions: Conceptualization: L.V., W.K., M.B., S.H., and H.M.F.; methodology: L.V., W.K., M.B., S.H., and H.M.F.; investigation: L.V., W.K., M.B., S.H., and H.M.F.; writing—original draft preparation: L.V., W.K., M.B., S.H., and H.M.F.; writing—review and editing: all authors. All authors have read and agreed to the published version of the manuscript.

Funding: This work was supported by a funding source through Portuguese Foundation for Science and Technology under the project UID04045/2020.

Institutional Review Board Statement: The study was conducted according to the guidelines of the Declaration of Helsinki, and approved by Ethics Committee of University of Trás-os-Montes and Alto Douro (CE.UTAD 27/2016).

Informed Consent Statement: Informed consent was obtained from all subjects involved in the study.

Data Availability Statement: The data presented in this study are available on request from the corresponding author. The data are not publicly available due to privacy and ethical restrictions.

Acknowledgments: The authors would like to thank all junior rugby players, coaching, and medical.

Conflicts of Interest: The authors did not report any potential conflict of interest.

References

1. Duthie, G.; Pyne, D.; Hooper, S. Applied physiology and game analysis of rugby union. *Sports Med.* **2003**, *33*, 973–991. [CrossRef]
2. Spamer, E. Talent identification and development in youth rugby players: A research review. *S. Afr. J. Res. Sport Phys. Educ. Recreat.* **2009**, *31*, 109–118. [CrossRef]
3. Maud, P.J. Physiological and anthropometric parameters that describe a rugby union team. *Br. J. Sports Med.* **1983**, *17*, 16–23. [CrossRef]
4. Fontana, F.Y.; Colosio, A.L.; De Roia, G.F.; Da Lozzo, G.; Pogliaghi, S. Anthropometrics of Italian Senior Male Rugby Union Players: From Elite to Second Division. *Int. J. Sports Physiol. Perform.* **2015**, *10*, 674–680. [CrossRef] [PubMed]
5. Posthumus, L.; MacGregor, C.; Winwood, P.; Tout, J.; Morton, L.; Driller, M.; Gill, N. The Physical Characteristics of Elite Female Rugby Union Players. *Int. J. Environ. Res. Public Health* **2020**, *17*, 6457. [CrossRef]
6. Smart, D.J.; Hopkins, W.G.; Gill, N.D. Differences and Changes in the Physical Characteristics of Professional and Amateur Rugby Union Players. *J. Strength Cond. Res.* **2013**, *27*, 3033–3044. [CrossRef] [PubMed]
7. Vaz, L.; Van Rooyen, M.; Sampaio, J. Rugby Game-Related Statistics that Discriminate Between Winning and Losing Teams in Irb and Super Twelve Close Games. *J. Sports Sci. Med.* **2010**, *9*, 51–55.
8. James, N.; Mellalieu, S.D.; Jones, N. The development of position-specific performance indicators in professional rugby union. *J. Sports Sci.* **2005**, *23*, 63–72. [CrossRef]
9. Nakamura, F.Y.; Pereira, L.A.; E Moraes, J.; Kobal, R.; Kitamura, K.; Abad, C.C.C.; Vaz, L.M.T.; LoTurco, I. Physical and physiological differences of backs and forwards from the Brazilian National rugby union team. *J. Sports Med. Phys. Fit.* **2016**, *57*, 1549–1556.
10. Adendorff, L.; Pienaar, A.E.; Malan, D.D.J.; Hare, E. Physical and motor abilities, rugby skills and anthropometric characteristics: A follow-up investigation of suc-cessful and less successful rugby players. *J. Hum. Mov. Stud.* **2004**, *46*, 441–457.

11. Barr, M.J.; Newton, R.U.; Sheppard, J.M. Were Height and Mass Related to Performance at the 2007 and 2011 Rugby World Cups? *Int. J. Sports Sci. Coach.* **2014**, *9*, 671–680. [CrossRef]
12. Sedeaud, A.; Marc, A.; Schipman, J.; Tafflet, M.; Hager, J.-P.; Toussaint, J.-F. How they won Rugby World Cup through height, mass and collective experience. *Br. J. Sports Med.* **2012**, *46*, 580–584. [CrossRef]
13. Smart, D.; Hopkins, W.G.; Quarrie, K.L.; Gill, N. The relationship between physical fitness and game behaviours in rugby union players. *Eur. J. Sport Sci.* **2011**, *14*, S8–S17. [CrossRef]
14. Hartwig, T.B.; Naughton, G.A.; Searl, J. Motion Analyses of Adolescent Rugby Union Players: A Comparison of Training and Game Demands. *J. Strength Cond. Res.* **2011**, *25*, 966–972. [CrossRef]
15. Brazier, J.; Antrobus, M.; Stebbings, G.K.; Day, S.H.; Callus, P.; Erskine, R.M.; Bennett, M.A.; Kilduff, L.P.; Williams, A.G. Anthropometric and Physiological Characteristics of Elite Male Rugby Athletes. *J. Strength Cond. Res.* **2020**, *34*, 1790–1801. [CrossRef] [PubMed]
16. Chiwaridzo, M.; Oorschot, S.; Dambi, J.M.; Ferguson, G.D.; Bonney, E.; Mudawarima, T.; Tadyanemhandu, C.; Smits-Engelsman, B. A systematic review investigating measurement properties of physiological tests in rugby. *BMC Sports Sci. Med. Rehabil.* **2017**, *9*, 24. [CrossRef]
17. LeSuer, D.A.; McCormick, J.H.; Mayhew, J.L.; Wasserstein, R.L.; Arnold, M.D. The Accuracy of Prediction Equations for Estimating 1-RM Performance in the Bench Press, Squat, and Deadlift. *J. Strength Cond. Res.* **1997**, *11*, 211–213.
18. Owen, C.; Till, K.; Weakley, J.J.S.; Jones, B. Testing methods and physical qualities of male age grade rugby union players: A systematic review. *PLoS ONE* **2020**, *15*, e0233796. [CrossRef]
19. Moore, A.; Murphy, A. Development of an anaerobic capacity test for field sport athletes. *J. Sci. Med. Sport* **2003**, *6*, 275–284. [CrossRef]
20. Tabacchi, G.; Sánchez, G.F.L.; Sahin, F.N.; Kizilyalli, M.; Genchi, R.; Basile, M.; Kirkar, M.; Silva, C.; Loureiro, N.; Teixeira, E.; et al. Field-Based Tests for the Assessment of Physical Fitness in Children and Adolescents Practicing Sport: A Systematic Review within the ESA Program. *Sustainability* **2019**, *11*, 7187. [CrossRef]
21. Green, B.S.; Blake, C.; Caulfield, B.M. A Valid Field Test Protocol of Linear Speed and Agility in Rugby Union. *J. Strength Cond. Res.* **2011**, *25*, 1256–1262. [CrossRef]
22. Lambert, M. Physiological Testing for the Athlete: Hype or Help? *Int. J. Sports Sci. Coach.* **2006**, *1*, 199–208. [CrossRef]
23. Wenger, H.A.; Green, H.J. *Physiological Testing of the High Performance Athlete*; Human Kinetics: Champaign, IL, USA, 1991; pp. 223–308.
24. Vaz, L.M.T.; Morais, T.; Rocha, H.; James, N. Fitness Profiles of Elite Portuguese Rugby Union Players. *J. Hum. Kinet.* **2014**, *41*, 235–244. [CrossRef] [PubMed]
25. Hopkins, W.G.; Schabort, E.J.; Hawley, J.A. Reliability of Power in Physical Performance Tests. *Sports Med.* **2001**, *31*, 211–234. [CrossRef]
26. Chen, P.Y.; Krauss, A.D. *The SAGE Encyclopedia of Social Science Research Methods*; SAGE Publications, Inc.: Thousand Oaks, CA, USA, 2004.
27. Darrall-Jones, J.D.; Jones, B.; Till, K. Anthropometric and Physical Profiles of English Academy Rugby Union Players. *J. Strength Cond. Res.* **2015**, *29*, 2086–2096. [CrossRef] [PubMed]
28. Argus, C.K.; Gill, N.D.; Keogh, J.W.L. Characterization of the Differences in Strength and Power Between Different Levels of Competition in Rugby Union Athletes. *J. Strength Cond. Res.* **2012**, *26*, 2698–2704. [CrossRef]
29. Quarrie, K.L.; Hopkins, W.G.; Anthony, M.J.; Gill, N.D. Positional demands of international rugby union: Evaluation of player actions and movements. *J. Sci. Med. Sport* **2013**, *16*, 353–359. [CrossRef] [PubMed]
30. Roberts, S.P.; Trewartha, G.; Higgitt, R.J.; El-Abd, J.; Stokes, K.A. The physical demands of elite English rugby union. *J. Sports Sci.* **2008**, *26*, 825–833. [CrossRef]
31. Pasin, F.; Caroli, B.; Spigoni, V.; Dei Cas, A.; Volpi, R.; Galli, C.; Passeri, G. Performance and antrhropometric characteristics of Elite Rugby Players. *Acta Biomed.* **2017**, *88*, 172–177.
32. Gabbett, T.; Jenkins, D.G.; Abernethy, B. Physiological and Anthropometric Correlates of Tackling Ability in Junior Elite and Subelite Rugby League Players. *J. Strength Cond. Res.* **2010**, *24*, 2989–2995. [CrossRef] [PubMed]
33. Fernandes, H.M.; Batista, M.; Vaz, L. Relationships between psychological skills and European U19 rugby union tournament outcomes and per-formance indicators. *J. Hum. Sport Exec.* **2019**, *14*, 1246–1249.
34. Vaz, L.M.T.; Leite, N.; João, P.V.; Gonçalves, B.; Sampaio, J. Differences between Experienced and Novice Rugby Union Players during Small-Sided Games. *Percept. Mot. Ski.* **2012**, *115*, 594–604. [CrossRef] [PubMed]

International Journal of
Environmental Research and Public Health

Article

An 8-Week Program of Plyometrics and Sprints with Changes of Direction Improved Anaerobic Fitness in Young Male Soccer Players

Ghaith Aloui [1], Souhail Hermassi [2,*], Aymen Khemiri [1], Thomas Bartels [3], Lawrence D. Hayes [4], El Ghali Bouhafs [5], Mohamed Souhaiel Chelly [1,†] and René Schwesig [6,†]

1   Research Unit (UR17JS01) Sport Performance, Health & Society, Higher Institute of Sport and Physical Education, University of La Manouba, Ksar-Saîd, Tunis 2010, Tunisia; gaithaloui@hotmail.fr (G.A.); aymenkha3@gmail.com (A.K.); mohamedsouhaiel.chelly@issep.uma.tn (M.S.C.)
2   Physical Education Department, College of Education, Qatar University, Doha 2713, Qatar
3   Center of Joint Surgery, Sports Clinic Halle, 06108 Halle (Saale), Germany; Thomas.Bartels@sportklinik-halle.de
4   School of Health and Life Sciences, University of the West of Scotland, Glasgow G72 0LH, UK; lawrence.hayes@uws.ac.uk
5   Department of Sports Science, Martin-Luther-University Halle-Wittenberg, Von-Seckendorff-Platz 2, 06120 Halle (Saale), Germany; bouhafs.elghali@gmail.com
6   Department of Orthopaedic and Trauma Surgery, Martin-Luther-University Halle-Wittenberg, Ernst-Grube-Str. 40, 06120 Halle (Saale), Germany; rene.schwesig@uk-halle.de
*   Correspondence: shermassi@qu.edu.qa
†   The last two authors are co-last authors.

check for **updates**

Citation: Aloui, G.; Hermassi, S.; Khemiri, A.; Bartels, T.; Hayes, L.D.; Bouhafs, E.G.; Souhaiel Chelly, M.; Schwesig, R. An 8-Week Program of Plyometrics and Sprints with Changes of Direction Improved Anaerobic Fitness in Young Male Soccer Players. *Int. J. Environ. Res. Public Health* **2021**, *18*, 10446. https://doi.org/10.3390/ijerph181910446

Academic Editors: Bruno Gonçalves, Jorge Bravo, Hugo Folgado and Paul B. Tchounwou

Received: 16 August 2021
Accepted: 29 September 2021
Published: 4 October 2021

Publisher's Note: MDPI stays neutral with regard to jurisdictional claims in published maps and institutional affiliations.

Abstract: This study examined the effects of 8 weeks of twice-weekly combined plyometric and sprint with change-of-direction (CPSCoD) training into habitual training regimes of young male soccer players. Participants were randomly allocated to an experimental group ($n = 17$, age: 14.6 ± 0.44 years, body mass: 61.2 ± 7.34 kg, height: 1.67 ± 0.09 m, body fat: $11.2 \pm 1.56\%$) and a control group ($n = 16$, age: 14.6 ± 0.39 years, body mass: 61.1 ± 3.96 kg, height: 1.67 ± 0.05 m, body fat: $11.8 \pm 1.47\%$). Measures obtained pre- and post intervention included vertical and horizontal jump performance (i.e., squat jump (SJ), countermovement jump (CMJ), and standing long jump (SLJ)), and sprint performance (i.e., 5 m and 20 m sprint). In addition, Measures obtained pre- and post-intervention included change-of-direction ability (4×5 m sprint test (S 4×5 m) and sprint 9–3–6–3–9 m with backward and forward running (SBF)), repeated change of direction (RCoD), and static balance performance (stork balance test). The training group experienced superior jump (all $p < 0.05$; d ≥ 0.61), sprint (all $p < 0.05$; d ≥ 0.58), change-of-direction (CoD) ability (all $p < 0.05$; d ≥ 0.58), RCoD (all parameters except the fatigue index $p < 0.01$; effect size (d) ≥ 0.71), and static balance (all $p < 0.05$; d ≥ 0.66) improvement. Adding twice-weekly CPSCoD training to standard training improves the anaerobic performance of U15 male soccer players.

Keywords: team sports; muscle power; complex performance diagnostic; training intervention

1. Introduction

Fitness is an important determinant of soccer performance [1]. During a 90 min match, elite young soccer players (13–18 years) complete intermittent running and often cover more than 6 km, emphasizing the importance of the aerobic metabolic pathway [2]. In a similar vein, players commonly perform high-speed actions over short distances (i.e., sprinting, acceleration, and deceleration), and their particular movements are associated with soccer-specific actions such as tackling, defending, or creating space during possession, with sprints being the most common action before scoring a goal [3–5]. The distance of six locomotor categories (CATs) and the total distance (TD) were chosen as performance results [6]. The following CATs were analyzed: standing (0.0–0.7 km/h), walking (0.7–7.2 km/h), jogging (7.2–14.4 km/h), running (14.4–19.8 km/h), high-speed running (19.8–25.2 km/h), and sprinting (>25.2 km/h). To quantify high-to-moderate accelerations, the number of peaks above a threshold of 2 ms^{-2} was determined [7,8]. In this context, Di Salvo et al. [9] investigated the physical demands of Spanish Premier League football players in relation to their playing position. These researchers reported that wide (11,990 m) and central (12,027 m) midfielders covered the greatest distance. In addition, they reported that wide midfielders covered the greatest distances in high-intensity speed zones (19.1–23.0 km/h: 738 m; >23.0 km/h: 446 m), as well as central midfielders had the highest volume in the mid-intensity zones (11.1 to 14.0 km/h: 1965 m; 14.1 to 19.0 km/h: 2116 m). Furthermore, soccer players complete an average of 50 turns per match [10] and complete a total of 723 $\pm$ 203 turns and dodges in a match [11].

Analysis of physical matches has shown that in elite soccer matches, players perform a significant number of high-intensity changes of direction (CoDs) using a broad spectrum of turning angles [11,12]. Some authors have proposed that the number and quality of CoDs in a match influence the outcome of the match in professional soccer [13,14]. Possibly as a result of this association, the ability to perform CoDs in a match is considered an important fitness component in all soccer abilities, and levels of competitions [14–16]. Recently, Lloyd et al. [17] reported that the application of optimal CoD training stimulus during the duration of athlete development was crucial for effective programming and improvement of CoD performance in youth. Several studies have reported CoD training improved CoD ability, [11,13]. However, more encouraging for practitioners is the efficacious nature of CoD training on other parameters such as jumping performance, sprint performance, and balance ability in young soccer players [11,13].

Plyometric training is effective at improving jump performance [14] (the mean relative improvement varied between 14% and 29% on vertical jump), given the specific nature of the training method [14–16]. However, much similar to CoD, the cross-training effects on CoD ability and sprint performance are encouraging for practitioners and athletes alike [14–16]. In addition to performance benefits, plyometric training also exerts healthogenic effects such as increased bone mineral density, improved neuromuscular function, body mass control, improved psychosocial well-being, improved cardiovascular risk profile, and decreased risk of [18–20]. A recent systematic review [21] indicated that a twice-weekly plyometric training for 8 to 10 weeks with a 72 h rest period between training sessions improves high-intensity physical abilities (e.g., jump, sprint, agility) in the young soccer player population aged 10 to 17 years, in both sexes.

Despite the increasingly abundant literature on combined training in several sport disciplines and different age categories, there is little research related to applications of this training method in young male soccer players [22–25]. Recently, Aloui et al. [22] studied the effects of 8 weeks bi-weekly combined plyometric and sprint with change-of-direction (CPSCoD) training in elite under-17 (U17) soccer players. Jumping (Δ19% on a squat jump; Δ21% on countermovement jump), sprinting (Δ−11% on 5 m; Δ−8% on 20 m) and changing-direction ability (Δ−9% sprint 4 × 5 m test), repeated sprinting ability performance (Δ−7% on fastest time; Δ−8% on mean time), and balance performance improved as a result of this training method. Additionally, Beato et al. [23] noted that CPSCoD training improved jump and sprint performance in elite U19 soccer players. Additionally, Jlid et al. [26] reported multidirectional plyometric training improved jump performance, CoD ability, and dynamic postural control in prepubertal soccer players.

Despite the efficacious nature of CPSCoD training to enhance anaerobic fitness in soccer, there are only a handful of studies that have examined this phenomenon, and mostly in adolescents >15 years of age. Therefore, the purpose of this investigation was to examine the effects of 8 weeks of twice-weekly CPSCoD training on anaerobic performance (i.e., jumping, speed, and CoD ability) and balance ability in elite male U15 soccer players during the competitive season. Particular tests of interest were vertical and horizontal jump performance (i.e., squat jump (SJ), countermovement-jump (CMJA), standing long jump test (SLJ)), sprint performance (i.e., 5 m and 20 m sprint), and CoD ability (4 × 5 m sprint test (S 4 × 5 m) and sprint 9–3–6–3–9 m with backward and forward running (SBF). In addition, repeated CoD (RCoD) ability and static balance performance (stork balance test) were examined. We hypothesized a priori that CPSCoD would improve physical fitness, particularly jump performance, sprint performance, and CoD ability, when compared with a control group, who maintained regular training during the season.

2. Materials and Methods

2.1. Participants

The 33 participants of all playing positions from a single male soccer team in the first national division took part in this study. Participants had 4.6 ± 0.9 years of systematic soccer training background comprising of five training sessions per week. Participants were examined by the team physician to ensure players were healthy enough to complete CPSCoD training. Participants were randomly assigned between an experimental group (n = 17, age: 14.6 ± 0.44 years, body mass: 61.2 ± 7.3 kg, height: 1.67 ± 0.09 m, body fat: 11.2 ± 1.6%) and a control group (n = 16, age: 14.6 ± 0.4 years, body mass: 61.1 ± 4.0 kg, height: 1.67 ± 0.05 m, body fat: 11.8 ± 1.5%). Participants were free from injury during the 6-month period before the beginning of the study and throughout the study. No baseline inter-group differences existed for anthropometric characteristics between groups ($p \geq 0.05$).

Prior to the study, participants had completed a 6-week preparation period (pre-season training) (5–6 sessions per week). The first 3 of these weeks were to focus on improving aerobic capacity by following low-to-moderate intensity interval training, and on improving resistance and muscle strength by following low-to-moderate intensity circuit training resistance. The last 3 weeks have been focused on improving aerobic power by following high-intensity interval training (HIIT) and small-sided games, and improving muscle power by following high-intensity circuit training, supplemented participation in a non-competitive training match.

Procedures were approved by the University Institutional Review Committee for ethical human experimentation (reference number: KS000002020 and date of approval 10 December 2020). Participants (and their guardians, in the case of minors) provided informed written consent. Two familiarization sessions were conducted 2 weeks before testing, which was 2 months into the competitive season.

2.2. Experimental Design

Participants' only physical training during the experimental phase was associated with the soccer team and one weekly 60 min school physical education session. The usual micro-training cycle for both groups consists of five sessions per week (approximately 90 min each session), with a competitive game each weekend. During the first three sessions of micro-cycle training each week, approximately 60% of the total time of the session was focused on technical–tactical training, as well as 40% of the time was focused on physical training. In addition, during the last two training sessions of each week, approximately 75% of the micro-cycle of training was focused on technical–tactical training, as well as 25% of the time was focused on physical training (Table 1).

The control group maintained its normal training program throughout the intervention of 8 weeks. While every Tuesday and Thursday, the experimental group replaced the technical–tactical part of their standard training program with a combined plyometric and sprint with CoD training. Indeed, the first session of the micro-cycle training of each week was focused on the development of aerobic capacity through skill drills/circuit training with medium volume and low intensity. The second training session included HIIT and small-sided games aimed at developing maximum aerobic power [27]. The third training session included dynamic exercises based on bodyweight only and jumping and sprinting exercises aimed at maximum anaerobic power. The fourth and fifth training sessions of each week included agility and vivacity training, respectively.

Table 1. Details of general training routine during the week performed by both control and experimental groups over the 8-week intervention.

Days	Objectives
Mondays	Rest
Tuesdays	Aerobic capacity training and defensive tactics training
Wednesdays	Aerobic power training and defensive tactics training
Thursdays	Power anaerobic training and defensive and offensive tactics training
Fridays	Agility training and technical training and offensive tactics training
Saturdays	Vivacity training and technical training and offensive tactics training
Sunday	Official matches

The investigation was completed during a local soccer season in 2020/2021.

2.3. Details of Combined Plyometric and Short Sprints with Change-of-Direction Training

The CPSCoD training program consisted of four drills, with increasing repetitions over the training period, twice per week (as visualized in Figure 1 and quantified in Table 2).

Each CPSCoD drill began with plyometric exercises and finished with a sprint with CoD. It should be noted that participants were used to applying plyometrics and CoD drills and had achieved good technical competency through training activities before starting the investigation.

Plyometric training variables were based on recommendations for volume and intensity from Bedoya et al. [21]. Furthermore, CoD angles were based on previously published recommendations [23].

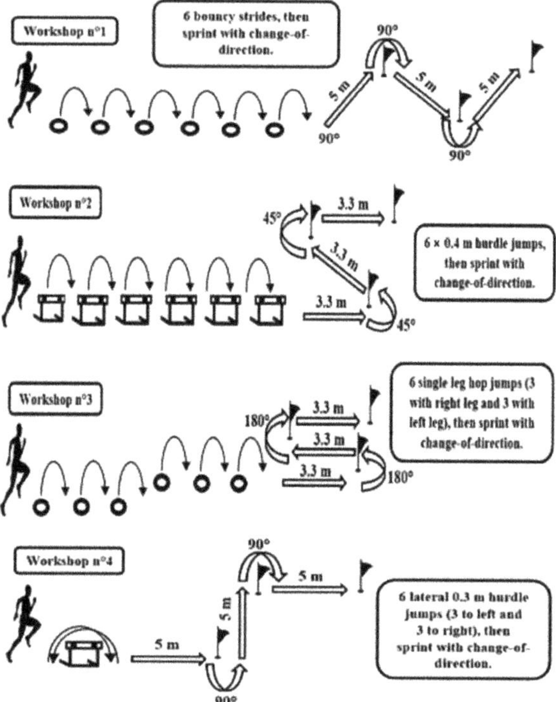

Figure 1. Exercise used in combined plyometric and sprint with a change-of-direction training program.

Table 2. Plyometric components were introduced into the program of the experimental group.

Week	Drill 1	Drill 2	Drill 3	Drill 4	Total (Contact)
1	3 Repetitions	3 Repetitions	3 Repetitions	3 Repetitions	72
2	3 Repetitions	3 Repetitions	3 Repetitions	3 Repetitions	72
3	4 Repetitions	4 Repetitions	4 Repetitions	4 Repetitions	96
4	4 Repetitions	4 Repetitions	4 Repetitions	4 Repetitions	96
5	5 Repetitions	5 Repetitions	5 Repetitions	5 Repetitions	120
6	5 Repetitions	5 Repetitions	5 Repetitions	5 Repetitions	120
7	6 Repetitions	6 Repetitions	6 Repetitions	6 Repetitions	144
8	6 Repetitions	6 Repetitions	6 Repetitions	6 Repetitions	144

All repetitions and sessions were separated by 90 s recovery intervals.

2.4. Testing Schedule

Tests were conducted at least 3 days after the last competitive match and 5–9 days after the previous training session. Testing was conducted on a tartan surface, which was part of the players' weekly training schedule. Standardized warm-up preceded each test. Tests were completed over three separate testing days in the following order: anthropometric assessment, squat jump (SJ), countermovement jump (CMJ), sprint 4×5 m test (S 4×5 m) (all day 1); the stork balance test, 5 and 20 m sprint, and sprint 9–3–6–3–9 m with backward and forward running (SBF) (all day 2); the standing long jump (SLJ) and the repeated CoD ability (RCoD) (both day 3).

The 5 and 20 m sprint performance, S 4×5 m, SBF, SJ, CMJ, SLJ, stork balance test, and the RCoD have all been previously described in detail [28–30] and therefore are not detailed here to avoid self-plagiarism. Anthropometric characteristics (body mass and body fat percentage) were evaluated after an overnight fast, in the morning (7–8 am), bare foot, with the bioelectrical impedance analysis (BIA) method (BC-602, Tanita Co., Tokyo, Japan) [31].

2.5. Statistical Analyses

All analysis was conducted on SPSS version 25.0 for Windows (IBM, Armonk, NY, USA). Normal distribution was tested using the Shapiro–Wilk test. Homogeneity of variance was determined using Levene's test. Independent samples *t*-tests examined between-group differences at baseline. An effect of training was considered by a mixed method of two-way (time $\times$ group) analysis of variance (ANOVA) with repeated measures. Subsequently, dependent samples *t*-tests tested for within-group training changes from pre- to post intervention were examined with Tukey's post hoc procedure applied to avoid type I error. Alpha level is reported as exact *p* values, as suggested by the American Statistical Association [31]. The effect size for paired comparisons is reported as Cohen's d [32] and interpreted as trivial (<0.35), small ($\geq$0.35–0.80), moderate ($\geq$0.80–1.50), and large ($\geq$1.50), based on the recommendations of Rhea [33] for recreationally trained subjects. Percentage changes were calculated as ((post-training value−pre-training value)/pre-training value) $\times$100. Reliability was evaluated using intraclass correlation coefficients (ICC) [34] and the coefficients of variation (CV) over consecutive pairs of intra-participant trials [35]. All performance measures had an ICC > 0.80 and a CV < 5% (Table 3). Data are reported as mean $\pm$ standard deviation (SD).

Table 3. Interclass correlation coefficient (ICC, 95% confidence intervals (95% CI)) and coefficient of variation (CV) showing acceptable reliability for all performance. ICC > 0.75 and CV < 10% marked in bold.

	ICC (95% CI)	CV (95%CI) [%]
	Sprint times	
5 m (s)	**0.93** (0.90–0.97)	**2.0** (1.7–2.4)
20 m (s)	**0.94** (0.91–0.97)	**1.9** (1.5–2.3)
	Change of direction	
Sprint 4 × 5 m (s)	**0.89** (0.84–0.93)	**2.2** (1.8–2.7)
SBF (s)	**0.87** (0.83–0.91)	**2.1** (1.7–2.5)
	Vertical jump	
SJ (cm)	**0.95** (0.90–0.98)	**2.4** (1.9–2.8)
CMJ (cm)	**0.94** (0.89–0.98)	**2.8** (2.4–3.2)
	Horizontal jump	
SLJ (m)	**0.91** (0.87–0.95)	**3.8** (3.4–4.3)
	Stork balance test	
Right leg (s)	**0.82** (0.72–0.88)	**4.7** (4.3–5.4)
Left leg (s)	**0.80** (0.70–0.87)	**4.9** (4.4–5.5)

ICC: Intraclass Correlation Coefficients; 95% CI: 95% Confidence Intervals; CV: Coefficients of Variation; SBF: Sprint 9–3–6–3–9 m with backward and forward running; SJ: Squat Jump; SLJ: Standing Long Jump.

3. Results

3.1. Normal Distribution and Homogeneity of Variance

The results of the Shapiro–Wilk exhibited approximately half the parameters (14/29, 48%) were not normally distributed. The following parameters were not normally distributed: height ($p = 0.002$), body fat ($p = 0.030$), SJ 1 ($p = 0.036$), SJ 2 ($p = 0.001$), SLJ 2: $p = 0.014$, sprint 20 m 2 ($p = 0.033$), RSA$_{best}$ 2 ($p = 0.011$), RSA$_{mean}$ 2 ($p = 0.020$), RSA fatigue index 1 ($p = 0.037$), RSA fatigue index 2 ($p = 0.003$), and all stork balance parameters ($p < 0.001$).

In total, 33% (4/12) of variance tests indicated heterogeneity of variance (SJ: $p = 0.003$, CMJ: $p = 0.048$, RSA Fatigue index: $p = 0.005$, and stork balance of left leg: $p = 0.040$).

3.2. Reliability

Measured reliability is displayed in Table 3. There were no baseline between-group differences.

3.3. Effect of Training on Jump Performance

In terms of vertical and horizontal jump performance, the present results showed significant intervention effects (group × time interaction), in the experimental group compared to the control, ($\Delta 22\%$; $p < 0.05$; d = 0.61 on squat jump $\Delta 20\%$; $p < 0.05$; d = 0.71 on countermovement jump and $\Delta 15\%$; $p < 0.05$; d = 0.66 on standing long jump test, respectively, for the experimental group) (Table 4).

3.4. Effect of Training on Sprint Performance

There was a group × time interaction in sprint performance with the experimental group (EG) improving more than the control group (CG) over distances of 5 ($\Delta -3\%$; $p < 0.001$; d = 0.39 and $\Delta -11\%$; $p < 0.001$; d = 1.81) and 20 m ($\Delta -3\%$; $p < 0.001$; d = 0.48 and $\Delta -10\%$; $p < 0.001$; d = 1.51), respectively, for the EG (Table 4).

3.5. Effect of Training on Change-of-Direction Ability

The sprint 4×5 m test displayed the largest group $\times$ time interaction effect ($p = 0.002$, $d = 0.82$) for all jump and change-of-direction performance parameters (Table 4). Especially, the EG ($p < 0.001$, $d = 2.12$) benefited from the intervention compared with the CG ($p < 0.001$, $d = 0.68$).

The SBF test showed an intervention effect with the EG improving more than CG ($\Delta-3\%$; $p < 0.001$; $d = 0.62$ and $\Delta-8\%$; $p < 0.001$; $d = 1.65$, respectively, for the EG) (Table 4).

3.6. Effect of Training on Repeated Change-of-Direction Ability

The RCoD test showed group $\times$ time interactions for $RCoD_{fastest}$ ($p = 0.006$, $d = 0.72$) and $RCoD_{mean}$ ($p = 0.006$, $d = 0.71$). Over the time, the improvement of performance was much higher in the EG ($d_{fastest} = 2.01$, $d_{mean} = 2.02$) than in the CG ($d_{fastest} = 0.42$, $d_{mean} = 0.43$) (Table 5).

3.7. Effect of Training on Balance Performance

Balance performance was improved more in the experimental group, compared with the control group (right leg: $\Delta68\%$; $p < 0.05$; $d = 0.66$; left leg: $\Delta79\%$; $p < 0.01$; $d = 0.72$, respectively for the experimental group) (Table 5).

Table 4. Vertical and horizontal jump tests, sprint times, and change-of-direction performance performances in experimental and control groups before and after the 8-week intervention.

	Experimental (*n* = 17)			Paired *t*-Test		Control (*n* = 16)			Paired *t*-Test		ANOVA (Group × Time)	
	Pre	Post	% Δ	*p*	ES	Pre	Post	% Δ	*p*	ES	*p*	ES
					Vertical jump							
SJ (cm)	25.1 ± 4.00	30.1 ± 4.75	20.2 ± 1.22	<0.001	1.15	24.5 ± 2.06	25.4 ± 2.16	3.70 ± 1.21	<0.001	0.43	0.018	0.61 (small)
CMJ (cm)	27.7 ± 3.66	33.2 ± 3.90	20.1 ± 2.26	<0.001	1.46	27.5 ± 2.50	28.6 ± 2.55	3.94 ± 1.42	<0.001	0.43	0.007	0.71 (small)
					Horizontal jump							
SLJ (m)	1.81 ± 0.13	2.08 ± 0.11	14.8 ± 2.78	<0.001	2.28	1.78 ± 0.18	1.85 ± 0.17	4.31 ± 1.68	<0.001	0.43	0.012	0.66 (small)
					Sprint							
5 m (s)	1.28 ± 0.10	1.12 ± 0.06	−11.3 ± 2.35	<0.001	1.81	1.24 ± 0.09	1.21 ± 0.09	−2.82 ± 0.71	<0.001	0.39	0.013	0.65 (small)
20 m (s)	3.61 ± 0.26	3.26 ± 0.19	−9.52 ± 1.30	<0.001	1.51	3.59 ± 0.21	3.49 ± 0.21	−2.81 ± 0.61	<0.001	0.48	0.027	0.58 (small)
					Change of direction Performance							
Sprint 4 × 5 m (s)	6.60 ± 0.32	5.97 ± 0.27	−9.39 ± 0.83	<0.001	2.12	6.60 ± 0.26	6.42 ± 0.27	−2.71 ± 0.72	<0.001	0.68	0.002	0.82 (moderate)
SBF (s)	9.09 ± 0.46	8.34 ± 0.44	−8.19 ± 0.78	<0.001	1.65	9.09 ± 0.42	8.83 ± 0.41	−2.68 ± 0.68	<0.001	0.62	0.026	0.58 (small)

ES: effect size; ANOVA: Analyze of Variance; CMJ: Countermovement Jump; Δ: percentage difference; *t*-test: *t* is the calculated difference in units of the standard error.

Table 5. Repeated change-of-direction ability and stork balance test performances in experimental and control groups before and after the 8-week intervention.

	Experimental (*n* = 17)			Paired *t*-Test		Control (*n* = 16)			Paired *t*-Test		ANOVA (Group × Time)	
	Pre	Post	% Δ	*p*	ES	Pre	Post	% Δ	*p*	ES	*p*	ES
				Repeated Change of Direction Ability parameters								
Fastest time (s)	7.14 ± 0.29	6.57 ± 0.40	−7.85 ± 0.67	<0.001	2.01	7.15 ± 0.31	7.01 ± 0.33	−1.93 ± 0.56	<0.001	0.42	0.006	0.72 (small)
Mean time (s)	7.28 ± 0.31	6.69 ± 0.41	−7.85 ± 0.65	<0.001	2.02	7.26 ± 0.32	7.12 ± 0.34	−1.94 ± 0.58	<0.001	0.43	0.006	0.71 (small)
Fatigue index (%)	1.97 ± 0.79	1.93 ± 0.40	1.96 ± 13.8	0.408	0.10	1.51 ± 0.32	1.49 ± 0.24	0.47 ± 9.91	0.844	0.04	0.844	0.06 (trivial)
				Stork balance test								
Right leg (s)	11.9 ± 4.39	19.6 ± 6.12	67.6 ± 10.8	<0.001	1.45	12.9 ± 4.04	14.6 ± 4.06	14.4 ± 7.07	<0.001	0.42	0.012	0.66 (small)
Left leg (s)	10.4 ± 4.28	18.2 ± 6.00	78.6 ± 14.3	<0.001	1.50	10.5 ± 3.03	12.3 ± 3.17	18.0 ± 8.11	<0.001	0.58	0.006	0.72 (small)

4. Discussion

This investigation examined the effect of 8 weeks of CPSCoD training on anaerobic performance and balance ability in elite U15 soccer players. We observed that adding CPSCoD to habitual training enhanced jump performance, sprint performance, CoD ability, RCoD ability, and balance performance.

4.1. Effect of Training on Jump Performance

In soccer, vertical jumps are performed when a player jumps toward a ball in an attempt to head direct the ball for an attacking (scoring/passing) or defensive (clearance/interception) task and is decisive action during matches [32]. We observed intervention effects in the experimental group, compared with the control, on vertical and horizontal jump performance (Table 4). The positive effects of CPSCoD in the present investigation are in line with Aloui et al. [22], who studied the effects of 8 weeks of CPSCoD training on U17 male soccer players and found this type of training improved vertical and horizontal jump performance. In addition, Beato et al. [23] reported significant increases in horizontal jump performance following CPSCoD training, in male young soccer players (U18). Although mechanisms for adaptation cannot be fully determined in the present study, it is likely improved jump performance was mediated by neural adaptation such as motor unit recruitment and the Hoffman reflex [20,33]. We suggest eccentric muscular contractions necessitated by the plyometric training are able to produce greater force and thus may provide a large stimulus above and beyond concentric contractions alone [34].

4.2. Effect of Training on Sprint Performance

In professional soccer match play, sprints in a straight line (45%), followed by vertical jumping (16%) were the two most commonly observed actions immediately prior to a goal being scored [35]. It is well known that muscle strength and contraction speed (i.e., muscle power) are determining factors in sprint speed [36–38], and the improved sprint ability evoked by CPSCoD training in the present study may translate to enhanced match play abilities and preferential match outcomes. We observed intervention effects in the EG, compared with the control for CoD ability (Table 4). Our results are aligned with those of Hammami et al. [28], who reported improved sprint performance over 5 and 20 m following CPSCoD training in young male handball players. Similarly, Michailidis et al. [25] observed improved 10 m, but not 30 m, sprint performance after CPSCoD training in prepubertal male soccer players. Improvement in sprinting performance after CoD training could be due to improved leg extensor power and the ability to produce lower limb strength more effectively after training [39,40]. In addition, temporal sequencing of muscle activation for more efficient movement, preferential recruitment of the fastest motor units, or increased nerve conduction speed promoted by plyometric training may be responsible for the improved sprint times [41].

4.3. Effect of Training on Change-of-Direction Ability

The CoD improvement observed herein following CPSCoD training is encouraging for trainers, as CoD ability is an important trait in competitive soccer [42,43]. Our data, especially the moderate interaction (d = 0.82) and the large (d = 2.12) effect for the EG regarding sprint 4 × 5 m, are in accordance with the existing literature, which has demonstrated improved CoD ability after several programs, including CPSCoD training [22–24], plyometric training [28,44], and CoD training [45,46]. Recently, Michailidis et al. [25] and Makhlouf et al. [24] reported improved CoD ability following CPSCoD training in prepubertal males. Improvements may have been caused by reduced contact times as a result of increased muscle strength and power, and efficiency of movement, as this has been noted previously following plyometric [20,47]. Similarly, enhanced eccentric lower limb strength could enhance the ability to accelerate and decelerate during CoDs [48]. In this context, Chaalali et al. [45] suggested the observed CoD enhancement may be related to improved CoD technique and increased strength and power of the leg extensors, after CoD training.

4.4. Effect of Training on Repeated Change-of-Direction Ability

Soccer players must complete high-intensity repeated efforts and/or sprints to compete at a high level [49]. We reported improved RCoD ability in the EG, compared with the CG for all parameters except the fatigue index. Reinforcing our results, Aloui et al. [22] observed improvement in RCoD ability following CPSCoD training in male U17 soccer players. Conversely, Hammami et al. [28] reported no improvement in RCoD ability following CPSCoD training in U15 male handball players. Improved RCoD observed herein was likely resultant from improved anaerobic fitness such as CoD ability, sprint ability, and jumping ability [50,51]. Particularly, RCoD ability improvements were most likely the result of improvements in stretch-shortening cycle efficiency, motor unit synchronization, or tendon stiffness [52].

4.5. Effect of Training on Balance Performance

Balance is the ability to maintain postural control during an and is classified can be classified as dynamic and static [53]. In soccer, players benefit from the ability to regain postural control to attenuate injury risk while landing from a dynamic movement, such as a pronounced deceleration, cut, or jump [54]. The EG in this study improved the static balance, compared with the control group, which confirms the results of Aloui et al. [22], who observed improved static balance after CPSCoD training in U17 male soccer players. Recently, Makhlouf et al. [18] reported improved static and dynamic balance following CPSCoD training in prepubertal male soccer players. These adaptations may be caused by neural factors such as improved reaction time or precision of movement [55]. Another possible reason for improved balance performance could be the increase in muscle strength of the lower limbs, the increase in muscle coordination, and the facilitation of fast-twitch motor units [56]. Ondra et al. [57] reported that a high level of postural stability is believed to be the factor reducing the risk of lower limb injury, emphasizing the importance of balance in this cohort.

4.6. Limitations

Results of this study suggest CPSCoD training confers advantages in anaerobic capacity and balance performance in elite U15 soccer players. These mostly small adaptations were observed after only 8 weeks, alongside other in-season training, and should be related to the higher risk of injury from performing plyometric movements. Thus, CPSCoD training during the in-season period may promote reduced injury risk in addition to improved anaerobic fitness. However, before we can promote CPSCoD training for all soccer squads, a larger confirmatory study may be required to elucidate if these effects are manifest across all age groups, players levels, and sexes. Moreover, as our participants were already well-trained and well-versed in the type of training included in the CPSCoD program, caution should be exercised when inducting untrained or novice athletes onto a CPSCoD training program.

5. Conclusions

CPSCoD training enhanced balance performance and anaerobic capacity (jumping, sprinting, CoD ability, and RCoD ability) during the competition season. However, 83% (10/12) of the interaction effects were small. Only for sprint 4 × 5 m, a moderate effect was observed. Note that during the intervention period, no injury was reported. Therefore, CPSCoD training can be considered safe for well-trained and well-versed young male soccer players provided progression guidelines are followed.

Author Contributions: Conceptualization, G.A. and A.K.; methodology, G.A.; software, M.S.C.; validation, G.A., M.S.C., A.K. and R.S.; formal analysis, M.S.C.; investigation, G.A.; resources, G.A. and A.K.; data curation, M.S.C.; writing—original draft preparation, G.A., and S.H.; writing—review and editing, L.D.H. and S.H.; visualization, A.K. and T.B.; supervision, E.G.B.; project administration, G.A. and A.K.; funding acquisition, R.S. All authors have read and agreed to the published version of the manuscript.

Funding: The authors thank the "Ministry of Higher Education and Scientific Research, Tunis, Tunisia" for financial support.

Institutional Review Board Statement: The study was conducted according to the guidelines of the Declaration of Helsinki and approved by the Institutional Review Board of Research Unit (UR17JS01) Sport Performance, Health & Society, Higher Institute of Sport and Physical Education, Ksar-Saîd, University.

Informed Consent Statement: Informed consent was obtained from all subjects involved in the study.

Data Availability Statement: The raw data supporting the conclusions of this article will be made available by the authors without undue reservation.

Conflicts of Interest: The authors declare no conflict of interest.

References

1. Howard, N.; Stavrianeas, S. In-season high-intensity interval training improves conditioning in high school soccer players. *Int. J. Exerc. Sci.* **2017**, *10*, 713–720.
2. Buchheit, M.; Mendez-Villanueva, A.; Simpson, B.M.; Bourdon, P.C. Match running performance and fitness in youth soccer. *Int. J. Exerc. Sci.* **2010**, *31*, 818–825. [CrossRef]
3. Buchheit, M.; Al Haddad, H.; Simpson, B.M.; Palazzi, D.; Bourdon, P.C.; Di Salvo, V.; Mendez-Villanueva, A. Monitoring accelerations with GPS in football: Time to slow down? *Int. J. Sports Physiol. Perform.* **2014**, *9*, 442–445. [CrossRef] [PubMed]
4. Cochrane, D.J.; Monaghan, D. Using sprint velocity decrement to enhance acute sprint performance. *J. Strength Cond. Res.* **2021**, *35*, 442–448. [CrossRef]
5. Mara, J.; Thompson, K.G.; Pumpa, K.L.; Morgan, S. The acceleration and deceleration profiles of elite female soccer players during competitive matches. *J. Sci. Med. Sports* **2017**, *20*, 867–872. [CrossRef] [PubMed]
6. Rampinini, E.; Bishop, D.; Marcora, S.M.; Bravo, D.F.; Sassi, R.; Impellizzeri, F.M. Validity of simple field tests as indicators of match-related physical performance in top-level professional soccer players. *Int. J. Sports Med.* **2007**, *28*, 228–235. [CrossRef] [PubMed]
7. Sæterbakken, A.; Haug, V.; Fransson, D.; Grendstad, H.N.; Gundersen, H.S.; Moe, V.F.; Ylvisaker, E.; Shaw, M.; Riiser, A.; Andersen, V. Match running performance on three different competitive standards in norwegian soccer. *Sports Med. Int. Open* **2019**, *3*, E82–E88. [CrossRef]
8. Akenhead, R.; Hayes, P.; Thompson, K.; French, D. Diminutions of acceleration and deceleration output during professional football match play. *J. Sci. Med. Sport* **2013**, *16*, 556–561. [CrossRef] [PubMed]
9. Di Salvo, V.; Baron, R.; Tschan, H.; Montero, F.C.; Bachl, N.; Pigozzi, F. Performance characteristics according to playing position in elite soccer. *Int. J. Sports Med.* **2007**, *28*, 222–227. [CrossRef]
10. Withers, R.; Maricic, Z.; Wasilewski, S.; Kelly, L. Match analysis of Australian professional soccer players. *J. Hum. Mov. Stud.* **1982**, *8*, 159–176.
11. Bloomfield, J.; Polman, R.; O'Donoghue, P. Deceleration movements performed during FA premier league soccer matches. *J. Sports Sci. Med.* **2007**, *6*, 63–70.
12. Bloomfield, J.; Polman, R.; O'Donoghue, P. Turning movements performed during FA premier league soccer matches. *J. Sports Sci. Med.* **2007**, *6*, 9–10.
13. Faude, O.; Koch, T.; Meyer, T. Straight sprinting is the most frequent action in goal situations in professional football. *J. Sports Sci.* **2012**, *30*, 625–631. [CrossRef]
14. Stølen, T.; Chamari, K.; Castagna, C.; Wisløff, U. Physiology of soccer: An update. *Sports Med.* **2005**, *35*, 501–536. [CrossRef]
15. Reilly, T.; Bangsbo, J.; Franks, A.L. Anthropometric and physiological predispositions for elite soccer. *J. Sports Sci.* **2000**, *18*, 669–683. [CrossRef] [PubMed]
16. Reilly, T.; Williams, A.M.; Nevill, A.; Franks, A. A multidisciplinary approach to talent identification in soccer. *J. Sports Sci.* **2000**, *18*, 695–702. [CrossRef]
17. Lloyd, R.S.; Oliver, J.L.; Faigenbaum, A.D.; Howard, R.; De Ste Croix, M.B.; Williams, C.A.; Best, T.M.; Alvar, B.A.; Micheli, L.J.; Thomas, D.P.; et al. Long-Term Athletic Development- Part 1: A pathway for all youth. *J. Strength Cond. Res.* **2015**, *29*, 1439–1450. [CrossRef]
18. Behm, D.G.; Faigenbaum, A.D.; Falk, B.; Klentrou, P. Canadian society for exercise physiology position paper: Resistance training in children and adolescents. *Appl. Physiol. Nutr. Metab.* **2008**, *33*, 547–561. [CrossRef] [PubMed]
19. Lloyd, R.; Meyers, R.; Oliver, J. The natural development and trainability of plyometric ability during child-hood. *Strength Cond. J.* **2011**, *33*, 23–30. [CrossRef]
20. Markovic, G.; Mikulic, P. Neuro-musculoskeletal and performance adaptations to lower-extremity plyometric training. *Sports Med.* **2010**, *40*, 859–895. [CrossRef]
21. Bedoya, A.A.; Miltenberger, M.R.; Lopez, R. Plyometric training effects on athletic performance in youth soccer athletes: A systematic review. *J. Strength Cond. Res.* **2015**, *29*, 2351–2360. [CrossRef] [PubMed]

22. Aloui, G.; Hermassi, S.; Hayes, L.; Hayes, N.S.; Bouhafs, E.; Chelly, M.; Schwesig, R. Effects of plyometric and short sprint with change-of-direction training in male U17 soccer players. *Appl. Sci.* **2021**, *11*, 4767. [CrossRef]
23. Beato, M.; Bianchi, M.; Coratella, G.; Merlini, M.; Drust, B. Effects of plyometric and directional training on speed and jump performance in elite youth soccer players. *J. Strength Cond. Res.* **2018**, *32*, 289–296. [CrossRef]
24. Makhlouf, I.; Chaouachi, A.; Chaouachi, M.; Ben Othman, A.; Granacher, U.; Behm, D.G. Combination of agility and plyometric training provides similar training benefits as combined balance and plyometric training in young soccer players. *Front. Physiol.* **2018**, *9*, 1611. [CrossRef]
25. Michailidis, Y.; Tabouris, A.; Metaxas, T. Effects of plyometric and directional training on physical fitness parameters in youth soccer players. *Int. J. Sports Physiol. Perform.* **2019**, *14*, 392–398. [CrossRef]
26. Jlid, M.C.; Racil, G.; Coquart, J.; Paillard, T.; Bisciotti, G.N.; Chamari, K. Multidirectional plyometric training: Very efficient way to improve vertical jump performance, change of direction performance and dynamic postural control in young soccer players. *Front. Physiol.* **2019**, *10*, 1462. [CrossRef] [PubMed]
27. Karahan, M. Effect of skill-based training vs. small-sided games on physical performance improvement in young soccer players. *Biol. Sport* **2020**, *37*, 305–312. [CrossRef]
28. Hammami, M.; Gaamouri, N.; Aloui, G.; Shephard, R.J.; Chelly, M.S. Effects of combined plyometric and short sprint with change-of-direction training on athletic performance of male u15 handball players. *J. Strength Cond. Res.* **2019**, *33*, 662–675. [CrossRef]
29. Negra, Y.; Chaabene, H.; Sammoud, S.; Prieske, O.; Moran, J.; Ramirez-Campillo, R.; Nejmaoui, A.; Granacher, U. The increased effectiveness of loaded versus unloaded plyometric jump training in improving muscle power, speed, change of direction, and kicking-distance performance in prepubertal male soccer players. *Int. J. Sports Physiol. Perform.* **2020**, *15*, 189–195. [CrossRef]
30. Sporis, G.; Jukic, I.; Milanovic, L.; Vucetic, V. Reliability and factorial validity of agility tests for soccer players. *J. Strength Cond. Res.* **2010**, *24*, 679–686. [CrossRef]
31. Sitko, S.; Cirer-Sastre, R.; Corbi, F.; Laval, I.L. Effects of a low-carbohydrate diet on body composition and performance in road cycling: A randomized, controlled trial. *Nutr. Hosp.* **2020**, *37*, 1022–1027. [CrossRef]
32. Filter, A.; Jabalera, J.O.; Molina-Molina, A.; Suárez-Arrones, L.; Robles-Rodríguez, J.; Dos'Santos, T.; Loturco, I.; Requena, B.; Santalla, A. Effect of ball inclusion on jump performance in soccer players: A biomechanical approach. *Sci. Med. Football* **2021**, 1–7. [CrossRef]
33. Hammami, M.; Negra, Y.; Shephard, R.J.; Chelly, M.S. The effect of standard strength vs. contrast strength training on the development of sprint, agility, repeated change of direction, and jump in junior male soccer players. *J. Strength Cond. Res.* **2017**, *31*, 901–912. [CrossRef]
34. Douglas, J.; Pearson, S.; Ross, A.; McGuigan, M. Eccentric exercise: Physiological characteristics and acute responses. *Sports Med.* **2017**, *47*, 663–675. [CrossRef] [PubMed]
35. Faude, O.; Roth, R.; Di Giovine, D.; Zahner, L.; Donath, L. Combined strength and power training in high-level amateur football during the competitive season: A randomised-controlled trial. *J. Sports Sci.* **2013**, *31*, 1460–1467. [CrossRef] [PubMed]
36. Comfort, P.; Stewart, A.; Bloom, L.; Clarkson, B. Relationships between strength, sprint, and jump performance in well-trained youth soccer players. *J. Strength Cond. Res.* **2014**, *28*, 173–177. [CrossRef] [PubMed]
37. Lockie, R.G.; Stage, A.A.; Stokes, J.J.; Orjalo, A.J.; Davis, D.L.; Giuliano, D.V.; Moreno, M.R.; Risso, F.G.; Lazar, A.; Birmingham-Babauta, S.A.; et al. Relationships and predictive capabilities of jump assessments to soccer-specific field test performance in division i collegiate players. *Sports* **2016**, *4*, 56. [CrossRef] [PubMed]
38. McFarland, I.T.; Dawes, J.J.; Elder, C.L.; Lockie, R.G. Relationship of two vertical jumping tests to sprint and change of direction speed among male and female collegiate soccer players. *Sports* **2016**, *4*, 11. [CrossRef]
39. Baker, D.A. Comparison of running speed and quickness between elite professional and young rugby league players. *Strength Cond. Coach* **1999**, *7*, 3–7.
40. Sayers, M. Running techniques for field sport players. *Sports Coach Autumn* **2000**, *23*, 26–27.
41. De Villarreal, E.S.-S.; Kellis, E.; Kraemer, W.J.; Izquierdo, M. Determining variables of plyometric training for improving vertical jump height performance: A meta-analysis. *J. Strength Cond. Res.* **2009**, *23*, 495–506. [CrossRef]
42. Göral, K. Examination of agility performances of soccer players according to their playing positions. *Sport J.* **2015**, *6*, 1–9. [CrossRef]
43. Muniroglou, S.; Subak, E. A comparison of 5, 10, 30 Meters sprint, modified t-test, ar rowhead and illinois agility tests on football referees. *J. Educ. Train. Stud.* **2018**, *6*, 70–76. [CrossRef]
44. Rædergård, H.G.; Falch, H.N.; Tillaar, R.V.D. Effects of strength vs. plyometric training on change of direction performance in experienced soccer players. *Sports* **2020**, *8*, 144. [CrossRef]
45. Chaalali, A.; Rouissi, M.; Chtara, M.; Owen, A.; Bragazzi, N.L.; Moalla, W.; Chaouachi, A.; Amri, M.; Chamari, K. Agility training in young elite soccer players: Promising results compared to change of direction drills. *Biol. Sport* **2016**, *33*, 345–351. [CrossRef] [PubMed]
46. Sariati, D.; Hammami, R.; Zouhal, H.; Clark, C.C.T.; Nebigh, A.; Chtara, M.; Chortane, S.G.; Hackney, A.C.; Souissi, N.; Granacher, U.; et al. Improvement of physical performance following a 6 week change-of-direction training program in elite youth soccer players of different maturity levels. *Front. Physiol.* **2021**, *12*, 668437. [CrossRef]

47. Asadi, A. Plyometric type neuromuscular exercise is a treatment to postural control deficits of volleyball players: A case study. *Rev. Andal. Med. Deport.* **2016**, *9*, 75–79. [CrossRef]
48. Markovic, G.; Newton, R.U. Does plyometric training improve vertical jump height? A meta-analytical review. *Br. J. Sports Med.* **2007**, *41*, 349–355. [CrossRef]
49. Buchheit, M.; Mendez-Villanueva, A. Changes in repeated-sprint performance in relation to change in locomotor profile in highly-trained young soccer players. *J. Sports Sci.* **2014**, *32*, 1309–1317. [CrossRef]
50. Buchheit, M.; Mendez-Villanueva, A.; Delhomel, G.; Brughelli, M.; Ahmaidi, S. Improving repeated sprint ability in young elite soccer players: Repeated shuttle sprints vs. explosive strength training. *J. Strength Cond. Res.* **2010**, *24*, 2715–2722. [CrossRef]
51. Negra, Y.; Chaabene, H.; Fernandez-Fernandez, J.; Sammoud, S.; Bouguezzi, R.; Prieske, O.; Granacher, U. Short-term plyometric jump training improves repeated-sprint ability in prepuberal male soccer players. *J. Strength Cond. Res.* **2020**, *34*, 3241–3249. [CrossRef] [PubMed]
52. Harris, N.; Cronin, J.; Keogh, J. Contraction force specificity and its relationship to functional performance. *J. Sports Sci.* **2007**, *25*, 201–212. [CrossRef] [PubMed]
53. Pollock, A.S.; Durward, B.R.; Rowe, P.J.; Paul, J.P. What is balance? *Clin. Rehabil.* **2000**, *14*, 402–406. [CrossRef]
54. Azevedo, R.R.; da Rocha, E.; Franco, P.S.; Carpes, F.P. Plantar pressure asymmetry and risk of stress injuries in the foot of young soccer players. *Phys. Ther. Sport* **2017**, *24*, 39–43. [CrossRef] [PubMed]
55. Manolopoulos, K.; Gissis, I.; Galazoulas, C.; Manolopoulos, E.; Patikas, D.; Gollhofer, A.; Kotzamanidis, C. Effect of combined sensorimotor-resistance training on strength, balance, and jumping performance of soccer players. *J. Strength Cond. Res.* **2016**, *30*, 53–59. [CrossRef]
56. Young, W.; Metzl, J.D. Strength training for the young athlete. *Pediatr. Ann.* **2010**, *39*, 293–299. [CrossRef]
57. Ondra, L.; Nátěsta, P.; Bizovská, L.; Kuboňová, E.; Svoboda, Z. Effect of in-season neuromuscular and proprioceptive training on postural stability in male youth basketball players. *Acta Gymnica* **2017**, *47*, 144–149. [CrossRef]

International Journal of
Environmental Research and Public Health

MDPI

Article

Different Pitch Configurations Constrain the Playing Tactics and the Creation of Goal Scoring Opportunities during Small Sided Games in Youth Soccer Players

Joaquín González-Rodenas [1,*], Rodrigo Aranda-Malavés [2,3], Andrés Tudela-Desantes [2], Pedro de Matías-Cid [2] and Rafael Aranda [2,*]

[1] Centre for Sport Studies, Rey Juan Carlos University, 28942 Madrid, Spain
[2] Department of Physical Education and Sports, University of Valencia, 46010 Valencia, Spain; Rodrigo.Aranda@uv.es (R.A.-M.); Andres.Tudela@uv.es (A.T.-D.); Pedro.Matias@uv.es (P.d.M.-C.)
[3] Doctoral School, Catholic University San Vicente Mártir, 46001 Valencia, Spain
* Correspondence: Joaquin.gonzalez@urjc.es (J.G.-R.); Rafael.Aranda@uv.es (R.A.)

Abstract: The aim of this study was to explore the tactical effects of different pitch configurations on the collective playing tactics and the creation of goal scoring opportunities (GSO) during small sided soccer games (SSG) in youth players. A total of 22 players performed a 7 vs. 7 + 1 floater (including goalkeepers) under three different pitch configurations ("Standard", 53 × 38 m; "Long", 63 × 32 m; and "Wide", 43 × 47 m). Eleven tactical indicators related to the development and the end of the team possessions were evaluated by systematic observation. Friedman tests (non-parametric ANOVA for repeated measures) revealed that the long and wide configurations produced more counterattacks ($p = 0.0028$; ES = 0.3), higher offensive penetration ($p = 0.007$; ES = 0.41), and more GSO ($p = 0.018$; ES = 0.30) than the standard format. Regarding the creation of GSO, the wide configuration produced more assists in the form of crosses than the long and standard formats ($p = 0.025$; ES = 0.31), more utilization of wide subspaces to assist the final player ($p = 0.022$; ES = 0.35), more number of headers as the final action ($p = 0.022$; ES = 0.32), and less assists in the form of passes in behind the defense ($p = 0.034$; ES = 0.28), than the long configuration. The modulation of the pitch configuration during SSG produced different tactical demands, requiring players to implement different tactical solutions to create GSO.

Keywords: non-linear pedagogy; constraints led approach; sport pedagogy; practice design; football training

check for **updates**

Citation: González-Rodenas, J.; Aranda-Malavés, R.; Tudela-Desantes, A.; de Matías-Cid, P.; Aranda, R. Different Pitch Configurations Constrain the Playing Tactics and the Creation of Goal Scoring Opportunities during Small Sided Games in Youth Soccer Players. *Int. J. Environ. Res. Public Health* **2021**, *18*, 10500. https://doi.org/10.3390/ijerph181910500

Academic Editors: Jorge Bravo, Hugo Folgado and Bruno Gonçalves

Received: 31 August 2021
Accepted: 30 September 2021
Published: 6 October 2021

Publisher's Note: MDPI stays neutral with regard to jurisdictional claims in published maps and institutional affiliations.

1. Introduction

The key offensive aim of soccer is to disorder the defensive organization of the opposing team to achieve goal scoring opportunities (GSO) and goals. During that process, the last actions such as the assist and the final shot are decisive and both depend on the spatial–temporal relationship between the penultimate and the last player as well as the interaction between the offensive and defensive organizations [1–4]. Due to its complexity and importance, one of the most challenging duties of professional soccer clubs is to develop or recruit players who have the capacity to score or create high amounts of GSO. In fact, recent studies have observed that goal scorers and players with high quantity of assists demonstrated to have higher levels of cognitive functions such as creativity and working memory, compared with other players [5,6]. In addition, the study of Kempe and Memmet [7] observed that showing high creativity in the last two actions before the actual shot on goal proved to be the best predictor for game success in elite soccer.

Regarding the tactical development of GSO and goals, existing literature found that the actions that led to goal were predominantly originated in central and advanced areas of the opposing half, while the crosses have been shown to represent around 30–40% of the goal assists [8]. Furthermore, the majority of goals and GSO seem to be produced by collective plays, so that the individual actions produced less than 20% of them [9–11]. For instance, the study of González-Rodenas et al. [12] analyzed 1172 GSO from top European teams and observed that crosses comprised the 19.6% of penultimate actions prior to GSO, while passes comprised the 62.4% and individual actions the 9.1%.

Nevertheless, the tactical actions performed prior to goals or GSO are influenced by the interaction with the opposing team [2,3,13] In this regard, González-Rodenas et al. [4] observed that crosses were more frequent against organized defenses, while passes in behind the defense or actions as dribbling or running with the ball had a greater percentage of goals against circumstantial defenses. Regarding the final action, 70.1% of the goals were scored by using only one contact to the ball in organized defenses but 46.6% in circumstantial defenses. These tactical facts suggest that players should be able to adapt to the defensive context and interact with their teammates to choose the best solution in order to disrupt the opposing team and create a GSO or a goal.

However, despite the growing emergence of research about small sided games (SSG) in soccer [14], there is a lack of studies that specifically focus on how different types of SSG can influence the way GSO and goals are achieved in training. In recent years, the analysis of SSG in soccer has focused on physiological variables [15], motion analysis [16,17], collective behavior [18,19], and technical and tactical performance of players [20–22].

Consequently, it seems necessary to explore training methods to optimize the creation of GSO in youth soccer. In this sense, the constraints-led approach based on non-lineal pedagogy has emerged as an optimal framework to create representative learning designs in collective sports [23]. Under this framework, coaches should manipulate task constraints during the SSG such as space, time, rules, goals, or number of players in order to create learning environments where players can interact with their teammates and opponents to explore opportunities for action and adapt to the specific tactical context. This manipulation of task constrains encourage the player's co-adaptive and exploratory behavior to search for effective solutions, embracing problem-solving situations [24].

One of the easiest constraints for coaches to modify in SSG is the space. In this sense, the effect of the manipulation of the space size on physical and technical demands has been analyzed in multiple studies about SSG [25]. However, no study to date has analyzed the effect of modifying the pitch configuration on the playing tactics and creation of GSO in soccer. In this sense, Coutinho et al. [19] analyzed the effect of different pitch configuration on the physical demands and movement behavior in young players, concluding that using different pitch configurations might help players to improve their ability to identify the most relevant cues that support the emergence of functional behaviors. Likewise, Folgado et al. [26] observed that a longer pitch configuration (40 × 30 m) during a 4 vs. 4 + GKs game registered more distance covered at high intensities, more passes, and dribbles than a wider format (30 × 40 m), which had more lateral passes and shots and a wider team positioning.

According to this background, our hypothesis is that the design of SSG with different pitch configurations (i.e., wide pitch vs. long pitch vs. standard pitch) may modulate the emergence of collective tactical behaviors and encourage players to explore different ways to build and create GSO. Therefore, the aim of this study was to check the tactical effects of different pitch configurations on the collective playing tactics and the creation of GSO during SSG in youth players.

2. Materials and Methods

2.1. Sample

A total of 22 elite youth players (age: 13.80 ± 0.58, years of experience: 6.30 ± 0.95) from an American professional soccer club participated in the study. The participants, parents, and the club were informed about the research procedures and provided written informed consent. This study followed the ethical standards for study in humans as outlined in the Declaration of Helsinki. The sample comprised 296 offensive possessions according to the definition of Pollard and Reep [27] that teams performed during the small sided soccer games.

2.2. Small Sided Games

The players performed the same type of SSG (7 vs. 7 + 1 floater, including goalkeepers) under three different pitch configurations ("Standard", 53 × 38 m; "Long", 63 × 32 m; and "Wide", 43 × 47 m) (Figure 1). The SSG were performed two times per week for a period of 3 weeks as part of the normal routine training sessions of the U14 team [28], following a standardized warm-up protocol consisting of 5 min of dynamic mobility and 5 min of technical actions. All the training sessions took place in the spring season, during the same hours (7.00 p.m.), on the same artificial turf surface. The SSG were recorded with one digital camera (Panasonic HC-V180) from an aerial perspective (10 m above the ground) to capture entirely the collective movements of both teams, following the routine way of filming of the team's coaching staff.

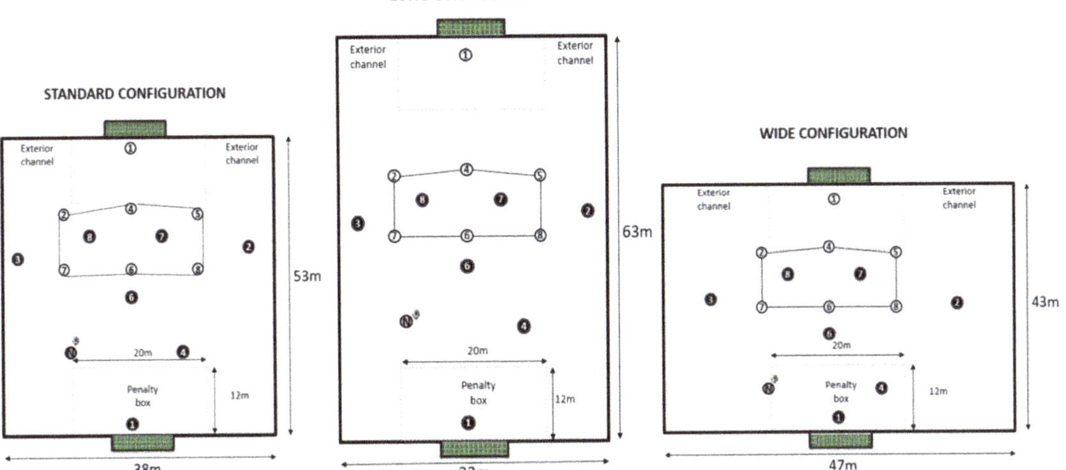

Figure 1. Different pitch configurations and field sizes of the small sided soccer games.

The order in which the SSG formats were performed during the course of the training sessions was randomized (order in week 1: standard, long, and wide; order in week 2: long, wide, and standard; order in week 3: wide, standard, and long).

Table 1 shows the main features of the SSG conducted in this study. The SSG design included official goals and goalkeepers, and the coach did not provide direct instructions about how to solve the tactical situations or to adapt to the different spatial constraints. In addition, the teams had identical tactical objectives and tactical formations during the SGG

The SSG implemented were part of the real tactical training of the team during the spring season. To contextualize, the team's game model was based on building-up from the back, having long ball possessions to disorder the opposing team and creating GSO. This style of play required high passing accuracy and speed but also high amount of patience to decide when and where to break lines of the defensive team to reach the opposing goal. Within this game model, the team's structure and dynamic encouraged the wingers (forwards in our design) to play in interior channels near the midfielders to create offensive superiority in central spaces. Meanwhile, the full backs were encouraged to advance through the wide channels to reach offensive zones, playing an important role in the attacking process to create GSO.

Table 1. Design of the small sided games.

Task Constraints	Description
Player number (including goalkeepers)	7 vs. 7
Floaters	1 (playing as a central back)
Area per player (m^2)	132 m^2
Time (work: passive recovery)	(5:2 min)
Tactical objective	Offensive: To create goal scoring opportunities and goals
	Defensive: To stay compact in a low block defense to protect the goal
Team systems	Offensive: 1.2.3.2
	Defensive: 1.3.3
Feedback	No direct instructions
Rules	Official soccer rules (including offside). The only exception is that all the restarts are taken as goal kicks (no throw ins, corner kicks, or free kicks)
Pitch configurations	Standard (53 × 38 m)
	Long (63 × 32 m)
	Wide (43 × 47 m)

Due to this tactical context, the objective of the coaching staff was to design representative learning designs [29] for players to experience different spatial scenarios where to dominate the ball possession and to create GSO, while they could interact with their teammates and opponents in their real positions.

Therefore, the team's system during the SSG (Figure 2B) tried to reproduce the actual game model in the attacking moment, having:

- One goalkeeper and one central back that should lead the process of building-up from the back.
- One midfielder that had the role of connecting the build-up with the finishing process.
- Two forwards that should occupy interior channels to create superiority in the build-up process and to play a relevant role in the finishing process.
- Two full backs that, in addition to build-up, had an important role reaching offensive zones and creating GSO.
- Additionally, one offensive in-floater was added to, according to the existing scientific literature, increase the passing possibilities and interactions of the attacking team [30,31] as well as to encourage the defensive team to stay compact and to prioritize the protection of the goal, [32], as we can observe in Figure 2A.

It is important to mention that although teams were structured in specific positions and roles, players were free to make their own decisions, movements, and actions to solve the different tactical situations during the SSG, in interaction with their teammates.

Regarding the pitch sizes and configurations selected, the coaching staff decided to compare the standard configuration with an extra-wide and an extra-long format, in order to expose players to different spatial constraints. The final design was based on the work of previous studies [19,26,33] that suggested that modifying the pitch configurations by changing the width and length affects team's spatial and temporal interaction. Although the pitch configuration was changed, the area per player was maintained stable across the three formats. This size (132 m^2 per player) was established considering previous literature on SSG [14,25] since most of the studies used pitch sizes that involved a range between 100 and 150 m^2 per player.

Furthermore, in order to promote the build-up of GSO from the back, the SSG did not have throw-ins, corner kicks, or free kicks, so that all the restarts were initiated as goal kicks by the team in possession of the ball.

Finally, the teams were balanced according to the technical and tactical ability of the players in order to have very competitive games and teams were maintained throughout the duration of the study.

2.3. Performance Analysis

The study was based on systematic observation [34]. Eleven tactical dimensions (Tables 2–4) related to the offensive team possessions were analyzed using the REOFUT theoretical framework [35] that provides a valid and reliable tool to analyze multiple tactical and technical dimensions related to the start, development, penultimate, and last action of teams' possessions as well as their association with achieving offensive performance [9,12].

For the analysis, a soccer coach/researcher experienced in match performance analyzed each possession post-event as many times as necessary. The Lince software [36] was used to code and register the data. The reliability of data was calculated by the intra and inter-observer agreement (Cohen's Kappa). For this purpose, the principal researcher, and another researcher with broad experience in soccer tactical analysis (UEFA Pro Soccer Coach and PhD in Sport Sciences) evaluated 100 random possessions (33% of the sample). This analysis showed good and very good level of reliability according to Altman criteria [37] (inter-observer kappa coefficient = 0.82–1.00; intra-observer kappa coefficient = 0.84–1.00).

Table 2. Description and categories for the dimensions related to the start and development of the team possession.

1-Possession type: way to start a team possession according to if the ball is in play or out of play. Two categories were considered:

(A) **Transition play:** when a player gains the possession of the ball by any means other than from a player of the same team with the ball in play.

(B) **Restart:** when a player initiates the team possession after the ball was out of play. In these SSG, all the restarts were taken as goal kicks by the goalkeeper.

2-Type of attack: degree of offensive directness and elaboration during the offensive process [38–41]. Two categories were considered:

(A) **Organized attack:** (a) the possession starts by winning the ball in play or restarting the game; (b) in this type of team possession the opposing team is organized defensively or is able to re-organize its collective defensive system during the possession.

(B) **Counterattack:** the possession starts by winning the ball in play; the opponent is not organized defensively and is not allowed to re-organize their collective defensive system during the team possession; the progression towards the goal attempts to utilize a degree of imbalance right from start to the end with high tempo [39]; the circulation of the ball takes place more in depth than in width, using a high percentage of penetrative passes. The intention of the team is to exploit the space left by the opponent when they were attacking.

3-Possession width: occupation of the interior and/or exterior channels within the space of defensive occupation of the opponent (SDO) [42] (Figure 2A). Three categories were considered:

(A) **Minimum width:** During the possession, the ball moves through one channel of the SDO.

Medium width: During the possession, the ball moves through two channels of the SDO© **Maximum width:** During the possession, the ball moves through the three channels of the SDO.

4-Passes per possession: number of passes performed by the offensive team during the possession.

Table 3. Description and categories for the dimensions related to the penultimate and last action of the possession.

5-Penultimate action: technical-tactical action performed immediately before the final action that allows the final player to have the opportunity of shooting at goal. This action may be performed by the same player that shoots at goal (individual action) or by a teammate that pass the ball to the final player (collective play). Four categories were considered:

(A) **Individual action:** the final player receives the ball without having a scoring opportunity but he succeeds in creating one by means of an individual action such as dribbling, running with the ball, collecting a free ball, or shooting from distance (the players shoot from outside the penalty box (Figure 1).

(B) **Collective play:** the penultimate player in the team possession performs a pass that allows the last player to have an immediate scoring opportunity. This category has three sub-categories.

(C) **b.1 Pass in behind the defense:** pass from central channels of the field that breaks the opposing defensive line and allows the receiver to have an immediate scoring opportunity in front of the goalkeeper.

(D) **b.2 Cross:** pass performed from the wide channels of the field in the opposing half (Figure 1) towards the penalty box [28] that allows the receiver to have an immediate scoring opportunity.

(E) **b.3. Goal pass:** the final player receives an assist in form of a pass (different from a pass in behind and cross) from a different player that allows him to have an immediate scoring opportunity.

6-Penultimate player: specific position of the player that performs the penultimate action. Five categories were considered **(A) Goalkeeper, (B) Central defender, (C) Full back, (D) Central midfielder, (EF) Forward**

7-Penultimate invasive subspace: area within the space of defensive occupation (SDO) [42] of the opponent where penultimate action is done (Figure 2A). Three categories were considered:

(A) Subspaces behind the defense (WBL, CB, WBR).
(B) Defensive subspace. (WDL, CD, WDL).
(C) Forward subspace (WFR, CF, WFL).

8-Last player: specific position of the player that performs the last action. Five categories were considered **(A) Goalkeeper, (B) Central defender, (C) Full back, (D) Central midfielder, EF) Forward.**

9-Last action: technical action performed by the last player who had the GSO. Three categories were considered:

(A) **Shoot with 1 contact:** the possession ends with a shot on goal by means of a single contact.
(B) **Shoot with 2 or more contacts:** the possession ends with a shot on goal by means of two or more contacts.
(C) **Header:** the final player shoots at goal by heading the ball.

10-Offensive performance: degree of offensive success of the possession, based on the degree of penetration over the opposing team and the achievement of GSO and goals. Three categories were considered:

(A) **Scoring opportunity:** the possession ends with a clear chance of scoring a goal during team possession (goals are included). This includes all shots and all the chances of shooting that one player has inside the penalty box (the player is facing the goal, there is not any opponents between him and the goal, and he has enough space and time to make a playing decision). The shots taken from outside the penalty box are considered GSO when the ball passes near the goal (2 m or less with respect to the goal).

(B) **Offensive penetration:** the team possession achieves to beat the forwards and midfielders' lines of the opponent and face directly the defensive line during the offensive sequence. However, the possession ends without creating any GSO. The player(s) facing the defensive line has/have enough time and space to perform intended actions on the ball at the moment of receiving the ball.

(C) **No offensive penetration:** the team possession does not achieve disorder and beat the forwards or midfielders' lines of the opposing team during the offensive sequence.

11-Last invasive zone: area within the space of defensive occupation (SDO) [42] of the opponent where last action is performed (Figure 2A). Three categories and nine sub-categories were considered:

(A) Subspaces behind the defense (WBL, CB, WBR).
(B) Defensive subspace. (WDL, CD, WDL).
(C) Forward subspace (WFR, CF, WFL).

Table 4. Comparison of playing tactics between the three different pitch configurations.

Category	Standard	Long	Wide	p *	ES #
	Mean ± SD (Median)	Mean ± SD (Median)	Mean ± SD (Median)		
Type of attack					
Counterattack	20.7 ± 13.4 (17.1)	34.6 ± 12.4 a (33.3)	27.4 ± 12.6 (26.8)	**0.028**	**0.30**
Positional attack	79.3 ± 13.4 (82.8)	65,3 ± 12.4 (67.7) [a]	72.6 ± 12.4 (73.2)	**0.028**	**0.30**
Passes per possession	5.0 ± 0.8 (4.9)	4.2 ± 1.2 (4.2)	4.53 ± 0.87 (4.6)	0.105	0.18
Transition play	45.7 ± 18.8 (46.4)	60.83 ± 15.95 (60.0) [a]	62.2 ± 12.4 (63.5) [a]	**0.010**	**0.39**
Restart	54.4 ± 18.8 (54.0)	39.16 ± 17.21 (40.0) [a]	37.7 ± 12.4 (36.4) [a]	**0.010**	**0.39**
Offensive width					
Maximum width	29.7 ± 11.8 (29.3)	35.5 ± 19.6 (36.6)	45.8 ± 17.0 (44.4)	0.083	0.21
Medium width	42.9 ± 17.7 (45.0)	37.8 ± 13.9 (38.7)	43.3 ± 18.1 (43.7)	0.517	0.06
Reduced width	27.3 ± 13.5 (25.0)	26.7 ± 17.8 (19.4)	10.9 ± 9.9 (12.5) [a]	**0.018**	**0.34**
Offensive penetration	49.1 ± 17.7 (50.0)	63.0 ± 15.2 (63.3) [a]	71.7 ± 17.4 (70.8) [a]	**0.007**	**0.41**
Scoring opportunity	26.9 ± 14.6 (28.6)	40.9 ± 21.0 (42.7) [a]	43.2 ± 17.8 (43.6) [a]	**0.026**	**0.30**

* Friedman Test. Values in bold indicate significant differences between pitch configurations. # Effect size calculated using the Kendall's W (coefficient of concordance; 0.1–0.3 = small effect; 0.3–0.69 = moderate effect; 0.70–1.0 = large effect size. [a] = significantly different ($p < 0.05$) from the "standard" configuration.

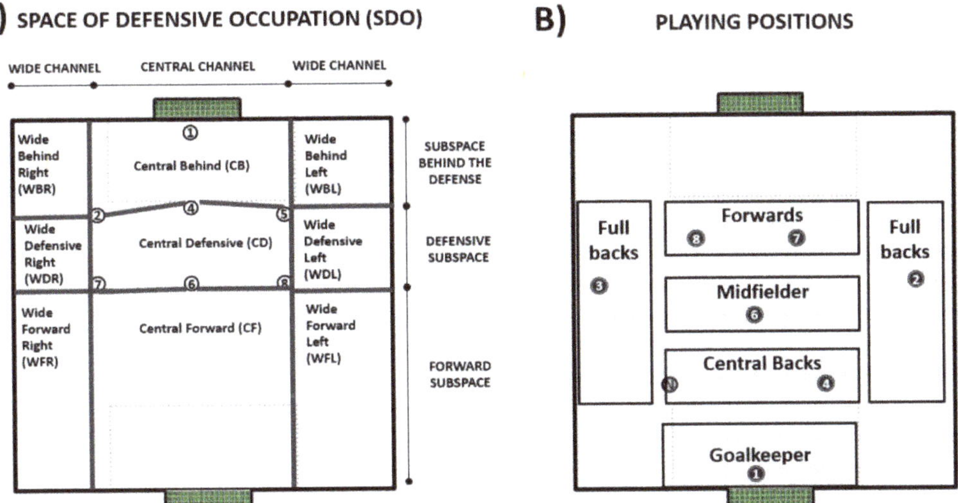

Figure 2. (A) Space of defensive occupation (SDO) of the defensive team [42]. This spatial organization is defined by Gréhaigne [43] as the "space that is constituted by the positions of the players located, at a given moment, in the periphery of a team in play, except the goalkeeper". This space is subdivided into 9 different subspaces that define the level of penetration and width in relation to the opponent (adapted from previous studies, [42–44]). These subspaces are dynamic and change every second depending on the positioning on the opposing players. (B) Playing positions of players considered in this study.

2.4. Statistical Analysis

Data was transcribed to a database created in the SPSS 20.0 program (SPSS, Chicago, IL, USA). All results are reported as mean, standard deviations (mean ± SD), and medians. Data represents the mean percentage of playing tactics or scoring opportunities implemented or created by the teams in each pitch configuration format. To show the differences between formats in the offensive penetration and creation of GSO, boxplot graphs were displayed to assess and compare the shape, central tendency, and variability of the sample.

The non-normality of the data was verified using the Shapiro–Wilk test. The Friedman test (non-parametric ANOVA for repeated measures) was used to detect tactical differences (dependent variables) between the three pitch configurations (independent variables). Dunn–Bonferroni post hoc tests were carried out to determine the differences between pairs. The confidence interval was set at 95%. The effect sizes were calculated using the Kendall's W (coefficient of concordance), (0.1–0.3 = small effect; 0.3–0.69 = moderate effect; 0.69–1.0 = large effect size).

3. Results

3.1. Collective Playing Tactics

Table 4 shows the playing tactics implemented by the teams during the SSG with different pitch configurations. The long configuration registered higher percentage of counterattacks (34.6 ± 12.4%) than the standard one (20.7 ± 13.4%) ($p < 0.05$) and both long and wide had more team possessions that started by means of transition play than the standard format (60.8 ± 15.9% and 62.2 ± 12.4% vs. 45.7 ± 18.8%, respectively) ($p < 0.05$). Additionally, the wide configuration registered less team possessions that achieved reduced width (10.9 ± 9.9%) in comparison with the standard format (27.3 ± 13.5%) ($p < 0.05$).

3.2. Offensive Performance

Figure 3 shows that both long (63.0 ± 15.2%) and wide (71.7 ± 17.4%) formats achieved more offensive penetration than the standard format (49.1 ± 17.7%) ($p < 0.05$). In addition, both long (40.9 ± 21.0%) and wide (43.2 ± 17.8%) were more effective in creating GSO than the standard configuration (26.9 ± 14.6%) ($p < 0.05$).

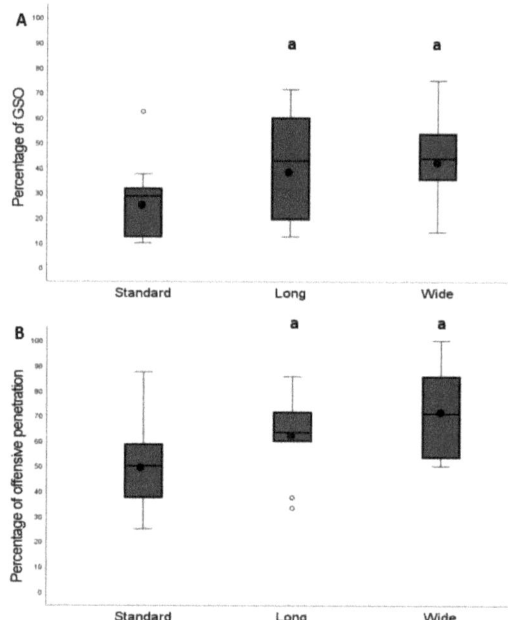

Figure 3. Box plot of the percentage of team possession that achieved (**A**) goal scoring opportunities and (**B**) offensive penetration during the different pitch configuration of the SSG. The box indicates the 25th and 75th quartiles and the central line is the median. The ends of the whiskers are the 2.5% and 97.5% values. Values outside the range of the whiskers are extreme values. Mean is represented with a black dot. a = significantly different ($p < 0.05$) from the "standard" configuration.

3.3. Goal Scoring Opportunities

Table 5 shows the playing tactics implemented when creating GSO. No significant differences were found for all the dimensions except for the offensive width, where the wide format registered lesser frequency of team possessions that had reduced width ($10.9 \pm 14.8\%$) than the long format ($30.1 \pm 31.8\%$) ($p < 0.05$).

Table 5. Comparison of playing tactics between the three different pitch configurations in the team possessions that led to GS.

Category	Standard	Long	Wide	p *	ES #
	Mean ± SD (Median)	Mean ± SD (Median)	Mean ± SD (Median)		
Type of attack					
Counterattack	35.3 ± 35.2 (35.9)	47.8 ± 33.5 (40.0)	25.8 ± 22.1 (33.3)	0.412	0.07
Positional attack	64.7 ± 35.18 (55.5)	52.2 ± 33.5 (60.0)	74.2 ± 22.1(67.2)	0.412	0.07
Pass number	4.5 ± 2.24 (4.9)	4.5 ± 2.7 (3.90)	4.5 ± 1.3 (4.5)	0.667	0.03
Type of start					
Transition play	47.8 ± 41.2 (53.3)	75.4 ± 20.9 (70.8)	55.1 ± 32.1 (60.0)	0.233	0.12
Restart	52.2 ± 41.2 (46.6)	24.6 ± 20.9 (29.1)	44.8 ± 32.1 (40.0)	0.233	0.12
Offensive width					
Maximum width	29.7 ± 37.8 (16.6)	43.7 ± 36.4 (40.0)	48.9 ± 30.73(45.5)	0.636	0.38
Medium width	54.4 ± 41.2 (58.3)	26.1 ± 28.9 (20.00)	40.1 ± 34.5 (40.0)	0.320	0.95
Reduced width	15.8 ± 24.5 (0.0)	30.1 ± 31.8 (22.50)	10.9 ± 14.8 (0.0) [b]	**0.018**	**0.36**

* Friedman Test. Values in bold indicate significant differences between pitch configurations. # Effect size calculated using the Kendall's W (coefficient of concordance; 0.1–0.3 = small effect; 0.3–0.69 = moderate effect; 0.70–1.0 = large effect size [b] = significantly different ($p < 0.05$) from the "long" configuration.

Regarding the penultimate action when creating GSO (Table 6), the wide configuration produced more assists in the form of crosses ($43.0 \pm 25.1\%$) than the long ($13.3 \pm 20.5\%$) and standard formats (16.6 ± 28.6) ($p < 0.05$). In addition, the wide configuration had more utilization of wide subspaces ($65.3 \pm 20.8\%$) to assist the final player than the long configuration ($29.2 \pm 22.0\%$) ($p < 0.05$). Finally, the long configuration registered more passes in behind the defense (30.8 ± 31.5 than the wide format ($12.8 \pm 24.0\%$) ($p < 0.05$).

Table 6. Comparison of the penultimate action between the three different pitch configurations in the team possessions that led to GSO.

Category	Standard	Long	Wide	p *	ES #
	Mean ± SD (Median)	Mean ± SD (Median)	Mean ± SD (Median)		
Penultimate action					
Individual play	55.0 ± 43.9 (53.3)	34.4 ± 30.4 (36.6)	37.6 ± 30.2 (33.3)	0.368	0.08
Cross	16.6 ± 28.6 (0)	13.3 ± 20.5 (0)	43.0 ± 25.1 (36.6) [a,b]	**0.025**	**0.31**
Pass in behind	11.4 ± 18.0 (0)	30.8 ± 31.5 (20.0)	12.8 ± 24.0 (0) [b]	0.034	0.28
Goal pass	16.9 ± 30.9 (0)	21.4 ± 30.2 (10.0)	6.5 ± 12.1 (0)	0.629	0.39
Assisting player					
Central Back	0 (0)	5.5 ± 16.6 (0.0)	10.6 ± 18.6 (0.0)	0.368	0.14
Full Back	56.7 ± 31.3 (50.0)	37.9 ± 42.4 (20.0)	65.9 ± 33.8 (66.7)	0.432	0.12
Midfielder	24.3 ± 38.2 (0.0)	20.00 ± 33.5 (0.0)	11.4 ± 30.3 (0.0)	0.444	0.12
Forward	19.0 ± 24.4 (0.0)	37.6 ± 43.3 (33.3)	12.1 ± 16.8 (0.0)	0.390	0.13
Assisting Space (width level)					
Central subspaces	56.1 ± 34.0 (50.0)	68.7 ± 20.1 (66.7)	34.7 ± 20.8 (33.3) [b]	**0.016**	**0.35**
Wide subspaces	43.9 ± 34.0 (50.0)	29.2 ± 22.0 (33.3)	65.3 ± 20.8 (66.7) [b]	**0.016**	**0.35**

Table 6. Cont.

Category	Standard	Long	Wide	p *	ES #
	Mean ± SD (Median)	Mean ± SD (Median)	Mean ± SD (Median)		
Assisting space (penetration level)					
Subspace behind the defense	19.4 ± 32.4 (0)	27.5 ± 31.7 (25.0)	26.5 ± 24.1 (29.1)	0.704	0.03
Defensive subspace	80.6 ± 32.4 (100)	72.5 ± 31.7 (75.0)	73.5 ± 24.1 (70.8)	0.704	0.03

* Friedman Test. Values in bold indicate significant differences between pitch configurations. # Effect size calculated using the Kendall's W (coefficient of concordance; 0.1–0.3 = small effect; 0.3–0.69 = moderate effect; 0.70–1.0 = large effect size. [a] = significantly different ($p < 0.05$) from the "standard" configuration. [b] = significantly different ($p < 0.05$) from the "long" configuration.

As for the last action when creating GSO, Table 7 shows that no significant differences were found for the final player and the final subspaces either in width or penetration level. As regards the last technical action, a greater number of headers was found in the wide configuration (16.6 ± 18.4%), in comparison with the long format (1.7 ± 5.8%).

Table 7. Comparison of the final action between the three different pitch configurations in the team possessions that led to GSO.

Category	Standard	Long	Wide	p *	ES #
	Mean ± SD (Median)	Mean ± SD (Median)	Mean ± SD (Median)		
Final player					
Central Back	5.5 ± 12.9 (0)	7.5 ± 11.7 (0)	2.1 ± 7.2 (0)	0.431	0.07
Full Back	18.3 ± 29.7 (0)	23.0 ± 30.7 (10.0)	7.9 ± 12.9 (0)	0.449	0.07
Midfielder	16.7 ± 38.9 (0)	8.9 ± 16.1 (0)	16.9 ± 22.8 (8.3)	0.446	0.07
Forward	59.4 ± 43.3 (73.3)	60.5 ± 29.8 (60.0)	71.0 ± 25.2 (70.8)	0.452	0.07
Final Space (width level)					
Central subspaces	74.7 ± 32.9 (10.0)	73.6 ± 30.9 (81.6)	74.4 ± 20.2 (75.0)	0.836	0.01
Wide subspaces	25.3 ± 32.9 (90.0)	26.5 ± 30.9 (18.8)	25.5 ± 20.2 (25.0)	0.836	0.01
Final space (penetration level)					
Behind the defense	60.0 ± 37.5 (50.0)	52.16 ± 33.38 (45.0)	53.5 ± 29.1 (50.0)	0.832	0.01
Defensive subspace	40.0 ± 37.6 (50.0)	47.91 ± 33.34 (55.0)	46.5 ± 29.1 (50.0)	0.832	0.01
Last action					
Shot (2 contacts)	76.9 ± 25.1 (83.3)	55.8 ± 38.6 (66.7)	47.2 ± 29.5 (50.0)	0.232	0.12
Shot (1contact)	18.8 ± 24.4 (0)	42.5 ± 40.1 (33.30)	36.1 ± 34.8 (29.1)	0.423	0.07
Header	4.2 ± 14.4 (0)	1.7 ± 5.8 (0)	16.6 ± 18.4 (12.5) [b]	**0.022**	**0.32**

* Friedman Test. Values in bold indicate significant differences between pitch configurations. # Effect size calculated using the Kendall's W (coefficient of concordance; 0.1–0.3 = small effect; 0.3–0.69 = moderate effect; 0.70–1.0 = large effect size. [b] = significantly different ($p < 0.05$) from the "long" configuration.

4. Discussion

The aim of this study was to explore the tactical effects of different pitch configurations on the collective playing tactics and the creation of goal scoring opportunities during SSG in youth soccer players. Our research observed that manipulating the pitch configuration during SSG to make the field "longer" or "wider" can modulate some of the technical and tactical actions performed by players and teams to create GSO.

To our knowledge, this study is the first to analyze the tactical creation of GSO in SSG with different pitch configurations by means of observational methodology. This fact makes it difficult to compare our findings with other studies, since previous studies evaluated the effects of different pitch shapes on physical and physiological variables [45], as well as on collective team behaviors [19,33] and technical actions [26] However, some of their findings may be useful for the interpretation of our results.

First of all, both the long and the wide configurations created more ball transitions between teams in open play, more counterattacks, more offensive penetration, and more GSO than the standard format. These tactical aspects may be due to the change in the space constraints experienced by the players. For instance, the study of Coutinho et al. [19] observed that SSG with standard condition registered higher collective movement synchronization in both longitudinal and lateral directions, in comparison to other formats such as a sided configuration. Although the methodology used in the research of Coutinho et al. [19] is very different from our study, their findings could help in the interpretation of our results. In this regard, a possible higher collective synchronization in the standard format would create a more defensively organized scenario in which penetrating and creating GSO could be more difficult than in the other formats, in which the different spatial constraints may reduce the collective organization and create a more open context to break lines of the opponent.

For example, the long configuration offers a tactical context where the reduced spatial width may provoke teams trying to advance the opposing goal with more verticality. This scenario may cause more ball losses and changes in the ball possessions between teams in open play, but also it could contribute to creating more "attempts" to break opposing lines, which would explain the higher offensive penetration and number of GSO than the standard format. In addition, the long configuration increases the distance that teams need to cover to reach the opposing goal or to move back to defend the own goal. In fact, the study of Folgado et al. [26] observed that increasing the field's length contributes to an increase in team's length and the distances between the team's centroids, which can create more space between the lines of the defensive team. This tactical constraint can increase the opportunities to perform counterattacks to exploit the space left behind by the opponent when trying to attack, which could lead to more offensive penetration and GSO than in the standard configuration.

As for the wide configuration, two main constraints can influence the modulation of the offensive process in comparison with other formats. On one hand, the wide format reduces the length of the field in comparison to the standard or long formats, which reduces the distance between goals and may make it easier to reach shooting areas, explaining the higher degree of offensive penetration and GSO. Our results agree with Folgado et al. [26] who observed more shots per player in a wider field (30 × 40 m) rather than in a standard one (40 × 30 m). On the other hand, this configuration allows the offensive team to have more space to progress in the wide channels, which makes it more demanding for the defensive team to move laterally and prevent the offensive penetration. In this sense, the study of Folgado et al. [26] observed that increasing the field's width contributes to increases in the team's width, which indicates that teams need to increase the distance between teammates to cover more space laterally, which can also help the offensive team penetrate through the interior subspaces of the defensive team. In addition to a higher team's width, Coutinho et al. [19] observed that a sided pitch did not lead to a higher time spent synchronized compared to the standard configuration. These findings defend the idea that in wider fields, the coordination between teammates decreases, which could lead to more opportunities for the offensive team to penetrate and create GSO.

Regarding the team possessions that led to GSO, the main findings of our study revealed that the wide configuration created more GSO by crossing than the rest of the formats, as well as more headers as the final action than the long configuration. Meanwhile, the long format had a higher frequency of penultimate actions in the form of passing in behind the defense than the wide format.

These results indicate that players explored and implemented different solutions to achieve GSO during the different SSG. For example, the long configuration offers a tactical context where the defensive team may not only have more length but they could also leave more space between their defensive line and the goalkeeper. Under these conditions, the offensive team may have more opportunities to make runs in behind the defensive line to try to exploit this subspace and create GSO. Interestingly, Folgado et al. [26] observed that more elongated pitch elicited more distance covered at high intensities, higher number of forward passes, and a larger distance between the goalkeeper and the last defender. In this tactical context, teams seem to play more vertical in order to gain advantage of the unoccupied space between the last defender and the goalkeeper, which makes players explore a different way of creating GSO than in a wider field, where the spatial constraints offer another tactical context.

In the wide configuration, offensive teams took advantage of the higher field width to perform more crosses from exterior channels of the field and wide subspaces of the opponent. Probably due to the higher frequency of crosses, more headers were found in this configuration. In this line, Frencken et al. [33] suggested that the availability of more lateral space at wider pitches offers players the opportunity to move into these regions, increasing teams' lateral displacement. In this manner, making the field wider seems to increase the importance of crossing and heading to create GSO.

Our study agrees with previous studies [19,26] in suggesting that altering the length and width of the pitch influences players' tendencies to explore along the goal-to-goal and lateral-to-lateral axes. This modulation of the pitch configuration adds variability to the SSG and promotes the players' tactical exploration and movement variability, which promotes the emergence of different tactical solutions to create GSO.

Limitations and Practical Applications

This study has several limitations. Firstly, the fact of only using observational methodology may not capture the entire complexity of soccer actions and interactions, as previous studies based on ecological models have claimed [46–48]. Secondly, this study did not measure collective and positional variables (team width and length, centroid distance, etc.), physical or physiological variables (heart rate, distance covered, accelerations, decelerations, etc.), or the accumulated training load, as other similar studies did [19,26,33,45,49]. Finally, this study only focused on offensive dimensions, while the possible effects of different pitch configurations on the defensive playing tactics were not analyzed.

Nevertheless, this study has important practical applications. Our findings suggest that soccer coaches should consider the manipulation of the pitch configuration to expose players to different spatial constraints and make them explore multiple solutions to create GSO.

5. Conclusions

In conclusion, the long and wide configurations produced more counterattacks, higher offensive penetration, and more GSO than the standard format. Regarding the creation of GSO, the wide configuration produced more assists in the form of crosses than the long and standard formats, more utilization of wide subspaces to assist the final player, greater number of headers as the final action, and less assists in the form of passes in behind the defense than the standard configuration.

Thus, the modulation of the spatial constraints by changing the pitch configuration during small sided soccer games produces different tactical demands, which requires players to adapt to the spatial context and implement different tactical solutions to create GSO.

Author Contributions: Conceptualization, J.G.-R. and R.A.; methodology, J.G.-R. and R.A.; validation, J.G.-R., R.A.-M., P.d.M.-C. and A.T.-D.; formal analysis, J.G.-R.; investigation, J.G.-R.; writing—original draft preparation, J.G.-R.; writing—review and editing, R.A., P.d.M.-C., A.T.-D. and R.A.-M.; supervision, R.A.; project administration, J.G.-R. and R.A. All authors have read and agreed to the published version of the manuscript.

Funding: This research received no external funding.

Institutional Review Board Statement: The study was conducted according to the guidelines of the Declaration of Helsinki. Ethical approval was not required for this study because the soccer exercises were part of the normal routine training sessions of the team and no invasive, individual, or identifiable measures were performed to obtain the data [28].

Informed Consent Statement: Informed consent was obtained from all subjects involved in the study.

Data Availability Statement: The data that support the findings of this study are available from the corresponding author, upon reasonable request.

Acknowledgments: The authors gratefully acknowledge the support of a Spanish government subproject mixed method approach on performance analysis (in training and competition) in elite and academy sport (PGC2018-098742-B-C33) (2019–2021) (del Ministerio de Ciencia, Innovación y Universidades (MCIU), la Agencia Estatal de Investigación (AEI) y el Fondo Europeo de Desarrollo Regional (FEDER)), that is part of the coordinated project new approach of research in physical activity and sport from mixed methods perspective (NARPAS_MM) (SPGC201800X098742CV0).

Conflicts of Interest: The authors declare no conflict of interest.

References

1. Shafizadeh, M.; Sproule, J.; Gray, G. The emergence of coordinative structures during offensive movement for goal-scoring in soccer. *Int. J. Perform. Anal. Sport* **2013**, *13*, 612–623. [CrossRef]
2. Vilar, L.; Araújo, D.; Davids, K.; Travassos, B.; Duarte, R.; Parreira, J. Interpersonal coordination tendencies supporting the creation/prevention of goal scoring opportunities in futsal. *Eur. J. Sport Sci.* **2014**, *14*, 28–35. [CrossRef]
3. Ramos, J.; Lopes, R.J.; Marques, P.; Araújo, D. Hypernetworks Reveal Compound Variables That Capture Cooperative and Competitive Interactions in a Soccer Match. *Front. Psychol.* **2017**, *8*, 1379. [CrossRef]
4. González-Ródenas, J.; López-Bondia, I.; Aranda-Malavés, R.; Tudela, A.; Sanz-Ramírez, E.; Aranda, R. Technical, tactical and spatial indicators related to goal scoring in European elite soccer. *J. Hum. Sport Exerc.* **2020**, *15*, 186–201. [CrossRef]
5. Vestberg, T.; Gustafson, R.; Maurex, L.; Ingvar, M.; Petrovic, P. Executive Functions Predict the Success of Top-Soccer Players. *PLoS ONE* **2012**, *7*, e34731. [CrossRef]
6. Vestberg, T.; Reinebo, G.; Maurex, L.; Ingvar, M.; Petrovic, P. Core executive functions are associated with success in young elite soccer players. *PLoS ONE* **2017**, *12*, e0170845. [CrossRef]
7. Kempe, M.; Memmert, D. «Good, better, creative»: The influence of creativity on goal scoring in elite soccer. *J. Sports Sci.* **2018**, *36*, 2419–2423. [CrossRef] [PubMed]
8. Rodenas, J.G.; Malaves, R.A.; Desantes, A.T.; Ramirez, E.S.; Hervas, J.C.; Malaves, R.A. Past, present and future of goal scoring analysis in professional soccer. *Retos* **2020**, *37*, 774–785.
9. Mitrotasios, M.; Gonzalez-Rodenas, J.; Armatas, V.; Aranda, R. The creation of goal scoring opportunities in professional soccer. Tactical differences between Spanish La Liga, English Premier League, German Bundesliga and Italian Serie A. *Int. J. Perform. Anal. Sport* **2019**, *19*, 452–465. [CrossRef]
10. Durlik, K.; Bieniek, P. Analysis of goals and assists diversity in English Premier League. *J. Health Sci.* **2014**, *4*, 047–056.
11. Armatas, V.; Yiannakos, Á. Analysis and evaluation of goal scored in 2006 world cup. *J. Sport Health Res.* **2010**, *2*, 119.
12. Gonzalez-Rodenas, J.; Mitrotasios, M.; Aranda, R.; Armatas, V. Combined effects of tactical, technical and contextual factors on shooting effectiveness in European professional soccer. *Int. J. Perform. Anal. Sport* **2020**, *20*, 280–293. [CrossRef]
13. Santos, R.; Duarte, R.; Davids, K.; Teoldo, I. Interpersonal Coordination in Soccer: Interpreting Literature to Enhance the Representativeness of Task Design, From Dyads to Teams. *Front. Psychol.* **2018**, *9*, 2550. [CrossRef] [PubMed]
14. Clemente, F.M.; Afonso, J.; Sarmento, H. Small-sided games: An umbrella review of systematic reviews and meta-analyses. *PLoS ONE* **2021**, *16*, e0247067. [CrossRef] [PubMed]
15. Halouani, J.; Chtourou, H.; Dellal, A.; Chaouachi, A.; Chamari, K. Soccer small-sided games in young players: Rule modification to induce higher physiological responses. *Biol. Sport* **2017**, *34*, 163–168. [CrossRef] [PubMed]
16. Hauer, R.; Störchle, P.; Karsten, B.; Tschan, H.; Baca, A. Internal, external and repeated-sprint demands in small-sided games: A comparison between bouts and age groups in elite youth soccer players. *PLoS ONE* **2021**, *16*, e0249906. [CrossRef] [PubMed]
17. Beenham, M.; Barron, D.J.; Fry, J.; Hurst, H.H.; Figueiredo, A.; Atkins, S. A Comparison of GPS Workload Demands in Match Play and Small-Sided Games by the Positional Role in Youth Soccer. *J. Hum. Kinet.* **2017**, *57*, 129–137. [CrossRef]
18. Olthof, B.; Frencken, W.G.; Lemmink, K.A. The older, the wider: On-field tactical behavior of elite-standard youth soccer players in small-sided games. *Hum. Mov. Sci.* **2015**, *41*, 92–102. [CrossRef] [PubMed]
19. Coutinho, D.; Gonçalves, G.; Santos, S.; Travassos, B.; Wong, D.P.; Sampaio, J. Effects of the pitch configuration design on players' physical performance and movement behaviour during soccer small-sided games. *Res. Sports Med.* **2019**, *27*, 298–313. [CrossRef] [PubMed]
20. Sánchez-Sánchez, J.; Hernández, D.; Casamichana, D.; Martínez-Salazar, C.; Ramirez-Campillo, R.; Sampaio, J. Heart Rate, Technical Performance, and Session-RPE in Elite Youth Soccer Small-Sided Games Played with Wildcard Players. *J. Strength Cond. Res.* **2017**, *31*, 2678–2685. [CrossRef] [PubMed]
21. Práxedes, A.; Moreno, A.; Gil-Arias, A.; Claver, F.; Del Villar, F. The effect of small-sided games with different levels of opposition on the tactical behaviour of young footballers with different levels of sport expertise. *PLoS ONE* **2018**, *13*, e0190157. [CrossRef] [PubMed]
22. Machado, J.C.; Barreira, D.; Teoldo, I.; Serra-Olivares, J.; Góes, A.; José Scaglia, A. Tactical Behaviour of Youth Soccer Players: Differences Depending on Task Constraint Modification, Age and Skill Level. *J. Hum. Kinet.* **2020**, *75*, 225–238. [CrossRef]

23. Ramos, A.; Coutinho, P.; Leitão, J.C.; Cortinhas, A.; Davids, K.; Mesquita, I. The constraint-led approach to enhancing team synergies in sport—What do we currently know and how can we move forward? A systematic review and meta-analyses. *Psychol. Sport Exerc.* **2020**, *50*, 101754. [CrossRef]
24. Davids, K.; Araújo, D.; Shuttleworth, R.; Button, C. Acquiring skill in sport: A constraints-led perspective. *Int. J. Comput. Sci. Sport* **2003**, *2*, 31–39.
25. Sarmento, H.; Clemente, F.M.; Harper, L.D.; Da Costa, I.T.; Owen, A.; Figueiredo, A.J. Small sided games in soccer—A systematic review. *Int. J. Perform. Anal. Sport* **2018**, *18*, 693–749. [CrossRef]
26. Folgado, H.; Bravo, J.; Pereira, P.; Sampaio, J. Towards the use of multidimensional performance indicators in football small-sided games: The effects of pitch orientation. *J. Sports Sci.* **2019**, *37*, 1064–1071. [CrossRef] [PubMed]
27. Pollard, R.; Reep, C. Measuring the effectiveness of playing strategies at soccer. *J. R. Stat. Soc. Ser. D* **1997**, *46*, 541–550. [CrossRef]
28. Winter, E.M.; Maughan, R.J. Requirements for ethics approvals. *J. Sports Sci.* **2009**, *27*, 985. [CrossRef]
29. Passos, P.J.; Davids, K. Learning design to facilitate interactive behaviours in Team Sports. *RICYDE* **2015**, *11*, 18–32.
30. Vilar, L.; Esteves, P.T.; Travassos, B.; Passos, P.; Lago-Peñas, C.; Davids, K. Varying Numbers of Players in Small-Sided Soccer Games Modifies Action Opportunities During Training. *Int. J. Sports Sci. Coach.* **2014**, *9*, 1007–1018. [CrossRef]
31. Praça, G.M.; Clemente, F.M.; Andrade, A.G.P.; De Morales, J.C.P.; Greco, P.J. Network Analysis in Small-Sided and Conditioned Soccer Games: The Influence of Additional Players and Playing Position. *Kinesiology* **2017**, *49*, 185–193. [CrossRef]
32. Padilha, M.B.; Guilherme, J.; Serra-Olivares, J.; Roca, A.; Teoldo, I. The influence of floaters on players' tactical behaviour in small-sided and conditioned soccer games. *Int. J. Perform. Anal. Sport* **2017**, *17*, 721–736. [CrossRef]
33. Frencken, W.; Van Der Plaats, J.; Visscher, C.; Lemmink, K. Size matters: Pitch dimensions constrain interactive team behaviour in soccer. *J. Syst. Sci. Complex.* **2013**, *26*, 85–93. [CrossRef]
34. Anguera, M.T.; Hernández-Mendo, A. La metodología observacional en el ámbito del deporte [Observational methodology in sport sciences]. *E-Balonmano.com Rev. De Cienc. Del Deporte* **2013**, *9*, 135–160.
35. Aranda, R.; González-Ródenas, J.; López-Bondia, I.; Aranda-Malavés, R.; Tudela-Desantes, A.; Anguera, M.T. "REOFUT" as an Observation Tool for Tactical Analysis on Offensive Performance in Soccer: Mixed Method Perspective. *Front. Psychol.* **2019**, *10*, 1476. [CrossRef]
36. Gabin, B.; Camerino, O.; Anguera, M.T.; Castañer, M. Lince: Multiplatform sport analysis software. *Procedia Soc. Behav. Sci.* **2012**, *46*, 4692–4694. [CrossRef]
37. Altman, D.G. Some common problems in medical research. In *Practical Statistics for Medical Research*; Altman, D.G., Ed.; Chapman y Hall: London, UK, 1991.
38. Bangsbo, J.; Peitersen, B. *Soccer Systems and Strategies*; Human Kinetics: Champaign, IL, USA, 2000.
39. Tenga, A.; Kanstad, D.; Ronglan, L.T.; Bahr, R. Developing a new method for team match performance analysis in professional soccer and testing its reliability. *Int. J. Perform. Anal. Sport* **2009**, *9*, 8–25. [CrossRef]
40. Sarmento, H.; Anguera, M.T.; Campaniço, J.; Leito, J. Development and validation of a notational system to study the offensive process in football. *Medicina* **2010**, *46*, 401–407. [CrossRef]
41. Lago-Ballesteros, J.; Lago, C.; Rey, E. The effect of playing tactics and situational variables on achieving score-box possessions in a professional soccer team. *J. Sports Sci.* **2012**, *30*, 1455–1461. [CrossRef]
42. Seabra, F.; Dantas, L. Space definition for match analysis in soccer. *Int. J. Perform. Anal. Sport* **2006**, *6*, 97–113. [CrossRef]
43. Gréhaigne, J. *La Organización del Juego en el Fútbol [Game Organization in Football]*; INDE: Barcelona, Spain, 2001.
44. Castellano, J.; Hernández Mendo, A. La observación de la acción de juego en fútbol. Contextualización de los acontecimientos [Game action observation in football. Events contextualization]. *El Entren. Español* **2001**, *90*, 15–20.
45. Casamichana, D.; Bradley, P.S.; Castellano, J. Influence of the Varied Pitch Shape on Soccer Players Physiological Responses and Time-Motion Characteristics during Small-Sided Games. *J. Hum. Kinet.* **2018**, *64*, 171–180. [CrossRef] [PubMed]
46. Araújo, D.; Davids, K.; Hristovski, R. The ecological dynamics of decision making in sport. *Psychol. Sport Exerc.* **2006**, *7*, 653–676. [CrossRef]
47. Vilar, L.; Araújo, D.; Davids, K.; Button, C. The Role of Ecological Dynamics in Analysing Performance in Team Sports. *Sports Med.* **2012**, *42*, 1–10. [CrossRef] [PubMed]
48. Travassos, B.; Davids, K.; Araújo, D.; Esteves, T.P. Performance analysis in team sports: Advances from an Ecological Dynamics approach. *Int. J. Perform. Anal. Sport* **2013**, *13*, 83–95. [CrossRef]
49. Teixeira, J.; Forte, P.; Ferraz, R.; Leal, M.; Ribeiro, J.; Silva, A.; Barbosa, T.; Monteiro, A. Monitoring accumulated training and match load in football: A systematic review. *Int. J. Environ. Res. Public Health* **2021**, *18*, 3906. [CrossRef] [PubMed]

International Journal of
Environmental Research and Public Health

Article

Substantiation of Methods for Predicting Success in Artistic Swimming

Olha Podrihalo [1], Leonid Podrigalo [2], Władysław Jagiełło [3], Sergii Iermakov [3] and Tetiana Yermakova [4,*]

[1] Department of Biological Science, Kharkiv State Academy of Physical Culture, 61022 Kharkiv, Ukraine; rovnayaolga77@ukr.net

[2] Department of Medical Science, Kharkiv State Academy of Physical Culture, 61022 Kharkiv, Ukraine; l.podrigalo@mail.ru

[3] Department of Sport, Gdansk University of Physical Education and Sports, 80-854 Gdansk, Poland; wjagiello1@wp.pl (W.J.); sportart@gmail.com (S.I.)

[4] Department of Pedagogy, Kharkiv State Academy of Design and Arts, 61002 Kharkiv, Ukraine

* Correspondence: yermakova2015@gmail.com

Abstract: To develop a methodology for predicting success in artistic swimming based on a set of morphofunctional indicators and indices, 30 schoolgirls, average age (12.00 ± 0.22), were divided into two groups. Group 1: 15 athletes, training experience 4–5 years. Group 2: 15 schoolgirls without training experience. For each participant, we determined the length and weight of the body, the circumference of the chest, vital lung capacity, and the circumference of the biceps in a tense and at rest. The Erisman index, biceps index, and the ratio of proper and actual vital lung capacity was calculated. Them, we conducted the Stange and Genchi hypoxic tests, and flexibility tests for "Split", "Crab position", and "Forward bend". Prediction was conducted using the Wald test with the calculation of predictive coefficients and their informativeness. A predictive table containing results of functional tests and indices of artistic swimming athletes is developed. It includes nine criteria, which informativeness varied in the range of 395.70–31.98. The content of the prediction consists of evaluating the results, determining the appropriate predictive coefficient, and summing these coefficients before reaching one of the predictive thresholds. The conducted research allowed us to substantiate and develop a method for predicting the success of female athletes with the use of morphofunctional indicators and indices.

Keywords: artistic swimming; prognosis; success; morphofunctional; indicators; indices; functional tests

Citation: Podrihalo, O.; Podrigalo, L.; Jagiełło, W.; Iermakov, S.; Yermakova, T. Substantiation of Methods for Predicting Success in Artistic Swimming. *Int. J. Environ. Res. Public Health* **2021**, *18*, 8739. https://doi.org/10.3390/ijerph18168739

Academic Editors: Paul B. Tchounwou, Bruno Gonçalves, Jorge Bravo and Hugo Folgado

Received: 3 June 2021
Accepted: 13 August 2021
Published: 19 August 2021

Publisher's Note: MDPI stays neutral with regard to jurisdictional claims in published maps and institutional affiliations.

1. Introduction

The selection and prediction of the growth of athletes' skills are the priority tasks of modern sports science [1–4]. Predicting, in a broad sense, means anticipatory reflection of the future and the identification of trends in the dynamics of a particular object based on an analysis of its condition in the past and present. The development of the prediction in the narrow sense is a special scientific research of concrete development prospects of any phenomenon. Predicting sports performance involves identifying predictors of success, analyzing their informativeness and making the dependent between them.

Sports prediction is an important part of sports statistics, the development of which is closely related to interdisciplinary integration. Huang and Shen [5] analyzed research on sports prediction and highlighted the main problems and shortcomings. It is concluded that the theory of sports predicting requires improvement.

Another study analyzed the possibility of applying different methods for sports prediction of results [6]. The authors concluded that the combination of methods significantly increases the probability of prediction. The study by Roberts et al. [7] was dedicated to identifying talented athletes based on predicting. The review and meta-analysis revealed a key topic of "coach instinct" as a main component of talent identification decisions.

Other authors suggest that the application of various mathematical methods allows us to obtain more accurate predictions in sports than subjective expert assessments [8]. It is argued that statistics and analytical methods are becoming increasingly important in basketball [9]. It is determined that players' performance prediction is a serious problem. The authors propose a methodology based on methods of processing rare and irregular data. The results demonstrate the competitiveness of the approach used.

Analyzing the features of sports prediction, Aldous [10] concludes that the use of special algorithms is important. Sports prediction should consider the rating system, answering questions arising in the process of model analysis. Other researchers developed a method for predicting an increase in the efficiency of training 400 m runners [11]. This method allows us to quickly assess the dynamics of physical fitness. It ensures the reliability and quality of prediction based on the training plan.

Other studies present approaches to assess the probability of talent selection in handball [12]. The system used included the general and special tests; assessments made by qualified professionals during training camp; and analysis of athletes' activities in the game based on video. The results of tests revealed the major coincidence of predictions.

Kalina et al. [13] justified the use of various tests to diagnose the abilities and capabilities of athletes. A conclusion is made about the effectiveness of hardware and simulation techniques. Another study considered predicting the results of an anaerobic sprint and 800 m running test at critical speed [14]. The ability of anaerobic distance was determined to be a significant predictor of anaerobic sprint test results and 800 m running results.

Thus, the available literature suggests the possibility of predicting success in sports based on the use of functional tests and the application of various mathematical methods. However, in artistic swimming (AS), this problem does not yet have a final solution. This determined the relevance of this study.

The aim of the study is to develop a method for predicting the success in artistic swimming based on a set of morphofunctional indicators and indices.

2. Materials and Methods

2.1. Ethics Statement and Participants

This study was approved by the Bioethics Committee for Clinical Research and conducted according to the Declaration of Helsinki (protocol of the Commission on Bioethics of the Kharkiv State Academy of Physical Culture No. 25).

The results of the survey of 30 schoolgirls, average age (12.00 $\pm$ 0.22) was used as the main materials of the study. The participants were divided into two groups. The first group included 15 AS athletes with 4–5 years training experience, and the second group included 15 schoolgirls who are not engaged in sports. All participants and their parents were informed about the purpose and objectives of the study, informed about the absence of possible harm to their health and gave the written consent to participate in the study.

2.2. The Design of the Study

The following parameters were determined to assess physical development: the body length (BL) and body weight (BW), chest circumference (CC) in the pause, vital lung capacity (VLC), and biceps circumference in a tense and at rest. The measurements were performed according to the requirements of the international unified methodology of anthropometric research [15].

The level and harmony of physical development was determined using the available age and gender standards of physical development of schoolchildren [16]. The regression scale method was used. The interval represented by body length determined the level of physical development. Development was considered harmonious if body weight and chest circumference were in the range $M \pm \delta_R$.

The following indices of physical development were calculated:

(1) Erisman index

$$ie = t - 0.5 \cdot 1 \tag{1}$$

where ie is the index, t is the chest circumference at rest (cm), and l is the body length (cm).

(2) The proper vital lung capacity (pvlc) was calculated using the formula:

$$pvlc = (l \cdot 0.041) - (b \cdot 0.018) - 3.7 \tag{2}$$

where l is the body length (cm) and b is age (years).

The ratio of pvlc to actual vlc was found.

(3) biceps index

$$ib = (lb1 - lb2)/lb2 \tag{3}$$

where lb1 is the biceps circumference in a tense condition (cm) and lb2 is the biceps circumference at rest (cm).

Hypoxic Stange (inhalation delay time, s) and Genchi (exhalation delay time, s) tests were used to assess the functional condition of the respiratory system.

The "Splits", "Crab position", and "Forward bend" tests were used as specific functional tests. These tests are used for selection in the AP according to the current training programs in this sport.

The "Splits" test: The participant is asked to perform the forward split. Behind her is a tripod, on which bars lie on the head. The distance from the floor to the inguinal region (cm) is measured. In the AS athletes, the leg is put forward, the heel resting on the gymnastic bench.

The "Crab position" test: The athlete is lying flat on her back. The athlete pulls her feet close to her buttocks, rests her hands at shoulder level and stretches upwards. The distance between the palms and heels (cm) is measured.

The "Forward bend" test: The athlete is in a standing position, feet together. The athlete bends forward while holding the leg grip. The bend time (s) is fixed.

2.3. Statistical Analysis

The analysis of the obtained data was performed using licensed MS Excel. The Wald test was used as a tool for solving the predictive problem [17,18]. The method is a predictive table, which includes predictive coefficients of signs and their informativeness. Predictive coefficients were calculated using the formula:

$$pc = 10 \cdot \log \{[p1 \cdot (d1/s)]/[p2 \cdot (d2/s)]\} \tag{4}$$

where pc is a predictive coefficient; s is the total number of people in the group; d1 is the number of test persons who had a value more than the average value in group 1; d2 is the number of test persons who had a value more than the average value in group 2; p1 is the probability of exceeding the average value in group 1; and p2 is the probability of exceeding the average value in group 2.

The predictive coefficients, in the case of the value being lower than average, were found similarly.

The informativeness is calculated by the formula:

$$i = pc \cdot 0.5 \cdot \{[p1 \cdot (d1/s)]/[p2 \cdot (d2/s)]\} \tag{5}$$

where i is the informativeness, and other designations are the same as in the previous Formula (4).

3. Results

The developed methodology for predicting success is given in Table 1.

Table 1. Methods for predicting the athletes' success in artistic swimming.

Indicators	Predictive Coefficients		Informativeness
	Availability	Absence	
The ratio of VLC to the PVLC is more than 1	10.8	−6.7	395.70
Biceps index more 6.7%	10.0	−4.5	300.00
Weight deficit of body in relation to age norms	9.5	−3.7	254.46
The time to perform the "Forward bend" test more than 10 s	6.7	−10.8	245.30
Splits less than 0 cm	6.7	−10.8	245.30
The result of the Stange test higher than the age norm, s	3.8	−5.2	88.72
The result of the Genchi test higher than the age norm, s	3.0	−4.8	60.21
The Erisman index less than 0 cm	2.4	−8.5	48.61
The "Crab position" test result is less than 60 cm	1.9	−7.8	31.98

The predictive table includes the indicator title, the value of its predictive coefficients and the value of informativeness. The indicators in the table are arranged in descending order of informativeness. This minimizes the number of steps in the predicting procedure and reduces the number of possible errors. In the case of the same value of informativeness, the order of location is determined randomly. The value of informativeness below 30.0 is considered insignificant. Indicators with the same or less informativeness are excluded in the table.

The individual prediction is made by successive summation of the values of the predictive coefficients peculiar to the inspected before reaching one of the thresholds. When the indicator specified in the table is performed, the availability indicator is summed. When the indicator specified in the table is not performed, the absence factor is summed. The value of the allowable error is 5%, which corresponds to the value of the threshold of 13 points. Upon reaching the threshold (+13), a conclusion is made about the high probability of success and growth of athletes' skills ($p < 0.05$). Upon reaching the threshold (−13), a conclusion is made about the low probability of success ($p < 0.05$). If, after completion of the table, none of the thresholds are reached, it is concluded that the prediction is uncertain and additional research is necessary.

4. Discussion

Artistic swimming is a unique sport with complex choreographic exercises performed both above the water surface and underwater. This sport requires athletes to have a high level of physical training, mastery of complex technical skills, and artistry. Viana et al. [19] emphasizes the high physical and physiological requirements for the athlete in this sport. The most significant are the need to adapt to long periods of apnea underwater during performing intense activities.

The development of a methodology for predicting success is most relevant at the stage of preliminary basic training. It is due to the specifics of this stage. Athletes had a certain level of morphofunctional indicator development, important for success. The level of mastering specific technical skills and abilities is still insufficient. Thus, predicting at this stage allows us to determine the feasibility of further training of athletes.

The Wald test is widely used in biomedical research for prediction [17,18]. The application of this method allowed us to develop methods for predicting the success of arm wrestling athletes [20] and kickboxing athletes [21]. The advantages of this method include the ability to choose the probability of prediction [17,18]. The probability of the prediction can vary in the range of 80–99.9%, depending on the selected threshold value (8–30 points). It allows us to significantly increase the efficiency of the prediction. To obtain a reliable prediction, 7–10 indicators in the prediction table are enough. The technique developed by us includes nine indicators. This allows us to consider it informative.

The sequential analysis procedure requires a comparison of two groups. In the context of our study, these are groups of AS athletes and non-athletes. Such a research design is common in sports science.

Li et al. [22] compared the condition of the elite AS athletes with the condition of non-athlete female students of the same age. A similar design was used in another study [23]. The authors compared anthropometric data of preschool children. It is a positive effect of the AS classes on the physical development of children. Another review presents data from the analysis of physiological parameters of the AS athletes and persons who did not engage in this sport [24]. The authors note the need to standardize the results.

The developed predictive table includes indices and results of functional tests, which determine the success in AS. This approach allows us to provide an integrated assessment of the condition of athletes, to increase the efficiency of the prediction. Similar data are available in the literature. Solana-Tramunt et al. [25] studied the possibility of monitoring the effect of training of AS athletes. The conclusion is made about the need for an integrated approach, with the use of various methods and indicators.

The available results confirm the legitimacy of including the results of functional tests in the predictive methodology. Escriva-Selles et al. [26] note the prospects of using functional tests to assess the effectiveness of training in the AS.

The index method is used for sports selection and prediction [27–29]. Harty et al. [30] used a fat mass index for selection in AS. The values of this index were significantly lower than in strength sport athletes. It is assessed as an illustration of the specifics of the impact of sport on the body of female athletes.

The functional state of the respiratory system should be recognized as a major factor in the success of AS athletes. The index of the ratio of actual and proper VLC and the results of Stange and Genchi tests are used for predicting. An increase in this index confirms the growth of functionality. The results of functional tests exceeding the age norm illustrate high resistance to hypoxia. It reflects the expansion of the functionality of athletes.

The data available in the literature confirm the validity of these assumptions. Garcia et al. [28] reported an increase in hypoxia resistance of the elite AS athletes compared to age standards.

Rovnaya et al. [31] confirmed the increased functionality of the external respiratory system of AS athletes. There was a direct correlation between the increase in functionality and the level of sportsmanship in the AS. The maximum value of the tidal volume and the minimum value of the respiratory rate were established at all stages of the hypoxic test in elite athletes of artistic swimming in comparison with beginners and sub-elite athletes.

The results of the biceps index obtained by us confirm the high level of development of shoulder muscles in AS athletes. It also illustrates the specifics of training in this sport. Exercises for the arm muscles are an important component of training; this quality is necessary to perform complex technical elements in such a sport as AS.

Similar results were obtained in another study [32]. The authors conclude that it is necessary to perform weightlifting exercises in the training of the AS athletes. Another study confirmed the importance of shoulder muscle balance in the AS athletes [33].

The deficit of body weight in AS athletes is due to the peculiarities of their morphological and nutritional status. Gracilization is one of the predictors of success in this sport. Its main manifestation is a deficit of body weight due to the reduction in the fat component. This coincides with the available literature data. It is determined that the increase in training experience in AS contributes to the increase in disharmony of physical development due to body weight deficit [34].

The assumption confirmed by the Erisman index, which is included in the predictive method. This index reflects the harmony of physical development due to the body muscles. The negative value of this indicator confirms the disharmony of development.

Other researchers provided similar data [35]. The authors determine changes in the homeostatic status of the AS athletes. This refers to changes in body weight under the influence of prolonged and intense training.

Similar results were obtained by Grznar et al. [36]. The authors concluded that the predictor of success in AS is low body weight (lower percentage of fat). In another study, the features of body composition and nutritional status in AS athletes were investigated [37].

A review by Robertson and Mountjoy [38] reported a high prevalence of specific energy deficiency syndrome in sports (RED-S) in AS athletes. This syndrome refers to impaired physiological function, including metabolic rate, menstrual function, bone health, immunity, protein synthesis, and cardiovascular health. It leads to psychological consequences that can either precede (due to restrictive dietary habits) or be the result of RED-S.

The motor tests "Crab position," "Forward bend", and "Splits" allow us to estimate the level of flexibility of AS athletes. The high level of flexibility allows athletes to perform difficult technical elements of such a sport as AS. The importance of this quality for success in AS has also been confirmed by a number of studies. Cho et al. [39] determine that synchronous swimmers had an increased range of motion in the joints of the spine and upper and lower extremities compared to swimming athletes. Other authors have studied the diagnostic validity of various functional tests in the AS athletes to assess functional disorders of the upper extremities [40]. The amplitude of movements of the shoulder and elbow joint can be used as tests of the strength of the upper extremities, criteria for the effectiveness of rehabilitation after injury.

5. Conclusions

Our research allowed us to develop a method for predicting success in the AS. The method is based on the Wald test and includes morphological parameters, results of functional tests, and indices based on them. The proposed method is a simple, informative, and objective tool for monitoring and managing the condition of AS athletes. Determining the indicators used is simple and accessible. It allows us to conclude the availability, clarity, and financial feasibility of the prediction.

The developed methodology can be used by coaches when selecting for artistic swimming, as well as a tool for controlling the fitness of athletes. In monitoring the state of artistic swimming athletes, it is necessary to use a set of indicators reflecting physical development (body weight deficit), indices of the state of the respiratory system (ratio of VLC to proper VLC), muscular system (Erisman index and biceps index), results of tests of the state of the respiratory system (Stange test and Genchi test) and flexibility (tests "Crab position", "Forward bend", and "Splits").

Author Contributions: Conceptualization, O.P., L.P., W.J., S.I. and T.Y.; methodology, L.P., W.J. and S.I.; validation, O.P., L.P. and S.I.; formal analysis, W.J. and T.Y.; investigation, L.P.; resources, O.P., L.P., W.J., S.I. and T.Y.; writing—original draft preparation, O.P., L.P., W.J., S.I. and T.Y.; visualization, W.J. and S.I.; supervision, S.I. and T.Y.; project administration, O.P. and L.P.; and funding acquisition, W.J. and S.I. All authors have read and agreed to the published version of the manuscript.

Funding: This research received no external funding.

Institutional Review Board Statement: This study was approved by the Bioethics Committee for Clinical Research and con-ducted according to the Declaration of Helsinki (protocol of the Commission on Bioethics of the Kharkiv State Academy of Physical Culture No. 25).

Informed Consent Statement: All participants and their parents were informed about the purpose and objectives of the study, informed about the absence of possible harm to their health and gave the written consent to participate in the study.

Conflicts of Interest: The authors declare no conflict of interest.

References

1. Osipov, A.Y.; Kudryavtsev, M.D.; Iermakov, S.S.; Jagiello, W. Criteria for Effective Sports Selection in Judo Schools–on Example of Sportsmanship's Progress of Young Judo Athletes in Russian Federation. *Arch. Budo* **2017**, *13*, 179–186.
2. Vinogradova, O.A.; Sovenko, S.P. Improving technical fitness of race walkers on the basis of special exercises to focus on key parameters of movements. *Pedagog. Phys. Cult. Sports* **2020**, *24*, 100–105. [CrossRef]
3. Vrublevskiy, E.; Skrypko, A.; Asienkiewicz, R. Individualization of selection and training of female athletes in speed-power athletics from the perspective of gender identity. *Phys. Educ. Stud.* **2020**, *24*, 227–234. [CrossRef]
4. Till, K.; Baker, J. Challenges and [Possible] Solutions to Optimizing Talent Identification and Development in Sport. *Front. Psychol.* **2020**, *11*, 664. [CrossRef] [PubMed]
5. Huang, C.; Shen, W. Characters and Development Tendency Analysis on Sports Prediction Scientific Research Papers in China. In Proceedings of the 2011 International Conference on Human Health and Biomedical Engineering, Jilin, China, 19–21 August 2011; pp. 814–819.
6. Spann, M.; Skiera, B. Sports Forecasting: A Comparison of the Forecast Accuracy of Prediction Markets, Betting Odds and Tipsters. *J. Forecast.* **2009**, *28*, 55–72. [CrossRef]
7. Roberts, A.H.; Greenwood, D.A.; Stanley, M.; Humberstone, C.; Iredale, F.; Raynor, A. Coach Knowledge in Talent Identification: A Systematic Review and Meta-Synthesis. *J. Sci. Med. Sport* **2019**, *22*, 1163–1172. [CrossRef]
8. Romanchenko, B.; Sperkach, M.; Zhdanova, O. Predicting of sports events results. *HAIT* **2019**, *2*, 278–287. [CrossRef]
9. Vinué, G.; Epifanio, I. Forecasting Basketball Players' Performance Using Sparse Functional Data. *Stat. Anal. Data Min. ASA Data Sci. J.* **2019**, *12*, 534–547. [CrossRef]
10. Aldous, D. Elo Ratings and the Sports Model: A Neglected Topic in Applied Probability? *Stat. Sci.* **2017**, *32*, 616–629. [CrossRef]
11. Bobkova, E.; Parfianovich, E. Neural networks for forecasting and modeling training in track-and-field athletics. *HSM* **2018**, *18*, 115–119. [CrossRef]
12. Schorer, J.; Rienhoff, R.; Fischer, L.; Baker, J. Long-Term Prognostic Validity of Talent Selections: Comparing National and Regional Coaches, Laypersons and Novices. *Front Psychol.* **2017**, *8*, 1146. [CrossRef] [PubMed]
13. Kalina, R.M.; Jagiełło, W. Non-apparatus, Quasi-apparatus and Simulations Tests in Diagnosis Positive Health and Survival Abilities. In *Advances in Human Factors in Sports, Injury Prevention and Outdoor Recreation*; Ahram, T., Ed.; Advances in Intelligent Systems and Computing; Springer International Publishing: Cham, Switzerland, 2018; Volume 603, pp. 121–128.
14. Arı, E.; Deliceoğlu, G. The Prediction of Repeated Sprint and Speed Endurance Performance by Parameters of Critical Velocity Models in Soccer. *Pedagog. Phys. Cult. Sports* **2021**, *25*, 132–143. [CrossRef]
15. Marfell-Jones, M.; Olds, T.; Stewart, A.; Lindsay Carter, L.E. *ISAK Manual, International Standards for Anthropometric Assessment*; International Society for the Advancement of Kinanthropometry; The University of South Australia Holbrooks Rd: Underdale, Australia, 2012.
16. Serdiuk, A.M. *Standards for Assessing the Physical Development of Schoolchildren*; Fairy Tale: Kiev, Ukraine, 2010. (In Ukrainian)
17. Antomonov, M.I. *Processing and Analysis of Biomedical Data*; Medinform: Kiev, Ukraine, 2018. (In Russian)
18. Gubler, E.V. *Information in Pathology, Clinical Medicine and Pediatry*; Kazan Medical: Leningrad, Russian, 1990. (In Russian)
19. Viana, E.; Bentley, D.J.; Logan-Sprenger, H.M. A Physiological Overview of the Demands, Characteristics, and Adaptations of Highly Trained Artistic Swimmers: A Literature Review. *Sports Med. Open* **2019**, *5*, 16. [CrossRef]
20. Podrigalo, L.V.; Galashko, M.N.; Iermakov, S.S.; Rovnaya, O.A.; Bulashev, A.Y. Prognostication of successfulness in arm-wrestling on the base of morphological functional indicators' analysis. *Phys. Educ. Stud.* **2017**, *21*, 46–51. [CrossRef]
21. Podrigalo, L.V.; Volodchenko, A.A.; Rovnaya, O.A.; Podavalenko, O.V.; Grynova, T.I. The Prediction of Success in Kickboxing Based on the Analysis of Morphofunctional, Physiological, Biomechanical and Psychophysiological Indicators. *Phys. Educ. Stud.* **2018**, *22*, 51–56. [CrossRef]
22. Li, F.; Bai, M.; Xu, J.; Zhu, L.; Liu, C.; Duan, R. Long-Term Exercise Alters the Profiles of Circulating Micro-RNAs in the Plasma of Young Women. *Front Physiol.* **2020**, *11*, 372. [CrossRef] [PubMed]
23. Domika, R.; Supsup, A.A.; Petric, V. valuation of the programme of synchronized swimming for pre-school children. *Acta Kinesiol.* **2018**, *12*, 41–45.
24. Ponciano, K.; Miranda, M.L.d.J.; Homma, M.; Miranda, J.M.Q.; Figueira Júnior, A.J.; Meira Júnior, C.D.M.; Bocalini, D.S. Physiological Responses during the Practice of Synchronized Swimming: A Systematic Review. *Clin. Physiol. Funct. Imaging* **2018**, *38*, 163–175. [CrossRef] [PubMed]
25. Solana-Tramunt, M.; Arboix-Alio, J.; Aguilera-Castells, J. Is Heart Rate Variablity a Suitable method for Monitoring the Effect of a Training Session in Synchronized Swimming? *Med. Sci. Sports Exerc.* **2019**, *51*, 28. [CrossRef]
26. Escrivá-Sellés, F.R.; González-Badillo, J.J. Efecto de Dos Periodos de Entrenamiento de Fuerza Sobre El Rendimiento En Los Ejercicios de Salto Vertical, Barracuda y Boost En Natación Sincronizada. *Apunts* **2020**, *4*, 35–45. [CrossRef]
27. Dorofieieva, O.; Yarymbash, K.; Skrypchenko, I.; Joksimović, M.; Mytsak, A. Complex Assessment of Athletes' Operative Status and Its Correction during Competitions, Based on the Body Impedance Analysis. *Pedagog. Phys. Cult. Sports* **2021**, *25*, 66–73. [CrossRef]
28. García, I.; Drobnic, F.; Pons, V.; Viscor, G. Changes in Lung Diffusing Capacity of Elite Artistic Swimmers during Training. *Int. J. Sports Med.* **2021**, *42*, 227–233. [CrossRef]

29. Müller, W.; Fürhapter-Rieger, A.; Ahammer, H.; Lohman, T.G.; Meyer, N.L.; Sardinha, L.B.; Stewart, A.D.; Maughan, R.J.; Sundgot-Borgen, J.; Müller, T.; et al. Relative Body Weight and Standardised Brightness-Mode Ultrasound Measurement of Subcutaneous Fat in Athletes: An International Multicentre Reliability Study, under the Auspices of the IOC Medical Commission. *Sports Med.* **2020**, *50*, 597–614. [CrossRef]
30. Harty, P.S.; Zabriskie, H.A.; Stecker, R.A.; Currier, B.S.; Moon, J.M.; Jagim, A.R.; Kerksick, C.M. Upper and Lower Thresholds of Fat-Free Mass Index in a Large Cohort of Female Collegiate Athletes. *J. Sport Sci.* **2019**, *37*, 2381–2388. [CrossRef]
31. Rovnaya, O.A.; Podrigalo, L.V.; Aghyppo, O.Y.; Cieslicka, M.; Stankiewicz, B. Study of Functional Potentials of Different Portsmanship Level Synchronous Swimming Sportswomen under Impact of Hypoxia. *Res. J. Pharm. Biol. Chem. Sci.* **2016**, *7*, 1210–1219.
32. Bellver, M.; Del Rio, L.; Jovell, E.; Drobnic, F.; Trilla, A. Bone Mineral Density and Bone Mineral Content among Female Elite Athletes. *Bone* **2019**, *127*, 393–400. [CrossRef]
33. Aguado-Henche, S.; Slocker de Arce, A.; Carrascosa-Sánchez, J.; Bosch-Martín, A.; Cristóbal-Aguado, S. Isokinetic Assessment of Shoulder Complex Strength in Adolescent Elite Synchronized Swimmers. *J. Bodyw. Mov. Ther.* **2018**, *22*, 968–971. [CrossRef] [PubMed]
34. Podrihalo, L.; Rovna, O.; Sokol, K.; Maksechko, O. Studying the peculiarities of physical development and functional state of female athletes engaged in synchronized swimming for educational process optimization. *Sci. Educ.* **2016**, *8*, 132–137. [CrossRef]
35. Carrasco-Marginet, M.; Castizo-Olier, J.; Rodríguez-Zamora, L.; Iglesias, X.; Rodríguez, F.A.; Chaverri, D.; Brotons, D.; Irurtia, A. Bioelectrical Impedance Vector Analysis (BIVA) for Measuring the Hydration Status in Young Elite Synchronized Swimmers. *PLoS ONE* **2017**, *12*, e0178819. [CrossRef]
36. Grznar, L.; Labudova, J.; Ryzkova, E.E.A. Anthropometry, body composition and ace genotype of elite female competitive swimmers and synchronized swimmers. In Proceedings of the 11th International Conference on Kinanthropology—Sport and Quality of Life, Brno, Czech Republic, 29 November–1 December 2017; pp. 162–171.
37. Costa, P.B.; Richmond, S.R.; Smith, C.R.; Currier, B.; Stecker, R.A.; Gieske, B.T.; Kemp, K.; Witherbee, K.E.; Kerksick, C.M. Physiologic, Metabolic and Nutritional Attributes of Collegiate Synchronized Swimmers. *Int. J. Sports Physiol. Perform.* **2019**, *14*, 658–664. [CrossRef] [PubMed]
38. Robertson, S.; Mountjoy, M. A Review of Prevention, Diagnosis and Treatment of Relative Energy Deficiency in Sport in Artistic (Synchronized) Swimming. *Int. J. Sport Nutr. Exerc.* **2018**, *28*, 375–384. [CrossRef] [PubMed]
39. Cho, N.M.Y.; Giorgi, H.P.; Liu, K.P.Y.; Bae, Y.-H.; Chung, L.M.Y.; Kaewkaen, K.; Fong, S.S.M. Proprioception and Flexibility Profiles of Elite Synchronized Swimmers. *Percept. Mot. Skills* **2017**, *124*, 1151–1163. [CrossRef] [PubMed]
40. Gaudet, S.; Begon, M.; Tremblay, J. Cluster Analysis Using Physical Performance and Self-Report Measures to Identify Shoulder Injury in Overhead Female Athletes. *J. Sci. Med. Sport* **2019**, *22*, 269–274. [CrossRef] [PubMed]

International Journal of
Environmental Research and Public Health

Article

The Acute Effects of a Swimming Session on the Shoulder Rotators Strength and Balance of Age Group Swimmers

Nuno Batalha [1,2,*], Jose A. Parraca [1,2], Daniel A. Marinho [3,4], Ana Conceição [4,5], Hugo Louro [4,5], António J. Silva [4,6] and Mário J. Costa [4,7]

1. Departamento de Desporto e Saúde, Escola de Saúde e Desenvolvimento Humano, Universidade de Évora, 7000-654 Évora, Portugal; jparraca@uevora.pt
2. Comprehensive Health Research Centre (CHRC), Universidade de Évora, 7000-654 Évora, Portugal
3. Department of Sport Sciences, University of Beira Interior, 6201-001 Covilhã, Portugal; marinho.d@gmail.com
4. Research Centre in Sports, Health and Human Development (CIDESD), 5000-801 Vila Real, Portugal; anaconceicao@esdrm.ipsantarem.pt (A.C.); hlouro@esdrm.ipsantarem.pt (H.L.); ajsilva@utad.pt (A.J.S.); mario.costa@ipg.pt (M.J.C.)
5. Department of Sport Sciences, Sport Sciences School of Rio Maior, 2040-413 Rio Maior, Portugal
6. Department of Sports Sciences, University of Trás-os-Montes and Alto Douro, 5000-801 Vila Real, Portugal
7. Department of Sports Sciences, Polytechnic Institute of Guarda, 6300-559 Guarda, Portugal
* Correspondence: nmpba@uevora.pt

Abstract: The purpose of this study was to analyze the acute effects of a standardized water training session on the shoulder rotators strength and balance in age group swimmers, in order to understand whether a muscle-strengthening workout immediately after the water training is appropriate. A repeated measures design was implemented with two measurements performed before and after a standardized swim session. 127 participants were assembled in male (n = 72; age: 16.28 ± 1.55 years, height: 174.15 ± 7.89 cm, weight: 63.97 ± 6.51 kg) and female (n = 55; age: 15.29 ± 1.28 years, height: 163.03 ± 7.19 cm, weight: 52.72 ± 5.48 kg) cohorts. The isometric torque of the shoulder internal (IR) and external (ER) rotators, as well as the ER/IR ratios, were assessed using a hand-held dynamometer. Paired sample t-tests and effect sizes (Cohen's d) were used ($p \leq 0.05$). No significant differences were found on the shoulder rotators strength or balance in males after training. Females exhibited unchanged strength values after practice, but there was a considerable decrease in the shoulder rotators balance of the non-dominant limb ($p < 0.01$ $d = 0.366$). This indicates that a single practice seems not to affect the shoulders strength and balance of adolescent swimmers, but this can be a gender specific phenomenon. While muscle-strengthening workout after the water session may be appropriate for males, it can be questionable regarding females. Swimming coaches should regularly assess shoulder strength levels in order to individually identify swimmers who may or may not be able to practice muscle strengthening after the water training.

Keywords: swimming; isometric strength; muscle balance; shoulder rotators

 check for **updates**

Citation: Batalha, N.; Parraca, J.A.; Marinho, D.A.; Conceição, A.; Louro, H.; Silva, A.J.; Costa, M.J. The Acute Effects of a Swimming Session on the Shoulder Rotators Strength and Balance of Age Group Swimmers. *Int. J. Environ. Res. Public Health* **2021**, *18*, 8109. https://doi.org/10.3390/ijerph18158109

Academic Editor: Paul B. Tchounwou

Received: 23 June 2021
Accepted: 28 July 2021
Published: 30 July 2021

Publisher's Note: MDPI stays neutral with regard to jurisdictional claims in published maps and institutional affiliations.

1. Introduction

The shoulder rotator muscles play a critical role in providing stability and mobility to the glenohumeral joint and shoulder joint complex [1]. In competitive swimming, the propulsive forces responsible for the total body displacement are produced mainly by the upper limbs through the arm adduction and shoulder internal rotation [2]. As such, the swimmers are classified as overhead athletes [3] because high levels of stress are installed in the upper body sections.

There is evidence that water training induces shoulder muscle imbalances [4]. The shoulder adductor and the internal rotator (IR) show a tendency to become proportionally stronger when compared to their antagonists. In addition, higher volumes of swimming training are associated with shoulder pain and injury [5]. Without the addition of preventive measures, this can lead to an acute or chronic injury process [6].

Shoulder strengthening programs are part of the training season and require detailed planning [3,7]. Evidence suggests that dryland workouts must be maintained throughout the entire sports season, otherwise the effects of detraining are felt [4]. Few studies have focused on shoulder rotator strengthening in competitive swimmers. Batalha et al. [8] found that a 10-week dryland training program reduced muscle imbalance and fatigue. Kluemper et al. [2] also reported increases in the shoulder rotators strength after a 6-week training program with consequent postural improvements. So, regular shoulder strength workouts are essential to maintain the integrity and longevity of the swimmers' glenohumeral joint [8].

Most times dryland strength training and in-water training are applied within the same training session. Coaches regularly assign dryland strength training before the swimming practice due to time-consuming issues [3,9]. However, it remains unclear when these workouts should be implemented and the impact an in-water session has on shoulder fatigue. It is not consensual if a single water session will compromise the shoulder joint and muscles, and what would be the degree of fatigue installed. The increasing fatigue in shoulder muscles, especially in the rotator cuff group, was identified as a possible cause of shoulder dysfunction [10] and was associated with performance decrements and a higher risk of injury [11]. Regarding the performance of muscle strengthening programs before the water training, Batalha et al. [9] evaluated the acute effects on the shoulder rotator strength. They concluded that shoulder rotator endurance and balance do not seem to be impaired after undergoing a shoulder rotator injury-prevention training program. However, to the best of our knowledge, there are no studies that assess the effects water training may have on the strength and balance of the shoulder rotators. Hence, it is crucial to understand whether it would be possible to implement a shoulder rotator strengthening program after swimming practice. Thus, the aim of this study was to analyze the acute effects of a standardized water training session on the strength and balance of the shoulder rotators in adolescent swimmers. It was hypothesized that a swimming training session would significantly reduce the shoulder rotator strength levels and muscle balance, limiting the performance of the dryland training afterward.

2. Materials and Methods

2.1. Participants

One hundred and twenty-seven national-level adolescent swimmers participated in this study. All participants were recruited from the clubs that agreed to participate in the study. Seventy two participants were males (age: 16.28 ± 1.55 years, height: 174.15 ± 7.89 cm, weight: 63.97 ± 6.51 kg, training/week: 6.75 ± 0.86 sessions, training time/day: 126 ± 26.39 min), and 55 were females (age: 15.29 ± 1.28 years, height: 163.03 ± 7.19 cm, weight: 52.72 ± 5.48 kg, training/week: 6.52 ± 0.57 sessions, training time/day: 115 ± 16.23 min) who met the following inclusion criteria: (i) do not have any clinical history of upper limb disorders; (ii) compete at the national level; and (iii) have a minimum of 10 h of training per week. The main goals of the study were explained to all participants and their legal guardians, who signed an informed consent allowing their participation. This research was approved by the ethics committee of the seeding institution. The research was undertaken in compliance with the Declaration of Helsinki and the international principles governing research on humans and animals.

2.2. Procedures

Before the data collection, all participants completed a questionnaire that included questions on hand dominance, shoulder injury, pain, and swimming training frequency. All subjects performed a 5-min shoulder warm-up with articular mobility and resistance tubing with the same directions used for testing. The IR and ER cuff strength data were collected during isometric actions performed with the microfet 2™ Digital Handheld Muscle Tester (Hoggan health, Draper, UT, USA), which is an accurate, portable Force Evaluation and Testing device, with a sample frequency of 10 sample/second. It is a modern adaptation of the time-tested art of hands-on manual muscle testing. The hand-held dynamometer (HHD) has wireless capability, is battery operated, and is ergonomically designed to fit comfortably in the palm of the hand. The reliability of this device to measure ER/IR ratios has already been documented [12]. An experienced tester with more than 10 years of experience performing muscle strength measurement with an HHD conducted all the tests.

Shoulder ER and IR strength were evaluated with the exact same methodology before and after a standardized swimming training. Tests were performed bilaterally in prone position with 90° of shoulder abduction and 0° of rotation with the elbow flexed to 90° (Figure 1). This position was considered the most suitable to replicate the arm position during the stroke, while also being repeatable [13,14]. Additionally, it has been used in previous studies within the same topic [3,15]. The order of testing position, sides (dominant, non-dominant), and motions (internal rotation, external rotation) was randomized. The tester stabilized the humerus distally against the stretcher, and the participants used the opposite arm to grasp it next to the test table for support. The HHD was placed just proximal to the ulnar styloid process on the anterior surface of the forearm to assess the IR strength. To assess the ER strength, the HHD was positioned using the same anatomical landmarks but on the posterior surface of the forearm. After positioning, the participants performed a warm-up protocol that consisted of 2 submaximal and 1 maximal isometric contraction in each direction, separated by 30 s of rest [14]. The maximal IR and ER isometric strength was evaluated with two repetitions for each shoulder and each rotation (2 × IR and 2 × ER), using a make-type test in which the participants were instructed to slowly produce and sustain a full isometric contraction (five seconds) of the involved muscle group until they were told to relax. The maximum value recorded from the two repetitions of each test session was used for analysis. In order to analyze muscle balance, the ER/IR ratio [(ER/IR) × 100] was calculated [16]. All tests had a resting period of 10 s between each repetition and 60 s between each strength test. During the entire period of force production, the tester verbally motivated the participants.

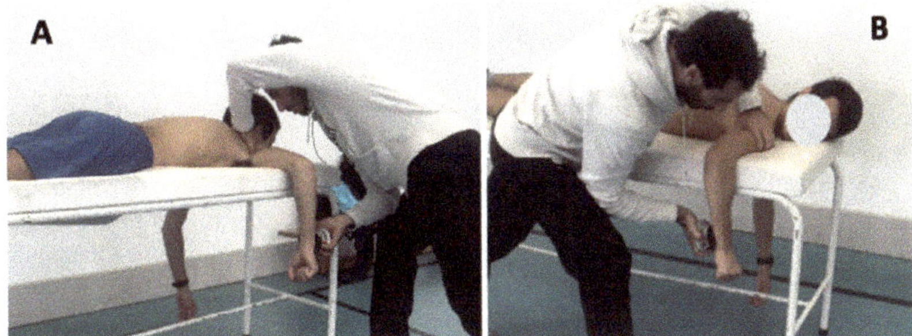

Figure 1. Shoulder external and internal rotation measurements. (**A**)—Shoulder external rotation measurement; (**B**)—Shoulder internal rotation measurement.

A "standard" swimming training session was performed between measurements. The training session was carried out according to the recommendations of the Portuguese Swimming Federation regarding the activities of the youth national squads participating in international competitions [17]. The total volume of the training session was 4600 m, including: (i) 900 m of warming-up tasks with low-intensity bouts; (ii) 800 m of technical training with technical drills; (iii) 400 m of velocity tasks with sprint bouts; (iv) 1000 m of aerobic capacity bouts; (v) 600 m of aerobic power bouts; and (vi) 900 m of recovery sets with low-intensity tasks. The session was previously sent to all coaches for approval and to be part of the training unit within each testing day.

2.3. Statistical Analysis

Assumptions of statistical tests such as normal distribution (Shapiro-Wilk test, $p > 0.05$) and sphericity (Mauchly test, $p > 0.05$) of data were checked as appropriate. All the parameters were normally distributed. Paired sample t-tests were conducted to compare the means of the shoulder strength, before and after the water training. In addition to the p values, the researchers provided detailed statistics, including the mean, the mean differences, and 95% confidence interval, to better depict the changes between the two testing points. Effect sizes were calculated using Cohen's d and interpreted as: null if <0.2, small if 0.2 to <0.6, medium if 0.6 to <1.2, and large if $\geq$1.2 [18]. All analyses were performed with SPSS (version 23.0; SPSS Inc, Chicago, IL, USA), and the significance level was set at $p \leq 0.05$ for all tests.

3. Results

Table 1 shows the strength and balance measures of the overall sample in pre and post-test. Although slight decreases in the strength and muscle balance are seen, those are not significant between both time points. Even though the result of the ER/RI ratio of the non-dominant shoulder was near the cut-off value for significance ($p = 0.050$; $d = -0.175$) this should be highlighted as a considerable reduction.

Table 1. Maximal isometric strength of internal and external, dominant and non-dominant shoulders, before and after training.

	Before Training N = 127	After Training N = 127	Mean Difference (95% CI)	p
IR_D (N)	169.40 ± 68.6	167.84 ± 45.3	−1.49 (−5.89, 2.92)	0.505
IR_ND (N)	172.60 ± 50.1	168.40 ± 50.9	−4.20 (−8.62, 0.48)	0.079
ER_D (N)	131.50 ± 35.3	130.63 ± 33.6	−0.87 (−4.08, 3.59)	0.901
ER_ND (N)	120.46 ± 28.88	121.10 ± 29.21	0.64 (−2.33, 3.61)	0.672
Ratio_D (%)	77.90 ± 11.71	78.50 ± 12.75	0.55 (−1.6, 2.71)	0.612
Ratio_ND (%)	78.65 ± 22.22	74.57 ± 13.84	−4.08 (−8.15, 0)	0.050

p-values for Paired sample *t*-tests; IR—Internal Rotation; ER—External Rotation; D—Dominant; ND—Non-Dominant.

Table 2 presents the shoulder rotators strength of male swimmers. Since there were no significant differences between testing points, there were no acute effects of aquatic training in this group. All the effect size values were under 0.2, indicating a null effect. Thus, there seems to be no issue carrying out shoulder-strengthening training programs following the water training with male swimmers.

Table 2. Maximal isometric strength of internal and external, dominant and non-dominant shoulders, before and after training.

	Before Training N = 72	After Training N = 72	Mean Difference (95% CI)	p
IR_D (N)	196.30 ± 37.5	191.69 ± 40.60	−4.61 (−11.27, 2.06)	0.173
IR_ND (N)	200.83 ± 42.88	196.43 ± 45.18	−4.4 (−11.22, 2.42)	0.202
ER_D (N)	151.15 ± 30.54	150.31 ± 32.98	−0.84 (−7.16, 5.48)	0.791
ER_ND (N)	137.73 ± 23.9	137.04 ± 24.63	−0.69 (−5.04, 3.67)	0.755
Ratio_D (%)	77.60 ± 11.72	79.50 ± 14.58	1.9 (−1.27, 5.07)	0.236
Ratio_ND (%)	70.22 ± 12.80	71.62 ± 13.83	1.4 (−1.69, 4.49)	0.369

p-values for Paired sample *t*-tests; IR—Internal Rotation; ER—External Rotation; D—Dominant; ND—Non-Dominant.

Table 3 shows the results for female swimmers, revealing some differences compared to their male counterparts. Although there were unchanged strength values between testing points, there was a considerable decrease in the balance of the shoulder rotators of the non-dominant limb after practice. These results suggest that there should be some caution when conducting strengthening programs after aquatic training with female swimmers.

Table 3. Maximal isometric strength of internal and external, dominant and non-dominant shoulders, before and after training.

	Before Training N = 55	After Training N = 55	Mean Difference (95% CI)	*p*
IR_D (N)	134.03 ± 25.39	136.62 ± 29.33	2.6 (−2.65, 7.85)	0.326
IR_ND (N)	134.02 ± 29.43	130.39 ± 29.01	−3.64 (−9.42, 2.14)	0.212
ER_D (N)	103.61 ± 19.24	104.16 ± 21.65	0.55 (−2.88, 3.97)	0.750
ER_ND (N)	97.86 ± 16.59	100.23 ± 20.08	2.37 (−1.55, 6.29)	0.231
Ratio_D (%)	78.29 ± 11.66	77.07 ± 9.64	−1.21 (−3.99, 1.57)	0.386
Ratio_ND (%)	89.69 ± 26.78	78.45 ± 12.98	−11.24 (−19.53, −2.96)	0.009

p-values for Paired sample *t*-tests; IR—Internal Rotation; ER—External Rotation; D—Dominant; ND—Non-Dominant.

The analysis of the different effect sizes in the distinct measurements is presented in Figure 2. The acute effect of the water training is related to a decrease in the ER/IR ratio mostly in the non-dominant shoulder of female swimmers ($p < 0.01$, $d = 0.366$).

Accute Water effect

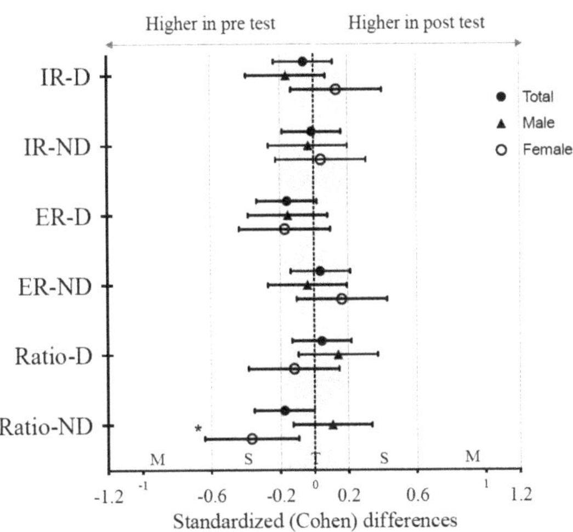

Figure 2. Cohen's d values (Mean ± SD) of pre and post water training. Paired-sample *t*-test, $^* p \leq 0.05$.

4. Discussion

The aim of this study was to analyze the acute effects of a standardized water training session on the shoulder rotator strength and balance in adolescent competitive swimmers. The results do not fully confirm the initial hypothesis that an in-water training session is enough to reduce the shoulder rotator strength levels and muscle balance. In fact, this seems to be a gender specific phenomenon. While the male swimmers revealed no acute effect after the water session, the female swimmers showed a significant increase in muscle imbalance in the non-dominant shoulder.

The comparative analysis of the strength levels between ER and IR is consistent in the total sample and between genders. This allows us to confirm that the muscle groups responsible for the internal rotation are in fact the strongest, and this is in line with previous studies that included swimmers [3,9], athletes from other sports [6,19], and sedentary people [20]. This can be explained by the size and number of the muscles around the glenohumeral joint. The muscular groups which produce the IR are not only greater in number, but are also anatomically larger and naturally stronger than their antagonists [21].

Additionally, in relation to IR and ER strength, it is a fact that shoulder muscles fatigue induces sensory motor perturbations that consequently alter the kinematics of the shoulder joint, overloading different structures in order to maintain the level of performance [22]. It is still unclear whether rotator cuff-specific muscle fatigue or general scapular muscle fatigue has the greatest influence on upper limb malfunction; however, it is a fact that there is a difference in the behavior and reaction regarding muscle fatigue around the shoulder joint, IRs appearing to be more fatigue-resistant than ERs in both physically-active people [23] and sedentary ones [20]. Our results do not fully support this statement since, in general, the IR strength values showed greater reductions compared to the ER, although with no significant values. In the analysis of the total sample results (Table 1), we can see that, despite there being no significant differences between the pre and post aquatic training, the reductions in the IRs strength levels are higher than those of the ERs (in the non-dominant shoulder there was no reduction). These results may point out that, contrary to what happens in other sports, swimming induces a greater degree of fatigue in the IR, which can be explained by the predominant internal-rotator forces that occur during pull-out and recovery [24]. During all stroke phases, the subscapularis muscle is always active, stabilizing the glenohumeral joint, because of the repetitive internal rotator forces [25]. It should be emphasized that IRs strength deficits may affect stroke dynamics and should be considered an important injury risk factor for swimmers [26].

When analyzing the results of the ER and IR strength values by gender, there was no acute effect in any of the variables after performing the aquatic training, but it showed some specific trends. In male swimmers, the strength values slightly decreased in the post-test, mostly in the IRs. Contrarily, in females, there was only a reduction in the IR values for the non-dominant shoulder, which can be seen as an alarming cue. Overall, the results for both genders are in agreement with the study by Matthews et al. [27]. The authors reported no significant strength differences between pre- and post-fatigue in the internal and external rotators of both shoulders. The changes were exclusively reductions in the stroke length, and in the IR and ER range of motion on both shoulders. More recently, Yoma et al. [24] observed that shoulder rotation isometric peak torque decreased immediately after a high-intensity training session but remained unchanged after a low-intensity session. These results highlight the importance of evaluating the strength values of the shoulder rotators, while controlling the training intensity. In this study, we did not perform an effective control of intensity, so no conclusions can be drawn. However, considering the characteristics of the swimming tasks (pointing to a medium/low intensity effort), we may argue that the intensity was not enough to induce fatigue as it happened in previous studies.

Regarding the shoulder rotators balance (ER/IR ratios), numerous studies reported that swimmers had greater IRs strength because of the repetitive concentric actions required during the propulsive phases of the swimming strokes [3,7]. In contrast, ER strength is weaker, which often leads to shoulder muscle imbalances. The ER/IR ratio assessment can be a useful measure to identify muscle imbalances in the swimmers' shoulders, and it is also associated with shoulder injuries [28], and scapular dyskinesis [29]. Previous normative data consider ER/IR ratios between 66 and 75% to be adequate [28,30]. Values below this threshold are commonly associated with muscle imbalance and instability in the glenohumeral joint [23]. In this study, the ER/IR ratios are within the considered normal standards for a healthy joint [28,30], varying between 70.22% ± 12.80% and 79.50% ± 14.58% in male swimmers, and between 77.07% ± 9.64% and 89.69 ± 26.78% in female swimmers in pre- and post-tests. In addition, these results are in line with previous studies that included swimmers [12], in which the authors carried out a reliability study, with the same evaluation technique, with ratios between 78.71% ± 9.36 and 81.81% ± 10.24%. The similarity of the results remained when compared to the study by Batalha et al. [9], but in this case the authors evaluated swimmers with an isokinetic dynamometer. The ratios were between 72.31% ± 15.66% and 77.37% ± 16.40%. On the other hand, Riemann et al. [14], in order to compare different evaluation positions with a handheld dynamometer, used a position and methodology similar to the one used in the current study. The ER/IR ratios presented were considerably higher (between 90% and 92% in boys, and 98 and 99% in girls). However, the sample was composed of healthy non-sports individuals, which may prove that swimmers have greater shoulder muscle imbalances when compared to non-sports individuals [20].

When we analyze the results of the ER/IR ratios, we can see that in male swimmers there was no acute effect of aquatic training on the shoulder rotators balance, since the ratio values are practically identical in pre- and post-training. In female swimmers, there was a slight reduction in the ratio in the dominant limb, although without a significant result. However, in the non-dominant shoulder there was a considerable acute effect of the water training ($p = 0.009$; $d = 0.366$). While this significant reduction is within the normal thresholds [30], coaches must be aware of it. Since the muscular strength balance between agonist and antagonist muscles is crucial for joint stability and ensures a dynamic centralization of the humeral head [29], this can be even more compromised if a strengthening training is to be implemented after a swim session.

Some limitations should be considered for further discussion: (i) no strict control of the training intensity may have compromised the strength response from swimmers; although the session was carefully planned, the effort during each training task was dependent on the athlete (ii) the assessment of strength deficits after the swimming practice was limited to the shoulder group, not allowing the study of those effects in other muscles in action.

5. Conclusions

The results of this study indicate that a single practice seems not to affect shoulders strength and balance of adolescent swimmers, but this can be a gender specific phenomenon. While a muscle-strengthening workout after the water session may be appropriate for males, this can be questionable for females, since the shoulder rotators' muscular balance, mostly of the non-dominant limbs, was compromised after training.

We believe these results have important practical implications for swimmers and coaches. In fact, swimming coaches should be careful if they intend to carry out shoulder-strengthening programs immediately after swim practice. It should be important to identify deficits in post swim rotation strength, serving as a practical way to reduce the athlete's susceptibility to shoulder injury. In addition, an individualized regular exercise program to improve shoulder rotation strength should be performed to minimize the post swimming adaptations. Finally, we believe that if swimming coaches regularly assess strength levels, they can easily find a method to check individual trends and see who will be able or not to maintain the strengthening workouts after the water training sessions.

Author Contributions: Conceptualization, N.B., J.A.P. and D.A.M.; methodology, N.B., A.C. and H.L.; software, M.J.C.; validation, N.B. and A.J.S.; formal analysis, N.B., J.A.P. and D.A.M.; investigation, N.B. and M.J.C.; resources, N.B., A.C. and H.L.; data curation, N.B. and A.J.S.; writing—original draft preparation, N.B., D.A.M. and M.J.C.; writing—review and editing, N.B., D.A.M. and M.J.C.; visualization, J.A.P.; supervision, N.B., M.J.C. and D.A.M.; project administration, N.B.; funding acquisition, N.B., J.A.P. and M.J.C. All authors have read and agreed to the published version of the manuscript.

Funding: The present publication was funded by Fundação Ciência e Tecnologia, IP national support through CHRC (UIDP/04923/2020). The English proofreading was funded by the project UIDB/04045/2020.

Institutional Review Board Statement: The study was conducted according with the guidelines of the Declaration of Helsinki and approved by the Institutional Ethics Committee of the University of Évora (proceeding 16019/2016).

Informed Consent Statement: Informed consent was obtained from all subjects involved in the study. Written informed consent has been obtained from the patient(s) to publish this paper.

Data Availability Statement: The data presented in this study are available on request from the corresponding author. The data are not publicly available due to privacy restrictions.

Acknowledgments: The authors would like to thank all clubs, coaches, and swimmers who were available to participate in this study.

Conflicts of Interest: The authors declare no conflict of interest.

References

1. Boettcher, C.E.; Cathers, I.; Ginn, K.A. The role of shoulder muscles is task specific. *J. Sci. Med. Sport* **2010**, *13*, 651–656. [CrossRef] [PubMed]
2. Kluemper, M.; Uhl, T.; Hazelrigg, H. Effect of stretching and strengthening shoulder muscles on forward shoulder posture in competitive swimmers. *J. Sport Rehabil.* **2006**, *15*, 58–70. [CrossRef]
3. Sawdon-Bea, J.; Benson, J. The Effects of a 6-Week Dry Land Exercise Program for High School Swimmers. *J. Phys. Educ. Sports Manag.* **2015**, *2*, 1–17. [CrossRef]
4. Batalha, N.M.; Raimundo, A.M.; Tomas-Carus, P.; Marques, M.A.; Silva, A.J. Does an in-season detraining period affect the shoulder rotator cuff strength and balance of young swimmers? *J. Strength Cond. Res.* **2014**, *28*, 2054–2062. [CrossRef] [PubMed]
5. Feijen, S.; Tate, A.; Kuppens, K.; Claes, A.; Struyf, F. Swim-Training Volume and Shoulder Pain Across the Life Span of the Competitive Swimmer: A Systematic Review. *J. Athl. Train.* **2020**, *55*, 32–41. [CrossRef]
6. Byram, I.R.; Bushnell, B.D.; Dugger, K.; Charron, K.; Harrell, F.E.; Noonan, T.J. Preseason Shoulder Strength Measurements in Professional Baseball Pitchers Identifying Players at Risk for Injury. *Am. J. Sports Med.* **2010**, *38*, 1375–1382. [CrossRef]
7. Batalha, N.; Marmeleira, J.; Garrido, N.; Silva, A.J. Does a water-training macrocycle really create imbalances in swimmers' shoulder rotator muscles? *Eur. J. Sport Sci.* **2015**, *15*, 167–172. [CrossRef] [PubMed]
8. Batalha, N.; Dias, S.; Marinho, D.A.; Parraca, J.A. The Effectiveness of Land and Water Based Resistance Training on Shoulder Rotator Cuff Strength and Balance of Youth Swimmers. *J. Hum. Kinet.* **2018**, *62*, 91–102. [CrossRef]
9. Batalha, N.; Paixao, C.; Silva, A.J.; Costa, M.J.; Mullen, J.; Barbosa, T.M. The Effectiveness of a Dry-Land Shoulder Rotators Strength Training Program in Injury Prevention in Competitive Swimmers. *J. Hum. Kinet.* **2020**, *71*, 11–20. [CrossRef] [PubMed]
10. Roy, J.S.; Ma, B.; Macdermid, J.C.; Woodhouse, L.J. Shoulder muscle endurance: The development of a standardized and reliable protocol. *Sports Med. Arthrosc. Rehabil. Ther. Technol.* **2011**, *3*, 1–14. [CrossRef]
11. Ebaugh, D.D.; McClure, P.W.; Karduna, A.R. Effects of shoulder muscle fatigue caused by repetitive overhead activities on scapulothoracic and glenohumeral kinematics. *J. Electromyogr. Kinesiol.* **2006**, *16*, 224–235. [CrossRef]
12. Conceicao, A.; Parraca, J.; Marinho, D.; Costa, M.; Louro, H.; Silva, A.; Batalha, N. Assessment of isometric strength of the shoulder rotators in swimmers using a handheld dynamometer: A reliability study. *Acta Bioeng. Biomech.* **2018**, *20*, 113–119. [CrossRef]
13. Cools, A.M.J.; Vanderstukken, F.; Vereecken, F.; Duprez, M.; Heyman, K.; Goethals, N.; Johansson, F. Eccentric and isometric shoulder rotator cuff strength testing using a hand-held dynamometer: Reference values for overhead athletes. *Knee Surg. Sports Traumatol. Arthrosc.* **2016**, *24*, 3838–3847. [CrossRef] [PubMed]
14. Riemann, B.L.; Davies, G.J.; Ludwig, L.; Gardenhour, H. Hand-held dynamometer testing of the internal and external rotator musculature based on selected positions to establish normative data and unilateral ratios. *J. Shoulder Elb. Surg.* **2010**, *19*, 1175–1183. [CrossRef] [PubMed]
15. Ramsi, M.; Swanik, K.A.; Swanik, C.B.; Straub, S.; Mattacola, C. Shoulder-rotator strength of high school swimmers over the course of a competitive season. *J. Sport Rehabil.* **2004**, *13*, 9–18. [CrossRef]
16. Wang, H.K.; Cochrane, T. Mobility impairment, muscle imbalance, muscle weakness, scapular asymmetry and shoulder injury in elite volleyball athletes. *J. Sports Med. Phys. Fit.* **2001**, *41*, 403–410.
17. Marinho, D.; Machado, J.; Silva, A. *Política Desportiva FPN–Natação Pura. Identificação E Desenvolvimento Do Talento Em Natação*; Federação Portuguesa de Natação: Lisboa, Portugal, 2020.
18. Cohen, J. *Statistical Power Analysis for the Behavioral-Sciences*, 2nd ed.; Academic Press: New York, NY, USA, 1988.
19. Pontaga, I. Shoulder external/internal rotation peak torques ratio side-asymmetry, mean work and power ratios balance worsening due to different fatigue resistance of the rotator muscles in male handball players. *MLTJ* **2018**, *8*, 513–519. [CrossRef]

20. Forthomme, B.; Dvir, Z.; Crielaard, J.M.; Croisier, J.L. Isokinetic assessment of the shoulder rotators: A study of optimal test position. *Clin. Physiol. Funct. Imaging* **2011**, *31*, 227–232. [CrossRef]
21. Dark, A.; Ginn, K.A.; Halaki, M. Shoulder muscle recruitment patterns during commonly used rotator cuff exercises: An electromyographic study. *Phys. Ther.* **2007**, *87*, 1039–1046. [CrossRef]
22. Stackhouse, S.K.; Stapleton, M.R.; Wagner, D.A.; McClure, P.W. Voluntary activation of the infraspinatus muscle in nonfatigued and fatigued states. *J. Shoulder Elb. Surg.* **2010**, *19*, 224–229. [CrossRef] [PubMed]
23. Ellenbecker, T.S.; Roetert, E.P. Testing isokinetic muscular fatigue of shoulder internal and external rotation in elite junior tennis players. *J. Orthop. Sports Phys. Ther.* **1999**, *29*, 275–281. [CrossRef] [PubMed]
24. Yoma, M.; Herrington, L.; Mackenzie, T.A.; Almond, T.A. Training Intensity and Shoulder Musculoskeletal Physical Quality Responses in Competitive Swimmers. *J. Athl. Train.* **2021**, *56*, 54–63. [CrossRef] [PubMed]
25. Pink, M.; Perry, J.; Browne, A.; Scovazzo, M.L.; Kerrigan, J. The normal shoulder during freestyle swimming. An electromyographic and cinematographic analysis of twelve muscles. *Am. J. Sports Med.* **1991**, *19*, 569–576. [CrossRef]
26. Tate, A.; Turner, G.N.; Knab, S.E.; Jorgensen, C.; Strittmatter, A.; Michener, L.A. Risk factors associated with shoulder pain and disability across the lifespan of competitive swimmers. *J. Athl. Train.* **2012**, *47*, 149–158. [CrossRef]
27. Matthews, M.J.; Green, D.; Matthews, H.; Swanwick, E. The effects of swimming fatigue on shoulder strength, range of motion, joint control, and performance in swimmers. *Phys. Ther. Sport* **2017**, *23*, 118–122. [CrossRef]
28. Ellenbecker, T.S.; Davies, G.J. The application of isokinetics in testing and rehabilitation of the shoulder complex. *J. Athl. Train.* **2000**, *35*, 338–350. [PubMed]
29. Madsen, P.H.; Bak, K.; Jensen, S.; Welter, U. Training induces scapular dyskinesis in pain-free competitive swimmers: A reliability and observational study. *Clin. J. Sport Med.* **2011**, *21*, 109–113. [CrossRef] [PubMed]
30. Van Cingel, R.; Kleinrensink, G.J.; Mulder, P.; De Bie, R.; Kuipers, H. Isokinetic strength values, conventional ratio and dynamic control ratio of shoulder rotator muscles in elite badminton players. *Isokinet. Exerc. Sci.* **2007**, *15*, 287–293. [CrossRef]

International Journal of
*Environmental Research
and Public Health*

MDPI

Review

Effects on Strength, Power and Speed Execution Using Exercise Balls, Semi-Sphere Balance Balls and Suspension Training Devices: A Systematic Review

Moisés Marquina [ID], Jorge Lorenzo-Calvo *[ID], Jesús Rivilla-García [ID], Abraham García-Aliaga [ID] and Ignacio Refoyo Román [ID]

Facultad de Ciencias de la Actividad Física y del Deporte (INEF—Sports Department), Universidad Politécnica de Madrid, 28040 Madrid, Spain; marquinascience@gmail.com (M.M.); jesus.rivilla@upm.es (J.R.-G.); abraham.garciaa@upm.es (A.G.-A.); ignacio.refoyo@upm.es (I.R.R.)
* Correspondence: jorge.lorenzo@upm.es; Tel.: +34-910-678-023

Abstract: Research in instability has focused on the analysis of muscle activation. The aim of this systematic review was to analyse the effects of unstable devices on speed, strength and muscle power measurements administered in the form of controlled trials to healthy individuals in adulthood. A computerized systematic literature search was performed through electronic databases. According to the criteria for preparing systematic reviews PRISMA, nine studies met the inclusion criteria. The quality of the selected studies was evaluated using STROBE. The average score was 14.3 points, and the highest scores were located in 'Introduction' (100%) and 'Discussion' (80%). There is great heterogeneity in terms of performance variables. However, instability seems to affect these variables negatively. The strength variable was affected to a greater degree, but with intensities near to the 1RM, no differences are observed. As for power, a greater number of repetitions seems to benefit the production of this variable in instability in the upper limb. Instability, in comparison to a stable condition, decreases the parameters of strength, power, and muscular speed in adults. The differences shown are quite significant in most situations although slight decreases can be seen in certain situations.

Keywords: instability; resistance training; exercise ball; suspension training; performance

Citation: Marquina, M.; Lorenzo-Calvo, J.; Rivilla-García, J.; García-Aliaga, A.; Refoyo Román, I. Effects on Strength, Power and Speed Execution Using Exercise Balls, Semi-Sphere Balance Balls and Suspension Training Devices: A Systematic Review. *Int. J. Environ. Res. Public Health* **2021**, *18*, 1026. https://doi.org/10.3390/ijerph 18031026

Academic Editors: Bruno Gonçalves, Hugo Folgado, Jorge Bravo and Pantelis T. Nikolaidis
Received: 26 November 2020
Accepted: 21 January 2021
Published: 24 January 2021

Publisher's Note: MDPI stays neutral with regard to jurisdictional claims in published maps and institutional affiliations.

1. Introduction

Strength training under unstable conditions, as well as destabilizing devices, have gained popularity in the last decade for athletes in order to strengthen core muscles, improve balance, proprioception and increase performance [1–4]. Muscle power is considered to be one of the main determinants of many short-term explosive sporting events [5]. Power has been defined by different authors throughout history. Bompa [6], defined muscle power as the ability to perform different actions, developing maximum strength in a short period. For this reason, power is a manifestation of the strength that most athletes in different disciplines consider to be of greater importance for the development of certain movements [7–12].

A potential reason for similar training-induced adaptations observed after unstable situations compared to stable ones could be related to similar or even higher levels of muscle activation in favour of unstable conditions [13,14]. In the case of the evidence shown concerning the use of unstable situations concerning strength, power and speed, maximum tests have been used to evaluate muscle strength; and tests to determine muscle power (i.e., abdominal power test, medicine ball throwing and different types of jumps) [15–20]. However, current results indicate that at least for non-elite athletes, there is a stress/strength and power training intensity pathway that is sufficient to induce positive training adaptations. In their review, Behm and Colado [21] reported that the average strength deficit in unstable situations compared to similar stable exercises was 29%. Furthermore, in healthy young adults, strength training with low loads compared to high loads is equally effective when improving muscle strength [22,23].

Because sport is not usually practised under stable conditions, such as throws, jumps, changes of direction, where the body must be stabilized while a specific action is being performed, training should try to represent the requirements of the specific sport [24–27]. Also, training under unstable conditions or unbalanced conditions can resemble training and daily activities, providing effective transference [1]. The use of unstable training has been proposed to improve movement specific effect through increased activation of stabilizers and core muscles [14,25,27].

Although initially unstable surface training was reserved for rehabilitation and prevention programs to reduce the rate of injury, due to the proprioceptive overload they provide [28–31] this type of training is now included in strength and conditioning programs [10,32–34]. Currently, the use of these devices has been incorporated into traditional exercises to promote neuromuscular coordination and recruitment but there is controversy regarding the effects of this combination on sports performance and core stability activation [27,35].

There are various ways of generating instability in the performance of exercises but the most common has been through the use of different devices [36–39]. The use of unstable devices in strength training has led to the development of numerous investigations focused on the analysis of muscle activation [40–46]. Nowadays, the use of different training methods that retain the stabilising capacity of athletes has become a common and frequent practice. The use of specific devices to create unstable environments, such as exercise or Swiss ball [47], the semi-sphere balance balls, like BOSU [48] and the suspension devices, like TRX [49] have been widely used in sports centres and are widely used throughout the population. Therefore, this review has focused on testing the use of these types of devices and not others, to clarify their influence on sports performance.

1.1. Swiss Balls or Exercise Balls

Swiss ball are air-filled balls covered with soft elastic material with a diameter of approximately 35 to 85 cm [47]. The use of unstable surfaces, such as the exercise ball, began to be used in strength and muscle conditioning training as a method of strengthening core, stabilizing muscles, that is the musculature with a deep location which is responsible for a good body posture both in our daily life and in the practice of sports [36,50–52]. The response of muscle activity to this unstable surface can be variable and depends on the type of exercise or muscles being tested. Patterns of muscle activation during bench presses have reported variable results based on muscle function. In the stabilising or core muscles, it has been shown that there is an increase in the activation of the internal obliques, the external obliques and the rectus abdominis [25,53–56] and spinal erectors [56] while having minimal effect on the rectus abdominis [13,56]. In the upper extremities, compared with a stable bench press, a greater increase in anterior deltoid, pectoralis major, triceps and serratus anterior activity has been demonstrated during execution on a stability ball [30,51,55,57–59]. However, improvement in muscle activation has not always been evident. In the case of the central or stabilising muscles, other studies have not shown any change in the oblique and internal spinal erectors [60]. In addition, other studies have shown higher data for the stable condition as opposed to the unstable condition, in the main motor musculature responsible for movement, as in the case of the pectorals and triceps in the bench press [43], or no main motor muscle in the shoulder press [59].

1.2. Semi-Sphere of Balance

The BOSU Balance Trainer ® (DW Fitness, LLC, Clifton, NJ, USA), or "both sides up" is an exercise device used to improve balance, core muscle or torso strength, and proprioception created for military service veterans [61]. The flat part of the device is a 25-inch platform with two built-in handles, and the other part is an inflatable rubber dome that rises about one foot above the ground. Each side can be used in different ways to create different situations depending on the exercise.

Different studies have analyzed the influence of the semi-sphere ball for training, providing a great variety of results. Authors found increased muscle activity in the rectus abdominis and external oblique in the performance of abdominal plates and gluteal bridge [53]. Anderson and Behm [14] reported increased EMG activity in the vast lateral, soleus and superficial trunk muscles, but not in the femoral biceps when comparing squatting with free weight on a stable versus an unstable surface. In bench press, greater activation of the internal oblique, spinal erector, soleus and biceps femoris was also evident [56]. However, Willardson et al. [35] compared EMG activity in the core performing 50% of 1RM in squats on a stable surface and in a semi-sphere ball and observed no differences between conditions. Authors examined the activity of the brachial triceps, spinal erector, rectus abdominis, internal oblique, and soleus while performing traditional and unstable bending in a single (hands or feet on the unstable surface) or dual (both hands and feet on the unstable surface) instability and found that the dual condition caused the highest percentage of change (>150%) for all muscles analyzed; compared with the other conditions [62].

1.3. Suspension Training Devices

A new method available to increase muscle activation is suspension training. This type of training uses the principles of body weight and strength boosting to improve motor unit recruitment [63]. The most applied suspension device is the TRX Home Suspension Training Kit (Fitness Anywhere LLC, San Francisco, CA, USA). In suspension training, a specific device is required to create an unstable condition. This method uses a system of straps with handles on the bottom and which are attached to a single anchor point [64]. Among the different strength training possibilities, suspension training is widely applied in various contexts. It is considered an effective technique for improving neuromuscular activation that precedes the use of heavy loads in traditional exercises [65]. Besides, improvements in speed and strength indicators have been found by the use of suspension training, suggesting increased recruitment of core/stabilizing muscles [66].

Regarding evidence from suspension devices, the effects of usage on both lower body muscle activity [41,67,68] and trunk stabilizing muscle [65,69–71] have been investigated. Clear evidence has been established regarding these devices that witnessed increased muscle activation in the stabilizing and synergistic muscles when performing exercises under these conditions [65,69,71,72]. Concerning lower body exercises, very high activation has been shown for the femoral and semitendinous biceps (>90% MVIC) [68], the hamstring, the gluteus maximus, the gluteus medius and the long adductor. Although no significant differences were found in the rectus femoris [41,67]. Also for the Bulgarian squat exercise, no difference in muscle activation was found between the stable and the suspended condition [64]. For the exercises of the upper part of the trunk and the stabilising muscles, they have been studied with the performance of the Push-up exercise. It has been shown that greater muscular activation in the core, rectus abdominis and external oblique muscles [69,73], however, the stable situation reported greater activation for the pectoral and deltoid muscles [65,69,71] the brachial triceps [65] and the clavicular portion of the pectoralis [71]. However, for the frontal plate exercise, Byrne et al. [70] reported no significant differences when studying exercise with suspension devices.

Research has focused on the analysis of muscle activation, determining different considerations, claiming their strengths and weaknesses. However, when the training objective is hypertrophy, or gain in muscle mass, strength, or power, it has not been recommended that the exercises be carried out using unstable situations [5].

As mentioned earlier, muscle activation has been shown to improve the stabilising muscles and reduce the main motor muscles involved in performing the task, but it is not known whether these activations can lead to improved performance.

In terms of the choice of devices, the 3 devices have been widely used for both rehabilitation, proprioception, and development of muscular capacity. With any one of them, there are numerous proposals for working on muscular strength, and all of them focus on the argument of greater activation of the central or core muscles. In fact, the use of suspension devices alone cannot bring about an improvement in strength or power by itself, since one always works with one's own body weight, without external loads that increase the intensity of the tasks to be carried out. Semi-spherical and exercise ball devices are not only an implement that increases instability, but there are concrete references that indicate that it could be a way to improve strength and power [30,43,59,74]. Although it could have certain limitations, therefore it has been decided to include it, to be able to evaluate all devices and tools that generate instability to a greater or lesser degree.

Therefore, a synthesis of the literature seems necessary to determine whether performing exercises with unstable material provides additional effects on measures of speed, strength and muscle power compared to stable execution. The existing controversy regarding the use and results provided by instability training, and the variability of surfaces, devices, instability positions, samples and exercises generated a heterogeneity of results that makes this review necessary. The aim of this systematic review was to provide a scientifically based study regarding the effects of unstable devices on speed, strength and muscle power measurements administered in the form of controlled trials to healthy individuals in adulthood, apart from muscular activation. It is hypothesized that unstable devices produce similar or not excessively inferior performance improvements to stable conditions because performance with instability is very demanding on the neuromuscular system (i.e., additional stability of joints and posture during exercise is required).

2. Materials and Methods

The present research was designed to qualitatively synthesize the available scientific evidence concerning the effect of instability in strength, power, and speed training. The stages of the review procedure and subsequent analysis of the original articles stayed within the guidelines set out in the Preferred Reporting Items for Systematic Reviews and Meta-Analysis (PRISMA) [75] checklist and the Population, Interventions, Comparisons, Outcomes and Study Design (PICOS) question model for the definition of inclusion criteria.

2.1. Study Selection and Eligibility Criteria

Primary and original studies to evaluate the strength, power or execution speed in instability were included. Furthermore, studies had to have been published in any language, in peer-reviewed journals with an impact factor included in the Journal Citation Reports of the Web of Science (JCR of WoS) or Scimago Journal & Country Rank (SJR of Scopus) until November 2020.

According to the 'PICOS' question model, the inclusion criteria were: (1) 'Population': physically active and healthy participants (both men and women) between 18 and 65 years. This age range includes all participants considered to be of adult age; (2) 'Intervention': acute training effects on strength, power and/or speed of execution using a Swiss ball, semi-sphere ball or suspension devices; (3) 'Comparison': differences in tasks multi-articular upper or lower limb between the execution of exercises in stable conditions and execution in unstable situations; (4) 'Outcomes': at least one strength, power and/or speed result had to be reported in the study; (5) 'Study Design': descriptive and quasi-experimental research based on a comparison between stable and unstable situations.

The exclusion criteria were: (1) the studies were for intervention periods, randomized control trials, and clinical trials; (2) they included patients or persons with disease or injury; (3) any data about muscle activation, because it is not a performance variable; (4) the subjects were not of adult age (under 18; e.g., children and adolescents and over 65's as the elderly) (5) the chronic effects of the situations under investigation were assessed; (6) Any other type of unstable device other than Swiss ball, semi-sphere ball or suspension devices, because these are the most frequently used devices; (7) any measurement that includes unilateral exercises or with different types of support such as exercises executed in a monopodal position.

2.2. Literature Search

A systematic computerized literature search of the Web of Science, PubMed and EBSCOhost with full text was conducted until November 2020 to capture all relevant articles investigating the effectiveness of instability versus stability. The following Boolean search strategy was applied using the operators 'AND', 'OR' and 'NOT': ('instability resistance training' OR 'instability strength training' OR 'free-weight training' OR 'suspension training' OR 'unstable devices') AND ('power' OR 'power performance' OR 'speed') AND ('strength' OR 'muscle strength' OR 'muscle power' OR 'muscular power') NOT ('natural surfaces') AND ('stability balls' OR 'bosu' OR 'suspension devices' OR 'unstable devices'). The unrestricted language search was limited to the human species and the availability of the full text of original articles reporting on a quasi-experimental trial in an academic journal. Also, we checked the reference lists of each included article and reviewed relevant review articles to identify additional studies suitable for inclusion in the database.

2.3. Systematic Review Protocol

The authors worked separately and independently to ensure the reliability of the process and the suitable eligibility of the studies. According to the criteria for preparing systematic reviews "Preferred Reporting Items for Systematic Reviews and Meta-Analysis"—PRISMA [75], the protocol carried out in the months of July, August and September 2020 and it was made up of four stages (Figure 1): (1) Identification: the first author (M.M.) found 167 studies in the four digital databases; (2) Screening: the first author (M.M.) eliminated the duplicate files ($n = 8$) and excluded those considered not relevant through a previous reading of the title, abstract and keywords ($n = 90$). Furthermore, the first author (M.M.), jointly with the second (J.L.C.) and third (J.R.G.), rejected the studies linked to the instability according to the exclusion criteria through a full-text reading ($n = 55$); (3) Eligibility: the first (M.M.), second (J.L.C.) and third author (J.R.G.) eliminated full-text studies from the selection process by the eligibility criteria ($n = 45$); (4) Inclusion: the remaining studies ($n = 8$) based on the relationship between the execution of the exercises in a stable condition and their execution in an unstable condition were finally considered. An additional article was identified from the reference lists of included papers and review articles already published [24,63,76,77].

2.4. Data Extraction and Management

A standardized form was used to extract data from the studies included in the review for assessment study quality and scientific evidence. Thus, information about (A) 'authors and year of publication', (B) 'sample experience' (C) 'sample size and sex' (number of players, sex), (D) 'sample characteristics' (age, height and weight), (E) 'variable measured' (strength, power and/or speed), (F) 'type of exercise' (G) 'number of situations' whether the tasks were executed by comparing only the stable condition vs. the unstable condition or whether more variations were included), (H) 'device' (unstable device implemented), (I) 'training volume' (number of sets/repeats/rest per exercise), (J) 'intensity' (percentage of one maximum repetition (1RM)), (K) 'strength results' (maximum strength (e.g., 1RM), mean strength), (L) 'power results' (maximum power, mean power and concentric phase power) and (M) 'speed results' (maximum and mean speed) were collected. The results data reflect the percentage of decrease or increase in instability concerning stability. If the included studies did not report the results (i.e., means and standard deviations) of the pre-and post-tests, the authors of those studies were contacted.

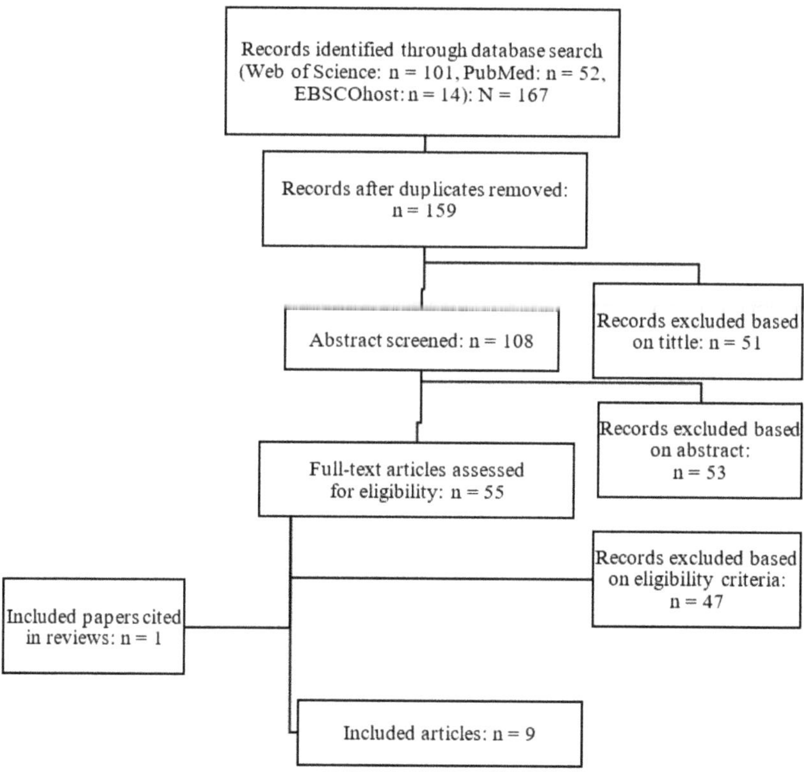

Figure 1. Flow chart illustrating the different phases of search and survey selection.

2.5. Study Quality Assessment

The quality of all eligible cross-sectional studies was evaluated using the criteria for strengthening the reporting of observational studies in epidemiology "STROBE" [78]. The following scale was used to rate the quality of studies: (a) good quality (>14 points, low risk of major or minor bias), (b) acceptable quality (7–4 points, moderate risk of major bias), and (c) poor quality (<7 points, high risk of major bias). The score was obtained through the 22 points on the STROBE checklist. Two independent reviewers (M.M. and J.L.) conducted study quality assessment. Rating disagreements were resolved by J.R. and inter-rater reliability calculated.

3. Results

3.1. Synthesis of Findings (Qualitative Analysis)

Scientific evidence on the sample characteristics (B, C, D), variables (E), exercise and variation (F, G) device used (H) volume and intensity training (I, J) and results for strength, power and speed is shown in Tables 1–3. Format and design, including the author and the year of publication, the sample characteristics (overall number, gender, age, height and weight), the variable measured (strength, power and/or speed), type and number of variations, the device used (exercise ball, semi-sphere ball or suspension device), training volume (number of sets/repeats/rest per exercise), intensity training (percentage of one maximum repetition (1RM)), strength results (maximum strength, mean strength), power results (maximum power, mean power and concentric phase power) and speed results (maximum and mean speed) are included.

Table 1. Scientific evidence on the sample characteristics (B, C, D) and variables (E).

Reference (A)	Sample			Variables (E)
	Experience (B)	Size and Sex (C)	Characteristics (D)	
Anderson et al. (2004) [13]	Trained in strength, Instability 1 year earlier	10 (M)	a: 26.2 ± 6.0 years h: 177.3 ± 6.0 cm w: 87.3 ± 12.2 kg	Strength
Behm et al. (2002) [79]	Trained	8 (M)	a: 24.3 ± 6.7 years h:178.1 ± 6.1 cm w: 82.3 ± 8.9 kg	Strength
Chulvi-Medrano (2010) [80]	Trained in strength Experience with instability	31 (M)	a: 24.29 ± 0.48 years h: 167.98 ± 8,11 cm w: 79.08 ± 2,37 kg	Strength
Goodman et al. (2008) [81]	Recreational	13 (10 M, 3 W)	a: 24.1 ± 1.6 years h: 176.7 ± 3.0 cm w: 76.0 ± 3.9 kg	Strength
Koshida et al. (2008) [82]	Trained	20 (M)	a: 21.3 ± 1.5 years h: 167.7 ± 7.7 cm w: 75.9 ± 17.5 kg	Strength Power Speed
Saeterbakken & Fimland (2013) [83]	Trained	15 (M)	a: 23.3 ± 2.7 h: 181 ± 0.09 cm w: 80.5 ± 8.5 kg	Strength
Sannicandro et al. (2015) [84]	No previous experience in strength or instability is indicated	24 (M)	a: 17.8 ± 0.8 years h: 179.1 ± 5.6 cm w: 73 ± 4.9 kg	Strength Power
Zemkova (2012) [85]	Trained in strength, no experience in instability	16 (M)	a: 23.4 ± 1.9 years h: 181.5 ± 6.1cm w: 75.1 ± 6.1 kg	Power
Zemkova et al. (2017) [86]	Trained in strength, no experience in instability	24 (M)	a: 22.1 ± 1.8 years h: 184.5 ± 8.3 cm w: 79.8 ± 9.4 kg	Power

M = men; W = women; a = age; h = height; w = weight; cm = centimetres; kg = kilograms.

3.2. Sample Characteristics

Table 1 shows scientific evidence on the sample characteristics (B, C, D) and variables (E). Format and design, including the author and the year of publication, the sample characteristics (overall number, gender, age, height, and weight), the variable measured (strength, power and/or speed).

Evaluation of the characteristics of the sample: (B) Experience. The experience of the sample was quite heterogeneous, with the participants standing out trained ($n = 3$; 33.34%); trained in strength without experience in instability ($n = 2$; 22.22%), trained in strength and instability 1 year earlier ($n = 2$; 22.22%); recreational ($n = 1$; 11.11%); no previous experience in strength or instability indicated ($n = 1$; 11.11%); (C) Sex. The distribution of the sample was very unbalanced with more male participants ($n = 158$; 98.14%) than female participants ($n = 3$; 1.86%); (D) Characteristics of the sample. The whole sample was identified as being between 18 (lower limit) and 25 (upper limit) years of age. The height range was identified as 167 cm to 185 cm. The weight range was identified as 79 kg to 88 kg.

Table 2. Scientific evidence on exercise and variation (F, G) device used (H) volume and intensity training (I, J).

Reference (A)	Tasks (F)	Situations (G)	Devices (H)	Volume Training (I)	Intensity Training (J)
Anderson et al. (2004) [13]	Bench Press	Stable and unstable device	Swiss ball	1 set 2 rps 2–3 min rest	75% 1 RM
Behm et al. (2002) [79]	Leg Extension Plantar Flexors	Stable and unstable device	Swiss ball	1 set 2–3 rps isometric 3 min rest	No external load
Chulvi-Medrano (2010) [80]	Deadweight	Stable and unstable device	Semi-sphere ball	1 set 6 rps 5 min rest	70% 1 RM
Goodman et al. (2008) [81]	Bench Press	Stable and unstable device	Swiss ball	1 set 3–6 rps 3 min rest	1 RM
Koshida et al. (2008) [82]	Bench Press	Stable and unstable device	Swiss ball	1 set 3 rps	50% 1 RM
Saeterbakken & Fimland (2013) [83]	Squat	Stable and unstable device	Semi-sphere ball	1 set Isometrics	20 kg
Sannicandro et al. (2015) [84]	Squat	Stable and unstable device	Suspension device	1 set 3 rps	No external load
Zemkova (2012) [85]	Bench Press Squat	Stable and unstable device	Swiss ball Semi-sphere ball	6 sets 8 rps	75% 1 RM
Zemkova et al. (2017) [86]	Bench Press Squat	Stable and unstable device	Swiss ball Semi-sphere ball	1 set 25 rps	75% 1 RM

rps = repetitions; min = minutes; RM = repetition maximum.

3.3. Tasks, Devices, and Training Parameters

Table 2 Shows scientific evidence on exercise and variation (F, G) device used (H) volume and intensity training (I, J). Format and design, including type and number of variations, the device used (exercise ball, semi-sphere ball or suspension device), training volume (number of sets/repeats/rest per exercise), intensity training (percentage of one maximum repetition (1RM)).

According to exercise (F): The most evaluated sports task was the "bench press" in five studies (41.67%) and "squat" in four (33.33%). "Deadweight", "plantar flexions" and "leg extension" were also evaluated (8.33% each of the exercises). (G) The number of situations. The number of comparisons between stable and unstable exercises was 100% of the situations that only compared the stable situation with an unstable one. (H) Type of device. The use of Swiss ball material was 54.55% (n = 6); the use of the semi-sphere ball was 36.36% of the studies analysed (n = 4). In only one study was a suspension device (TRX) used (n = 1; 9.09%). (I) Training volume. The number of series, repetitions and rest was quite heterogeneous. In the case of the series, only two studies are evident, comprising between three and six series. In the case of the repetitions, they varied from isometric execution to 25 repetitions, with the execution of 3–6 repetitions being the most used (60%). In terms of rest, they vary between 3 and 5 min. (J) Training intensity. In the case of the intensity of training, the percentage of load most used was 75% of 1RM (n = 3; 30%). 20% of the investigations did not use external load (n = 2). The rest of the investigations ranged from maximum repetition to 40% of 1RM.

Table 3 shows scientific evidence on strength results (K), power results (L) and speed results (M). Format and design, including strength results (maximum strength, mean strength), power results (maximum power, mean power and concentric phase power) and speed results (maximum and mean speed).

Table 3. Scientific evidence on strength results (K), power results (L) and speed results (M).

| | Performance Measures | | |
Reference (A)	Strength Results in Newtons (K)	Power Results in Watios (L)	Speed Results in cm/s (M)
Anderson et al. (2004) [13]	INS (S) = ↓59.4% MIVC		
Behm et al. (2002) [79]	INS (LE-S) = ↓75.4% MVC INS (PF-S) = ↓20.2% MVC		
Chulvi-Medrano (2010) [80] Goodman et al. (2008) [81] Koshida et al. (2008) [82] Saeterbakken & Fimland (2013) [83]	INS (B) = ↓10.2% MIVC INS (S) = No Differences MáxS INS (S) = ↓5.9% MS INS (B) = −19% MS	INS (S) = ↓9.9% MP	INS (S) = ↓9.1% MV
Sannicandro et al. (2015) [84]	INS (EF-LF-T) = ↓13.8 MS INS (EF-LF-T) = ↓46.8 MáxS INS (CF-LF-T) = ↓12.8 MS INS (CF-LF-T) = ↓12.6 MáxS INS (EF-RF-T) = ↓11.7 MS INS (EF-RF-T) = ↓42.9 MáxS INS (CF-RF-T) = ↓13.2 MS INS (CF-RF-T) = ↓11.9 MáxS		
Zemkova (2012) [85]		INS (BP-S) = ↓10.3% MP INS (BP-S) = ↓7.3% Pmáx INS (BP-S) = ↓11.5% CF INS (SQ-B) = ↓15.7% MP INS (SQ-B) = ↓17% Pmáx INS (SQ-B) = ↓15.1% CF	
Zemkova et al. (2017) [86]		INS (BP-S) = ↓12.9% MP (1–3 rps) INS (BP-S) = ↑5.6% MP (22–25 rps) INS (BP-S) = ↓6.9% MP (25 rps) INS (BP-S) = ↓13.8% CF (1–3 rps) INS (BP-S) = ↑13.2% CF (22–25 rps) INS (BP-S) = ↓4.6% CF (25 rps) INS (SQ-B) = ↓17.1% MP (1–3 rps) INS (SQ-B) = ↓21.4% MP (22–25 rps) INS (SQ-B) = ↓19.3% MP (25 rps) INS (SQ-B) = ↓16.2% CF (1–3 rps) INS (SQ-B) = ↓20.6% CF (22–25 rps) INS (SQ-B) = ↓18% CF (25 rps)	

INS = instability; S = Swiss ball; B = Semi-sphere ball; T = Suspension device; LE = leg extension; PF = plantar flexors; EF = eccentric phase; CF = concentric phase; MVIC = maximum voluntary isometric contractions; MVC = maximum voluntary contractions; MS = mean strength; MáxS = maximum strength; LF = left foot; RF = right foot; BP = bench press; SQ = squat; MP = mean power; PMáx = maximum power.

In performance measures, it can be seen how the use of instability decreases in some cases substantially in relation to the stable condition. Although with loads close to the RM no differences are appreciated. In terms of power, the difference seems to be slighter in stable and unstable condition and even at a higher number of repetitions the instability seems to improve power production. The execution speed also shows a lower production when instability is added.

3.4. Strength Results

Concerning the strength parameter with a Swiss ball, the bench press exercise showed 59.4% less isometric strength in instability [13], 5.9% [82], but no differences were found in 1RM [81]. For the lower limb exercises, 70.5% less was evidenced in the unstable condition in leg extension exercises than in the stable condition while the unstable force in plantar flexors was 20.2% less than the stable condition [79]. Also with the same exercise, a decrease with the execution with the semi-sphere ball concerning the stable condition of 19% was evidenced [83]. In the case of the deadweight exercise, the decrease in the maximum isometric contraction between the stable condition and the execution with semi-sphere ball was 10.2% [80]. Regarding suspension training squat exercise with bipodal execution, in the eccentric phase, peak and average force showed a decrease of 46.8% and 13.8% respectively for the lower left limb. In the concentric phase, the use of the suspension training tool caused a decrease of 12.6% in peak force and 12.8% in mean force. For the right lower limb, in the eccentric phase, during execution with the suspension training tool, the force decreased by 42.9% and the mean force by 11.7%. In the concentric phase, during execution with the suspension training tool, the peak and average force decreased respectively by 11.9% and 13.2%. During monopodal execution, the eccentric phase in the left limb, the peak force suffered a decrease of 41.8% and the average force a decrease of 18.1%. In the concentric phase, on the other hand, the use of the suspension training tool caused a decrease of 13.5% in peak force and 15.8% in average force. For the right limb during monopodal execution, in the eccentric phase, the force has decreased by 45.1% and the average force by 17.4%. In the concentric phase, the use of the suspension training tool caused a decrease of 12.4% in the force and 14.3% in the mean force [84].

3.5. Power Results

For the variable of power with a Swiss ball, a decrease in the unstable situation of 9.9% concerning the stable situation has been evidenced with the chest press exercise [82], and of 10.3% in the average power, 7.3% in the maximum power and 11.5% in the power exercised in the concentric phase [85]. For the average power, with the execution of 25 repetitions, a decrease of 6.9% was found in the unstable bench press, although in the last three repetitions the average power exercised in the unstable condition was 5.6% higher than the stable condition. On the contrary, among the first three repetitions, the unstable data was 12.9% lower than the stable condition. The power exercised in the concentric phase of the bench press was reported to be 4.6% lower in the unstable bench press, although in the last three repetitions the power exercised in the concentric phase of the unstable condition was 13.2% higher than the stable condition. On the contrary, among the first three repetitions, the unstable data was 13.8% lower than the stable condition [86]. To average power, 25 repetitions showed a decrease of 19.3% in the unstable squat, although in the last three repetitions the average power exercised in the unstable condition was 21.4% higher than the stable condition. On the contrary, among the first three repetitions, the unstable condition was 17.1% lower than the stable condition. The power exercised in the concentric phase of the squat was reported to be 18% lower in the unstable squat, although in the last three repetitions the power exercised in the concentric phase of the unstable condition was 20.6% lower than the stable condition. On the contrary, among the first 3 repetitions, the unstable condition was 16.2% lower than the stable condition [86].

3.6. Speed Results

Only one research study has been shown to consider the execution speed of strength exercises measured in instability. In the case of the speed variable with a Swiss ball, there has been a 9.1% decrease in the unstable condition concerning the stable banking press [82].

3.7. Study Selection and Assessment (Qualitative Analysis)

The quality analysis (STROBE' checklist) yielded the following results (Table 4): (a) The quality scores ranged from 13–16; (b) The average score was 14.3 points; (c) Of the 9 included studies, 5 (55.55%) were considered to 'fair quality' (13–14 points); and 4 (44.44%) were categorized as 'good quality' (15–16 points).

Table 4. The study quality analysis (STROBE' checklist).

Reference	Title and Abstract	Introduction		Methods										Results			Other Analysis	Discussion				Other Information	Strobe Points	Study Quality
	1	2	3	4	5	6	7	8	9	10	11	12	13	14	15	16	17	18	19	20	21	22		
Anderson et al. (2004) [13]	+	+	+	-	-	+	+	+	-	-	+	+	-	-	+	+	-	+	-	+	+	-	13	FAIR
Behm et al. (2002) [79]	+	+	+	-	-	+	+	+	-	-	+	+	-	-	+	+	+	+	-	+	+	-	14	FAIR
Chulvi-Medrano (2010) [80]	+	+	+	+	+	+	+	+	-	-	+	+	-	-	+	+	-	+	+	+	+	-	16	GOOD
Goodman et al. (2008) [81]	+	+	+	+	+	+	+	+	-	-	+	+	-	-	+	+	+	+	-	+	+	-	16	GOOD
Koshida et al. (2008) [82]	+	+	+	+	-	+	+	+	-	-	+	+	-	-	+	+	-	+	+	+	+	-	15	GOOD
Saeterbakken & Fimland (2013) [83]	+	+	+	+	-	+	+	+	-	-	+	+	-	-	+	+	-	+	-	+	+	-	13	FAIR
Sannicandro et al. (2015) [84]	+	+	+	-	-	+	+	+	-	-	+	+	-	-	+	+	+	+	-	+	+	+	15	GOOD
Zemkova (2012) [85]	+	+	+	-	-	+	+	+	-	-	+	+	-	-	+	+	-	+	+	+	+	+	14	FAIR
Zemkova et al. (2017) [86]	+	+	+	-	-	+	+	+	-	-	+	+	-	-	+	+	-	+	-	+	+	-	13	FAIR

By sections, the highest scores were located in 'introduction' (100%) and 'discussion' (80%) and among the highest quality studies, items no. 2 (background/rationale); no. 3 ('Objectives—State specific objectives and/or any pre-specified hypothesis'); no. 6 (participants); no. 7 (variables); no. 8 ('data source—procedure for determining performance measurement'), no. 11 ('descriptive results—the number (absolute frequency) or percentage (relative frequency) of participants found in each grouping category and subcategory'); no. 12 (statistical methods); no. 15 (outcome data); no. 16 (main results); no. 18 ('key results—a summary of key results concerning study objectives'); no. 20 (interpretation) and no. 21 (generalisability) were considered complete (100%), while the most commonly absent or incomplete item (0 points) was found in items no. 9, 10, 13 and 14 ('main results—a measure of effect size'). The lowest scores were found in the 'funding' section (20%).

4. Discussion

This is the first systematic review of the literature to examine the effects of instability on measures of muscle strength, power, and speed, administered in the form of quasi-experimental studies in healthy individuals during adulthood.

About the production of force, for the exercises of the upper limb, high decreases in values have been observed for the unstable condition. These differences have ranged from 20 to 75% loss in force development in unstable conditions. According to Kornecki, Kebel, and Siemieński [87], the stabilising function of the skeletal muscles is necessary for the coordinated performance of any voluntary movement, and it significantly influences muscle coordination patterns. Therefore, significant reductions in muscle production probably occurred because the muscles around the shoulder complex needed to give priority to stability over force production. Furthermore, under conditions of instability, the stiffness of the joints that act can limit gains in strength, power and speed of movement [88].

However, in the data evidenced in the study by Koshida et al. [82], the losses in force values are much lower than in the rest of the studies (5.9% loss in instability) compared with 59.4% loss in Anderson and Behm [13] and in the case of Goodman et al. [81], no significant differences are observed. This inconsistency between the previous research can be attributed to the type of muscle contraction, the degree of instability during the recorded task and the equipment. In Koshida et al. [82] the bench press movement was performed in a supine position with the Swiss ball placed in the thoracic area, which provided a broader support base than for other activities performed in a sitting or standing position. Therefore, the instability imposed on the trunk stabilising muscles would probably be less significant than in previous research. Besides, both studies used dynamic contractions with an Olympic bar with weight plates, while Anderson and Behm [13] used isometric contractions with two independent handles held by straps to force the transducers into the ground. The difference in equipment could impose different levels of instability on the shoulder joint and trunk muscles. Although bilateral contractions were performed in both studies, the independence of each hand in the study by Anderson and Behm [13] may have increased the effort required to maintain balance and the need for the muscles to stabilize during maximum isometric contractions, therefore reducing the net force output. In the case of Goodman et al. [81], where no differences were found, it could be due to the use of different loads, since 1RM was used while in Anderson and Behm [13] 75% of 1RM loads were used and in Goodman et al. [81] were used 50% of 1RM. These data could indicate that the percentage of external load can influence the effect of the instability in the training.

In the case of force production in the lower limbs, there have been notable decreases when comparing tasks performed in instability concerning stable conditions. These decreases ranged from 10% to 19%, so it seems that instability affects the upper body more than the lower. With the use of the semi-sphere ball, analysed in terms of strength, with the performance of a dominant hip exercise such as deadweight the decrease in strength was 10.2% [80] while with a dominant knee exercise such as squat it was 19% [83]. This could indicate that certain movements could be affected to a lesser extent depending on the instability. However, as detailed above the methods and loads used were very different.

It is noteworthy that many of the studies to check force production in the lower limb using isometric contractions. However, isometric contractions are not usually used in strength training. Despite this, results obtained under isometric contractions have reported that conditions are strongly correlated with dynamic mobility performance [89]. However, due to the isometric test mode, subjects could gradually build up strength while stabilizing and maintaining balance on different surfaces. During the 3 s of maximum effort, the subjects may have been able to stabilize the limbs and trunk and therefore be able to exert a considerable amount of force in unstable conditions. We only know of one study that investigates the production of maximum force in squats on a stable and unstable surface [3]. These researchers used an inflatable balance disk and reported a decrease of approximately 46% in peak force. Although there was a greater decrease in force in that study, it could be attributed to the lack of a familiarization session, which the rest of the studies did consider appropriate.

In terms of strength, there seems to be a differentiation in the data concerning the devices used. When the main movement involves the muscles of the upper body the Swiss ball has been used and in the case of the main movement being performed on the lower body, the semi-sphere ball has been used. The exceptional case was the execution of an exercise such as squatting where the instability with the device in suspension was placed in the upper body. In the case of the Swiss ball, it seems to have a greater influence on the decrease of the force values (differences of 20–75%) compared to the semi-sphere ball (differences between the and 10% and 19%) suspended device (detriments between 12–47%). However, in some cases, the Swiss ball did not produce any differences between the conditions. So, it does not seem to be a determining factor in the case of muscle strength.

Concerning the production of muscular power in the upper limb, decreases in the unstable condition have been observed. Decreases with the unstable condition ranged from 7% to 17%. The data found in the studies that analyzed the bench press with Swiss ball reported a very similar percentage decrease in terms of average power (10.3% [85]; 9.9% [82]; and 12.9% [86]). These small deviations found may again be due to the different percentage of load used (50% vs. 75%) and the different volume of training applied. However, in some situations, instability has produced better power data than a stable condition. These better data have been produced in the last repetitions of the executions with high numbers of repetitions in the exercise (22–25 repetitions). The improvements observed in average power were 5.6% and in the concentric phase 13.2%.

The increase in power observed could be due to previous evidence that has shown that producing a high power output with a light to moderate load would be more effective in developing maximum power than using a heavy load [90,91]. Thus, it appears that such a low rate of reduction may still allow muscle power to be gained from strength exercise in the unstable condition. The mechanism of energy production using the stretch-shortening cycle employs the energy storage capacity of several elastic components and the stimulation of the stretching reflex to facilitate muscle contraction for a minimum period. The concentric muscle action does not occur immediately after the eccentric, the stored energy is dissipated and lost in the form of heat and also the strengthening stretching reflex is not activated. Resistance to instability exercise can compromise the three phases of the stretching-shortening cycle, including the amortization phase. Around this turning point, where the eccentric phase becomes concentric, the maximum force is produced. At the same time, the subjects must stabilize the torso on an unstable surface to provide firm support for the contracted muscles. This additional task can compromise the contraction of the muscles acting on the bar. Their less intense contraction not only prolongs the change of direction of movement but, due to the lower maximum force, impairs the accumulation of elastic energy. The consequence is less speed and power in the subsequent concentric phase [92,93]. However, the subjects of the study by Zemkova et al. [86] were able to produce greater power during the executions on an unstable surface than on a stable one. This higher energy production can be attributed to the so-called ball bounce effect.

In terms of power, there seems to be a differentiation in the data about the devices used. The use of semi-sphere ball seems to have a greater influence on the decrease in power since the detriments with this device varied between 15% and 22%. In the case of the Swiss ball, the decrease in power oscillated between 7% and 14%, with better power data being found in unstable conditions with this device (5–13%). However, when the instability was placed where there was no movement, as in the case of the device in suspension and the squat, improvements of between 5% and 10% in power production were shown. Therefore, placing the instability where the main movement does not occur seems to be a good option for power improvement.

Finally, about the speed of execution, a decrease in the values in unstable conditions in comparison to stable conditions is observed, but the analysis of this variable has hardly been studied. According to Adkin et al. [94], a postural threat in a subject (fear of falling) will lead to a reduction in the magnitude and speed of voluntary movements. Thus, muscle stabilization seems to compromise gains in strength, power and speed of movement [95]. It should also be noted that new patterns of movement are generally learned at low speed, while sport-specific motor actions are executed at high speed [26].

The great heterogeneity in terms of volume and intensity of the load is remarkable. Also, instability seems to affect the force variable to a higher degree, but with intensities close to 1RM no differences are observed. As for power, a greater number of repetitions seems to benefit the production of this variable in instability in the upper limb. Finally, speed has barely been analysed and seems to show losses of speed in instability but not excessively so.

5. Implications for Practice

The great heterogeneity found is a limitation of the study, however, the results of this study can be applied in various ways. It would be interesting to include training in instability in athletes trained in this type of situation. All the information about this is with beginner athletes, and what is interesting about the application of instability is the individualization of the subjects. Unstable surfaces can be very interesting tools for optimising training, because although decreases in performance variables have been shown, this may not be the case for experienced athletes. As for integrated work, where besides strength, power and speed, other qualities such as balance can be analysed. Also, the angles with which the work is done in instability are different concerning stable conditions, so that other complementary muscles are worked. Finally, variety in environments, methods and exercises is one of the principles of training and these unstable situations provide it.

The complexity of execution in this type of unstable situation, where the technique can be affected, is noteworthy. For this reason, the level of experience of the athletes is important to be able to apply this type of training. Besides, the population requires the help of a qualified professional who can help and direct the sessions or tasks with this type of device, with an appropriate and individualised programme according to the different users.

6. Conclusions

The main findings of this review were that there is great heterogeneity in analysing the acute effects of instability on performance variables. Instability compared to a stable condition decreases the parameters of strength, power, and muscular speed in adults. The differences shown are quite significant in most situations although slight decreases can be seen in certain situations. However, for the upper limb, a greater number of repetitions seems to increase the power values in instability compared to the stable situation. The variables of force, power and speed seem to be affected when instability is implemented. However, it seems necessary to extend the investigation of instability with the performance variables because the results are very heterogeneous and there are no unified criteria to evaluate the different conditions, subjects, tasks and devices.

Author Contributions: Conceptualization, M.M. and J.L.-C.; methodology, M.M., J.L.-C. and J.R.-G.; software, M.M.; validation, J.L.-C. and I.R.R.; formal analysis, M.M.; investigation, M.M.; resources, M.M. and. J.L.-C.; data curation, M.M. and A.G.-A.; writing—original draft preparation, M.M.; writing—review and editing, J.L.-C., J.R.-G. and I.R.R.; visualization, M.M., J.L.-C., J.R.-G., A.G.-A. and I.R.R.; supervision, J.L.-C. and J.R.-G.; project administration, J.L.-C., J.R.-G. and I.R.R.; funding acquisition, A.G.-A. and I.R.R. All authors have read and agreed to the published version of the manuscript.

Funding: This research received no external funding.

Institutional Review Board Statement: Not applicable.

Informed Consent Statement: Not applicable.

Acknowledgments: We wish to thank all athletes who completed the experimental protocol. We would also like to thank the Faculty of Physical Activity and Sport Sciences (INEF) and the Universidad Politécnica de Madrid (UPM) for their support.

Conflicts of Interest: The authors declare no conflict of interest.

References

1. Ignjatovic, A.M.; Radovanovic, D.S.; Kocić, J. Effects of eight weeks of bench press and squat power training on stable and unstable surfaces on 1RM and peak power in different testing conditions. *Isokinet. Exerc. Sci.* **2019**, *27*, 203–212. [CrossRef]
2. Kohler, J.M.; Flanagan, S.P.; Whiting, W.C. Muscle activation patterns while lifting stable and unstable loads on stable and unstable surfaces. *J. Strength Cond. Res.* **2010**, *24*, 313–321. [CrossRef] [PubMed]
3. McBride, J.M.; Cormie, P.; Deane, R. Isometric squat force output and muscle activity in stable and unstable conditions. *J. Strength Cond. Res.* **2006**, *20*, 915–918. [CrossRef] [PubMed]
4. Nascimento, V.Y.S.; Torres, R.J.B.; Beltrão, N.B.; dos Santos, P.S.; Pirauá, A.L.T.; de Oliveira, V.M.A.; Pitangui, A.C.R.; de Araújo, R.C. Shoulder muscle activation levels during exercises with axial and rotational load on stable and unstable surfaces. *J. Appl. Biomech.* **2017**, *33*, 118–123. [CrossRef] [PubMed]
5. Wang, R.; Hoffman, J.R.; Sadres, E.; Bartolomei, S.; Muddle, T.W.; Fukuda, D.H.; Stout, J.R. Effects of different relative loads on power performance during the ballistic push-up. *J. Strength Cond. Res.* **2017**, *31*, 3411–3416. [CrossRef] [PubMed]
6. Bompa, T.O.; Haff, G.G. *Periodization: Theory and Methodology of Training*, 5th ed.; Human Kinetics: Champaign, IL, USA, 2009; ISBN 978-0-7360-7483-4.
7. Arriscado Alsina, D.; Martínez, J. Muscular strength training in young football players. *J. Sport Heal. Res.* **2017**, *9*, 329–338.
8. Billich, R.; Stvrtna, J.; Jelen, K. Optimal velocity to achieve maximum power. *KInanthropologica* **2014**, *50*, 37–46. [CrossRef]
9. Dallas, G.; Kirialanis, P.; Mellos, V. The acute effect of whole body vibration training on flexibility and explosive strenght of young gymnasts. *Biol. Sport* **2014**, *31*, 233–237. [CrossRef]
10. Górski, M.; Starczewski, M.; Pastuszak, A.; Mazur-Różycka, J.; Gajewski, J.; Buśko, K. Changes of strength and maximum power of lower extremities in adolescent handball players during a two-year training cycle. *J. Hum. Kinet.* **2018**, *63*, 95–103. [CrossRef]
11. Pereira, L.A.; Nimphius, S.; Kobal, R.; Kitamura, K.; Turisco, L.A.L.; Orsi, R.C.; Cal Abad, C.C.; Loturco, I. Relationship between change of direction, speed and power in male and female national olympic team handball athletes. *J. Strength Cond. Res.* **2018**, *63*, 1. [CrossRef]
12. Spieszny, M.; Zubik, M. Modification of strength training programs in handball players and its influence on power during the competitive period. *J. Hum. Kinet.* **2018**, *63*, 149–160. [CrossRef] [PubMed]
13. Anderson, K.G.; Behm, D.G. Maintenance of EMG activity and loss of force output with instability. *J. Strength Cond. Res.* **2004**, *18*, 637. [CrossRef] [PubMed]
14. Anderson, K.; Behm, D.G. Trunk Muscle Activity Increases With Unstable Squat Movements. *Can. J. Appl. Physiol.* **2005**, *30*, 33–45. [CrossRef] [PubMed]
15. Alagesan, J. Effect of Instability Resistance Training of Abdominal Muscles in Healthy Young Females-An Experimental Study. *Int. J. Pharm. Sci. Health Care* **2012**, *2*, 91–97.
16. Marinković, M.; Bratić, M.; Ignjatović, A.; Radovanović, D. Effects of 8-Week Instability Resistance Training on Maximal Strength in Inexperienced Young Individuals. *Serbian J. Sport. Sci.* **2012**, *6*, 17–21.
17. Sukalinggam, C.; Sukalinggam, G.; Kasim, F.; Yusof, A. Stability ball training on lower back strength has greater effect in untrained female compared to male. *J. Hum. Kinet.* **2012**, *33*, 133–141. [CrossRef]
18. Sparkes, R.; Behm, D.G. Training Adaptations Associated With an 8-Week Instability Resistance Training Program With Recreationally Active Individuals. *J. Strength Cond. Res.* **2010**, *24*, 1931–1941. [CrossRef]
19. Cowley, P.; Swensen, T.; Sforzo, G. Efficacy of Instability Resistance Training. *Int. J. Sports Med.* **2007**, *28*, 829–835. [CrossRef]
20. Maté-Muñoz, J.L.; Antón, A.J.M.; Jiménez, P.J.; Garnacho-Castaño, M.V. Effects of instability versus traditional resistance training on strength, power and velocity in untrained men. *J. Sport. Sci. Med.* **2014**, *13*, 460–468. [CrossRef]

21. Behm, D.G.; Colado Sanchez, J.C. Instability Resistance Training Across the Exercise Continuum. *Sports Health* **2013**, *5*, 500–503. [CrossRef]
22. Tanimoto, M.; Sanada, K.; Yamamoto, K.; Kawano, H.; Gando, Y.; Tabata, I.; Ishii, N.; Miyachi, M. Effects of whole-body low-intensity resistance training with slow movement and tonic force generation on muscular size and strength in young men. *J. Strength Cond. Res.* **2008**, *22*, 1926–1938. [CrossRef] [PubMed]
23. Kanehisa, H.; Nagareda, H.; Kawakami, Y.; Akima, H.; Masani, K.; Kouzaki, M.; Fukunaga, T. Effects of equivolume isometric training programs comprising medium or high resistance on muscle size and strength. *Eur. J. Appl. Physiol.* **2002**, *87*, 112–119. [CrossRef] [PubMed]
24. Behm, D.G.; Drinkwater, E.J.; Willardson, J.M.; Cowley, P.M. The use of instability to train the core musculature. *Appl. Physiol. Nutr. Metab.* **2010**, *35*, 91–108. [CrossRef]
25. Behm, D.G.; Leonard, A.M.; Young, W.B.; Bonsey, W.A.C.; MacKinnon, S.N. Trunk muscle electromyographic activity with unstable and unilateral exercises. *J. Strength Cond. Res.* **2005**, *19*, 193–201. [CrossRef] [PubMed]
26. Behm, D.G. Neuromuscular Implications and Applications of Resistance Training. *J. Strength Cond. Res.* **1995**, *9*, 264–274. [CrossRef]
27. Behm, D.G.; Anderson, K.G. The Role of Instability With Resistance Training. *J. Strength Cond. Res.* **2006**, *20*, 716. [CrossRef]
28. Chapman, D.W.; Needham, K.J.; Allison, G.T.; Lay, B.; Edwards, D.J. Effects of experience in a dynamic environment on postural control. *Br. J. Sports Med.* **2007**, *42*, 16–21. [CrossRef]
29. Drake, J.D.M.; Fischer, S.L.; Brown, S.H.M.; Callaghan, J.P. Do Exercise Balls Provide a Training Advantage for Trunk Extensor Exercises? A Biomechanical Evaluation. *J. Manipulative Physiol. Ther.* **2006**, *29*, 354–362. [CrossRef]
30. Nairn, B.C.; Sutherland, C.A.; Drake, J.D.M. Location of Instability During a Bench Press Alters Movement Patterns and Electromyographical Activity. *J. Strength Cond. Res.* **2015**, *29*, 3162–3170. [CrossRef]
31. Paillard, T.; Margnes, E.; Portet, M.; Breucq, A. Postural ability reflects the athletic skill level of surfers. *Eur. J. Appl. Physiol.* **2011**, *111*, 1619–1623. [CrossRef]
32. Hornsby, W.; Gentles, J.; MacDonald, C.; Mizuguchi, S.; Ramsey, M.; Stone, M. Maximum Strength, Rate of Force Development, Jump Height, and Peak Power Alterations in Weightlifters across Five Months of Training. *Sports* **2017**, *5*, 78. [CrossRef] [PubMed]
33. Prieske, O.; Muehlbauer, T.; Borde, R.; Gube, M.; Bruhn, S.; Behm, D.G.; Granacher, U. Neuromuscular and athletic performance following core strength training in elite youth soccer: Role of instability. *Scand. J. Med. Sci. Sport.* **2016**, *26*, 48–56. [CrossRef] [PubMed]
34. Dolezal, S.M.; Frese, D.L.; Llewellyn, T.L. The effects of eccentric, velocity-based training on strength and power in collegiate athletes. *Int. J. Exerc. Sci.* **2016**, *9*, 657–666.
35. Willardson, J.M.; Fontana, F.E.; Bressel, E. Effect of Surface Stability on Core Muscle Activity for Dynamic Resistance Exercises. *Int. J. Sports Physiol. Perform.* **2009**, *4*, 97–109. [CrossRef] [PubMed]
36. Escamilla, R.F.; Lewis, C.; Bell, D.; Bramblet, G.; Daffron, J.; Lambert, S.; Pecson, A.; Imamura, R.; Paulos, L.; Andrews, J.R. Core Muscle Activation During Swiss Ball and Traditional Abdominal Exercises. *J. Orthop. Sport. Phys. Ther.* **2010**, *40*, 265–276. [CrossRef] [PubMed]
37. Fredericson, M.; Moore, T. Muscular Balance, Core Stability, and Injury Prevention for Middle- and Long-Distance Runners. *Phys. Med. Rehabil. Clin. N. Am.* **2005**, *16*, 669–689. [CrossRef] [PubMed]
38. Kibler, W.B.; Press, J.; Sciascia, A. The Role of Core Stability in Athletic Function. *Sport. Med.* **2006**, *36*, 189–198. [CrossRef] [PubMed]
39. Schwanbeck, S.; Chilibeck, P.D.; Binsted, G. A Comparison of Free Weight Squat to Smith Machine Squat Using Electromyography. *J. Strength Cond. Res.* **2009**, *23*, 2588–2591. [CrossRef]
40. Saeterbakken, A.H.; Solstad, T.E.J.; Behm, D.G.; Stien, N.; Shaw, M.P.; Pedersen, H.; Andersen, V. Muscle activity in asymmetric bench press among resistance-trained individuals. *Eur. J. Appl. Physiol.* **2020**. [CrossRef]
41. Miller, W.M.; Barnes, J.T.; Sofo, S.S.; Wagganer, J.D. Comparison of Myoelectric Activity During a Suspension-Based and Traditional Split Squat. *J. Strength Cond. Res.* **2019**, *33*, 3236–3241. [CrossRef]
42. Saeterbakken, A.H.; Fimland, M.S. Muscle activity of the core during bilateral, unilateral, seated and standing resistance exercise. *Eur. J. Appl. Physiol.* **2012**, *112*, 1671–1678. [CrossRef]
43. Saeterbakken, A.H.; Fimland, M.S. Electromyographic activity and 6RM strength in bench press on stable and unstable surfaces. *J. Strength Cond. Res.* **2013**, *27*, 1101–1107. [CrossRef]
44. Saeterbakken, A.; Andersen, V.; Brudeseth, A.; Lund, H.; Fimland, M.S. The effect of performing bi- and unilateral row exercises on core muscle activation. *Int. J. Sports Med.* **2015**, *36*, 900–905. [CrossRef]
45. Dunnick, D.D.; Brown, L.E.; Coburn, J.W.; Lynn, S.K.; Barillas, S.R. Bench press upper-body muscle activation between stable and unstable loads. *J. Strength Cond. Res.* **2015**, *29*, 3279–3283. [CrossRef]
46. Gonzalo-Skok, O.; Tous-Fajardo, J.; Suarez-Arrones, L.; Arjol-Serrano, J.L.; Casajús, J.A.; Mendez-Villanueva, A. Single-leg power output and between-limbs imbalances in team-sport players: Unilateral versus bilateral combined resistance training. *Int. J. Sports Physiol. Perform.* **2016**, *12*, 106–114. [CrossRef]
47. Jakubek, M.D. Stability Balls: Reviewing the Literature Regarding Their Use and Effectiveness. *Strength Cond. J.* **2007**, *29*, 58–63. [CrossRef]
48. Ruiz, R.; Richardson, M.T. Using a Domed Device. *Strength Cond. J.* **2005**, *27*, 50–55. [CrossRef]

49. Bettendorf, B. *TRX Suspension Training Bodyweight Exercises: Scientific Foundations and Practical Applications*, 1st ed.; Fitness Anywhere Inc.: San Francisco, CA, USA, 2010.
50. Imai, K.; Keele, L.; Tingley, D. A general approach to causal mediation analysis. *Psychol. Methods* **2010**, *15*, 309–334. [CrossRef]
51. Lehman, G.J.; Gilas, D.; Patel, U. An unstable support surface does not increase scapulothoracic stabilizing muscle activity during push up and push up plus exercises. *Man. Ther.* **2008**, *13*, 500–506. [CrossRef]
52. Vera-Garcia, F.J.; Grenier, S.G.; McGill, S.M. Abdominal muscle response during curl-ups on both stable and labile surfaces. *Phys. Ther.* **2000**, *80*, 564–569. [CrossRef]
53. Czaprowski, D.; Afeltowicz, A.; Gebicka, A.; Pawłowska, P.; Kedra, A.; Barrios, C.; Hadała, M. Abdominal muscle EMG-activity during bridge exercises on stable and unstable surfaces. *Phys. Ther. Sport* **2014**, *15*, 162–168. [CrossRef] [PubMed]
54. Feldwieser, F.M.; Sheeran, L.; Meana-Esteban, A.; Sparkes, V. Electromyographic analysis of trunk-muscle activity during stable, unstable and unilateral bridging exercises in healthy individuals. *Eur. Spine J.* **2012**, *21*. [CrossRef] [PubMed]
55. Marshall, P.W.M.; Murphy, B.A. Increased Deltoid and Abdominal Muscle Activity During Swiss Ball Bench Press. *J. Strength Cond. Res.* **2006**, *20*, 745. [CrossRef]
56. Norwood, J.T.; Anderson, G.S.; Gaetz, M.B.; Twist, P.W. Electromyographic activity of the trunk stabilizers during stable and unstable bench press. *J. Strength Cond. Res.* **2007**, *21*, 343–347. [CrossRef] [PubMed]
57. Barros Beltrão, N.; Torres Pirauá, A.L. Analysis of muscle activity during the bench press exercise performed with the pre-activation method on stable and unstable surfaces. *Kinesiology* **2017**, *49*, 161–168. [CrossRef]
58. de Oliveira, A.S.; de Morais Carvalho, M.; de Brum, D.P.C. Activation of the shoulder and arm muscles during axial load exercises on a stable base of support and on a medicine ball. *J. Electromyogr. Kinesiol.* **2008**, *18*, 472–479. [CrossRef] [PubMed]
59. Uribe, B.P.; Coburn, J.W.; Brown, L.E.; Judelson, D.A.; Khamoui, A.V.; Nguyen, D. Muscle Activation When Performing the Chest Press and Shoulder Press on a Stable Bench vs. a Swiss Ball. *J. Strength Cond. Res.* **2010**, *24*, 1028–1033. [CrossRef]
60. Lehman, G.J.; Hoda, W.; Oliver, S. Trunk muscle activity during bridging exercises on and off a swissball. *Chiropr. Osteopat.* **2005**, *13*, 1–8. [CrossRef]
61. Brooks, D.; Brooks, C. *BOSU Balance Trainer: Integrated Balance Training*; DW Fitness LLC: Greater Manchester, UK, 2002.
62. Anderson, G.S.; Gaetz, M.; Holzmann, M.; Twist, P. Comparison of EMG activity during stable and unstable push-up protocols. *Eur. J. Sport Sci.* **2013**, *13*, 42–48. [CrossRef]
63. Aguilera-Castells, J.; Buscà, B.; Fort-Vanmeerhaeghe, A.; Montalvo, A.M.; Peña, J. Muscle activation in suspension training: A systematic review. *Sport. Biomech.* **2020**, *19*, 55–75. [CrossRef]
64. Aguilera-Castells, J.; Buscà, B.; Morales, J.; Solana-Tramunt, M.; Fort-Vanmeerhaeghe, A.; Rey-Abella, F.; Bantulà, J.; Peña, J. Muscle activity of Bulgarian squat. Effects of additional vibration, suspension and unstable surface. *PLoS ONE* **2019**, *14*, 1–20. [CrossRef] [PubMed]
65. Snarr, R.L.; Esco, M.R. Electromyographic comparison of traditional and suspension push-ups. *J. Hum. Kinet.* **2013**, *39*, 75–83. [CrossRef]
66. Coswig, V.S.; Dall'Agnol, C.; Del Vecchio, F.B. Anthropometric measurements usage to control the exercise intensity during the performance of suspension rowing and back squats. *Rev. Andaluza Med. Deport.* **2016**, *9*, 119–123. [CrossRef]
67. Krause, D.A.; Elliott, J.J.; Fraboni, D.F.; McWilliams, T.J.; Rebhan, R.L.; Hollman, J.H. Electromyography of the hip and thigh muscles during two variations of the lunge exercise: A cross-sectional study. *Int. J. Sports Phys. Ther.* **2018**, *13*, 137–142. [CrossRef] [PubMed]
68. Malliaropoulos, N.; Panagiotis, T.; Jurdan, M.; Vasilis, K.; Debasish, P.; Peter, M.; Tsapralis, K. Muscle and intensity based hamstring exercise classification in elite female track and field athletes: Implications for exercise selection during rehabilitation. *Open Access J. Sport. Med.* **2015**, 209. [CrossRef]
69. Borreani, S.; Calatayud, J.; Colado, J.C.; Moya-Nájera, D.; Triplett, N.T.; Martin, F. Muscle activation during push-ups performed under stable and unstable conditions. *J. Exerc. Sci. Fit.* **2015**, *13*, 94–98. [CrossRef]
70. Byrne, J.M.; Bishop, N.S.; Caines, A.M.; Crane, K.A.; Feaver, A.M.; Pearcey, G.E.P. Effect of Using a Suspension Training System on Muscle Activation During the Performance of a Front Plank Exercise. *J. Strength Cond. Res.* **2014**, *28*, 3049–3055. [CrossRef]
71. Calatayud, J.; Borreani, S.; Colado, J.C.; Martin, F.; Batalha, N.; Silva, A. Muscle activation differences between stable push-ups and push-ups with a unilateral v-shaped suspension system at different heights. *Motricidade* **2014**, *10*, 84–93. [CrossRef]
72. Snarr, R.L.; Hallmark, A.V.; Nickerson, B.S.; Esco, M.R. Electromyographical comparison of pike variations performed with and without instability devices. *J. Strength Cond. Res.* **2016**, *30*, 3436–3442. [CrossRef]
73. Cugliari, G.; Boccia, G. Core muscle activation in suspension training exercises. *J. Hum. Kinet.* **2017**, *56*, 61–71. [CrossRef]
74. Nairn, B.C.; Sutherland, C.A.; Drake, J.D.M. Motion and muscle activity are affected by instability location during a squat exercise. *J. Strength Cond. Res.* **2017**, *31*, 677–685. [CrossRef] [PubMed]
75. Moher, D.; Shamseer, L.; Clarke, M.; Ghersi, D.; Liberati, A.; Petticrew, M.; Shekelle, P.; Stewart, L.A. Preferred reporting items for systematic review and meta-analysis protocols (PRISMA-P) 2015 statement. *Syst. Rev.* **2015**, *4*, 1. [CrossRef] [PubMed]
76. Behm, D.G.; Muehlbauer, T.; Kibele, A.; Granacher, U. Effects of strength training using unstable surfaces on strength, power and balance performance across the lifespan: A systematic review and meta-analysis. *Sport. Med.* **2015**, *45*, 1645–1669. [CrossRef] [PubMed]
77. Tan, B. Variables to Optimize Maximum Strength in Men: A Review. *J. Strength Cond. Res.* **1999**, *13*, 289–304. [CrossRef]

78. Vandenbroucke, J.P.; Von Elm, E.; Altman, D.G.; Gøtzsche, P.C.; Mulrow, C.D.; Pocock, S.J.; Poole, C.; Schlesselman, J.J.; Egger, M. Strengthening the Reporting of Observational Studies in Epidemiology (STROBE): Explanation and elaboration. *PLoS Med.* **2007**, *4*, 1628–1654. [CrossRef]
79. Behm, D.; Anderson, K.G.; Curnew, R.S. Muscle force and activation under stable and unstable conditions. *J. Strength Cond. Res.* **2002**, *16*, 416–422. [CrossRef]
80. Chulvi-Medrano, I.; García-Massó, X.; Colado, J.C.; Pablos, C.; de Moraes, J.A.; Fuster, M.A. Deadlift muscle force and activation under stable and unstable conditions. *J. Strength Cond. Res.* **2010**, *24*, 2723–2730. [CrossRef]
81. Goodman, C.A.; Pearce, A.J.; Nicholes, C.J.; Gatt, B.M.; Fairweather, I.H. No difference in 1RM strength and muscle activation during the barbell chest press on a stable and unstable surface. *J. Strength Cond. Res.* **2008**, *22*, 88–94. [CrossRef]
82. Koshida, S.; Urabe, Y.; Miyashita, K.; Iwai, K.; Kagimori, A. Muscular outputs during dynamic bench press under stable versus unstable conditions. *J. Strength Cond. Res.* **2008**, *22*, 1584–1588. [CrossRef]
83. Saeterbakken, A.H.; Fimland, M.S. Muscle force output and electromyographic activity in squats with various unstable surfaces. *J. Strength Cond. Res.* **2013**, *27*, 130–136. [CrossRef]
84. Sannicandro, I.; Cofano, G.; Rosa, A.R. Strength and power analysis in half squat exercise with suspension training tools. *J. Phys. Educ. Sport* **2015**, *15*, 433–440. [CrossRef]
85. Zemková, E.; Jeleň, M.; Kováčiková, Z.; Ollé, G.; Vilman, T.; Hamar, D. Power outputs in the concentric phase of resistance exercises performed in the interval mode on stable and unstable surfaces. *J. Strength Cond. Res.* **2012**, *26*, 3230–3236. [CrossRef]
86. Zemkova, E.; Jelen, M.; Radman, I.; Svilar, L.; Hamar, D. L'effetto delle condizioni di sollevamento stabili e instabili sulla forza muscolare e sul tasso di affaticamento durante esercizi di resistenza. *Med. Dello Sport* **2017**, *70*, 36–49. [CrossRef]
87. Kornecki, S.; Kebel, A.; Siemieński, A. Muscular co-operation during joint stabilisation, as reflected by EMG. *Eur. J. Appl. Physiol.* **2001**, *84*, 453–461. [CrossRef]
88. Carpenter, M.; Frank, J.; Silcher, C.; Peysar, G. The influence of postural threat on the control of upright stance. *Exp. Brain Res.* **2001**, *138*, 210–218. [CrossRef]
89. Stone, M.H.; Sanborn, K.; O'Bryant, H.S.; Hartman, M.; Stone, M.E.; Proulx, C.; Ward, B.; Hruby, J. Maximum Strength-Power-Performance Relationships in Collegiate Throwers. *J. Strength Cond. Res.* **2003**, *17*, 739. [CrossRef]
90. Häkkinen, K. Neuromuscular and hormonal adaptations during strength and power training. A review. *J. Sports Med. Phys. Fitness* **1989**, *29*, 9–26.
91. Wilson, G.J.; Newton, R.U.; Murphy, A.J.; Humphries, B.J. The optimal training load for the development of dynamic athletic performance. *Med. Sci. Sports Exerc.* **1993**, *25*, 1279–1286. [CrossRef]
92. Komi, P.V.; Bosco, C. Utilization of stored elastic energy in leg extensor muscles by men and women. *Med. Sci. Sports* **1978**, *10*, 261–265.
93. González Badillo, J.J.; Gorostiaga Ayestarán, E. *Fundamentos del Entrenamiento de La Fuerza. Aplicación al Alto Rendimiento Deportivo*; INDE: Barcelona, Spain, 1995; ISBN 9788487330384.
94. Adkin, A.L.; Frank, J.S.; Carpenter, M.G.; Peysar, G.W. Fear of falling modifies anticipatory postural control. *Exp. Brain Res.* **2002**, *143*, 160–170. [CrossRef]
95. Kornecki, S.; Zschorlich, V. The nature of the stabilizing functions of skeletal muscles. *J. Biomech.* **1994**, *27*, 215–225. [CrossRef]

International Journal of
Environmental Research and Public Health

MDPI

Systematic Review

Exercise Intensity in Patients with Cardiovascular Diseases: Systematic Review with Meta-Analysis

Catarina Gonçalves [1,2], Armando Raimundo [1,2], Ana Abreu [3] and Jorge Bravo [1,2,*]

1   Departamento de Desporto e Saúde, Escola de Saúde e Desenvolvimento Humano, Universidade de Évora, Largo dos Colegiais, 7000 Évora, Portugal; cjg@uevora.pt (C.G.); ammr@uevora.pt (A.R.)
2   Comprehensive Health Research Centre (CHRC), Universidade de Évora, Largo dos Colegiais, 7000 Évora, Portugal
3   Servico de Cardiologia, Hospital Universitário de Santa Maria/Centro Hospitalar Universitário Lisboa Norte (CHULN), Centro Académico de Medicina de Lisboa (CAML), Centro Cardiovascular da Universidade de Lisboa (CCUL), Faculdade de Medicina, Universidade de Lisboa, 1649-028 Lisboa, Portugal; ananabreu@hotmail.com
*   Correspondence: jorgebravo@uevora.pt

Abstract: Exercise-induced improvements in the VO$_2$peak of cardiac rehabilitation participants are well documented. However, optimal exercise intensity remains doubtful. This study aimed to identify the optimal exercise intensity and program length to improve VO$_2$peak in patients with cardiovascular diseases (CVDs) following cardiac rehabilitation. Randomized controlled trials (RCTs) included a control group and at least one exercise group. RCTs assessed cardiorespiratory fitness (CRF) changes resulting from exercise interventions and reported exercise intensity, risk ratio, and confidence intervals (CIs). The primary outcome was CRF (VO$_2$peak or VO$_2$ at anaerobic threshold). Two hundred and twenty-one studies were found from the initial search (CENTRAL, MEDLINE, CINAHL and SPORTDiscus). Following inclusion criteria, 16 RCTs were considered. Meta-regression analyses revealed that VO$_2$peak significantly increased in all intensity categories. Moderate-intensity interventions were associated with a moderate increase in relative VO$_2$peak (SMD = 0.71 mL·kg^{-1}·min^{-1}; 95% CI = [0.27–1.15]; $p = 0.001$) with moderate heterogeneity ($I^2 = 45\%$). Moderate-to-vigorous-intensity and vigorous-intensity interventions were associated with a large increase in relative VO$_2$peak (SMD = 1.84 mL·kg^{-1}·min^{-1}; 95% CI = [1.18–2.50], $p < 0.001$ and SMD = 1.80 mL·kg^{-1}·min^{-1}; 95% CI = [0.82–2.78] $p = 0.001$, respectively), and were also highly heterogeneous with I^2 values of 91% and 95% ($p < 0.001$), respectively. Moderate-to-vigorous and vigorous-intensity interventions, conducted for 6–12 weeks, were more effective at improving CVD patients' CRF.

Keywords: cardiac rehabilitation; cardiorespiratory fitness; exercise therapy; heart diseases; high-intensity intermittent exercise

Citation: Gonçalves, C.; Raimundo, A.; Abreu, A.; Bravo, J. Exercise Intensity in Patients with Cardiovascular Diseases: Systematic Review with Meta-Analysis. *Int. J. Environ. Res. Public Health* **2021**, *18*, 3574. https://doi.org/10.3390/ijerph18073574

Academic Editor: Paul B. Tchounwou

Received: 23 February 2021
Accepted: 25 March 2021
Published: 30 March 2021

Publisher's Note: MDPI stays neutral with regard to jurisdictional claims in published maps and institutional affiliations.

1. Introduction

Cardiovascular diseases (CVDs) are the leading cause of mortality in today's society, being responsible for up to one-third of all deaths worldwide and 50% of all deaths in Europe, and this scenario is expected to worsen in the coming years [1].

The concept of cardiac rehabilitation (CR) has been defined as the effort towards cardiovascular risk factor reduction, designed to lessen the chance of a subsequent cardiac event, and to slow and perhaps stop the progression of the disease process. In the context of CR programs, exercise training has been recognized as one of the main components, combined with education, control, pharmacological adherence and lifestyle changes of cardiovascular risk factors [2]. Physical exercise inclusion in CR programs resulted in several beneficial effects on cardiovascular functional capacity, quality of life, risk factor modification, psychological profile, hospital readmissions, and mortality [3,4]. Such benefits can be justified by a 20% reduction in mortality from all causes and in the levels of cardiorespiratory fitness (CRF) for each metabolic equivalent improvement (MET) in CRF of patients with CVD [5].

Exercise programs for patients with CVD traditionally involve mostly low- to moderate-intensity continuous aerobic exercise training, with the consensus that one of the benefits of aerobic exercise is the increase in peak oxygen uptake (VO_2peak) [6–8]. Continuous aerobic exercise training implicates higher durations under moderate-intensity and nonvariable aerobic activity (60–80% of VO_2peak) [9–12], compared to high-intensity protocols, which consist of intermittent, short high-intensity work periods (85–100% of VO_2peak) with relative resting periods [13,14].

Exercise intensity appears to influence the number of cardioprotective benefits achieved from aerobic exercise [15,16]. The current consensus recommends that exercise intensity prescribed for patients with CVD should be approximately 60% of the maximal heart rate (MHR), 50% of the heart rate reserve (HRR), or 12–13 on the Borg scale. Intensities around 85% MHR, 80% HRR, or 15–16 on the Borg scale should represent the upper limits [6]. Additionally, high-intensity protocols (85–100% of VO_2peak) appear to be of particular interest to scientists, considering their application in patients with CVD based on the effects on the cardiorespiratory and muscle systems [7]. High-intensity protocols elicit a greater training stimulus than moderate continuous exercise in improving maximal aerobic capacity [8–19]. In addition, high-intensity exercise appears to improve the limiting factors of VO_2peak, and VO_2peak itself has been found to be more effective in improving cardiovascular risk factors than moderate-intensity exercise [17,19].

Training sessions based on moderate-intensity continuous exercise have shown improvements in HRR after eight weeks [20] and after 12 weeks [21,22]. Moderate- to high-intensity continuous exercise (6 and 12 MET, corresponding to 21 and 42 mL-kg^{-1}-min^{-1} of VO_2peak) has also been shown to reduce all-cause mortality in healthy individuals, independent of activity duration [7], and reduce the risk of heart disease [15], supporting the need to further investigate the potential health effects of protocols based on higher intensities. Therefore, during the last two decades, several studies have demonstrated that high-intensity exercise protocols induce more beneficial cardiovascular adaptations in patients with mild-to-severe heart disease when compared to moderate-intensity exercise protocols [8,17–19].

A recent meta-analysis [23] reported higher improvements in maximal aerobic capacity after high-intensity interval training (HIIT) programs compared to moderate-intensity programs. Nevertheless, the optimum exercise intensity prescription in patients with CVD is still a subject of debate. A recent systematic review on the topic [24] did not report optimal intensity prescription (e.g., the intensity interval that is most effective during exercise interventions to induce favorable changes in aerobic capacity). Thus, despite the literature being replete with studies showing that regular and structured exercise is beneficial for CVD patients, the optimal intensity and length of exercise interventions that bring about greater benefits remain equivocal. Hence, the objective of this systematic review with meta-analysis was to identify, through Randomized Controlled Trials (RCTs) of exercise-based CR, the most effective exercise intensity and intervention length to optimize VO_2peak in patients with CVD.

2. Materials and Methods

This systematic review was undertaken as detailed in the protocol registered with PROSPERO (Registration Number CRD42018097319).

2.1. Search Strategy

The search strategies were designed in accordance with the methods suggested by the Cochrane Handbook for Systematic Reviews of Interventions [25]. The following databases were searched from their inception to January 2021: Cochrane Central Register of Controlled Trials (CENTRAL), MEDLINE (Ovid), CINAHL (EBSCO) and SPORTDiscus. Data are provided as the risk difference (95% CI), based on RCTs published until January

2021, ensuring that all studies have been included if reporting data on established outcomes. Reference lists of eligible studies were also systematically searched.

We used the PICO model [26] to identify free text terms and controlled vocabulary terms to create our searches. The following key concepts were chosen: "Patients with cardiovascular diseases" AND "Cardiac Rehabilitation" AND "Exercise Intensity" AND "Cardiorespiratory Fitness". The search strategy for the MEDLINE (Ovid) database is available in the Supplementary Materials of this manuscript.

2.2. Inclusion Criteria

The inclusion criteria were full-length research articles published in peer-reviewed journals in the English language with no limits set on the date of first publication or gender. Only RCTs up to January 2021 were eligible. Studies included participants who were diagnosed with CVD, such as those involved in some exercise programs, assessed by analyzing expired air during a maximal cardiopulmonary exercise test at baseline and postintervention.

We included RCTs to compare aerobic capacity changes resulting from exercise interventions, with an exercise group (or groups), that described exercise intensities, including data for risk ratio and CI.

Studies were required to detail the exercise prescription in patients with CVD, including the frequency, intensity and duration of each session, mode of exercise and the overall length of intervention. The main authors of studies and experts in this field were asked for any missed, unreported, or ongoing trials. The quantitative synthesis included studies reporting sample size and the mean and standard deviations (SDs) for VO_2peak preintervention and postintervention.

2.3. Exclusion Criteria

Abstracts, conference presentations or posters, letters to editors or book chapters, unpublished papers, and retrospective design studies were excluded. In addition, studies were excluded if participants had documented heart failure (ejection fraction < 40%) or arrhythmia, they were targeting a specific comorbidity (e.g., diabetes, chronic obstructive pulmonary disease, or stroke) and they featured interventions involving resistance exercises only. We also excluded studies based on exercise prescriptions including testing food supplements and nutritional or pharmacological aids.

Studies were also excluded if baseline or postintervention data were not published, and the authors were not available for contact or did not wish to provide the missing data.

2.4. Study Selection and Data Extraction

All data were extracted by the principal investigator and their accuracy was assessed by the second author. The EndNote software (Clarivate Analytics, Philadelphia, PA, USA) was used to import, manage and remove duplicated articles for final review. After removing the duplicates, the two reviewers independently reviewed titles and abstracts against the inclusion/exclusion criteria. If in doubt, the full texts were evaluated to verify if they met the criteria. Subsequently, abstracts were selected for eligibility, and full manuscripts were retrieved for further evaluation of eligibility. Discrepancies were resolved between both authors, and a third expert, not involved in the previous procedures, was consulted to verify the ratings. The selection process was entered into a Preferred Reporting Items for Systematic Reviews and Meta-analysis (PRISMA) diagram [27] (Figure 1).

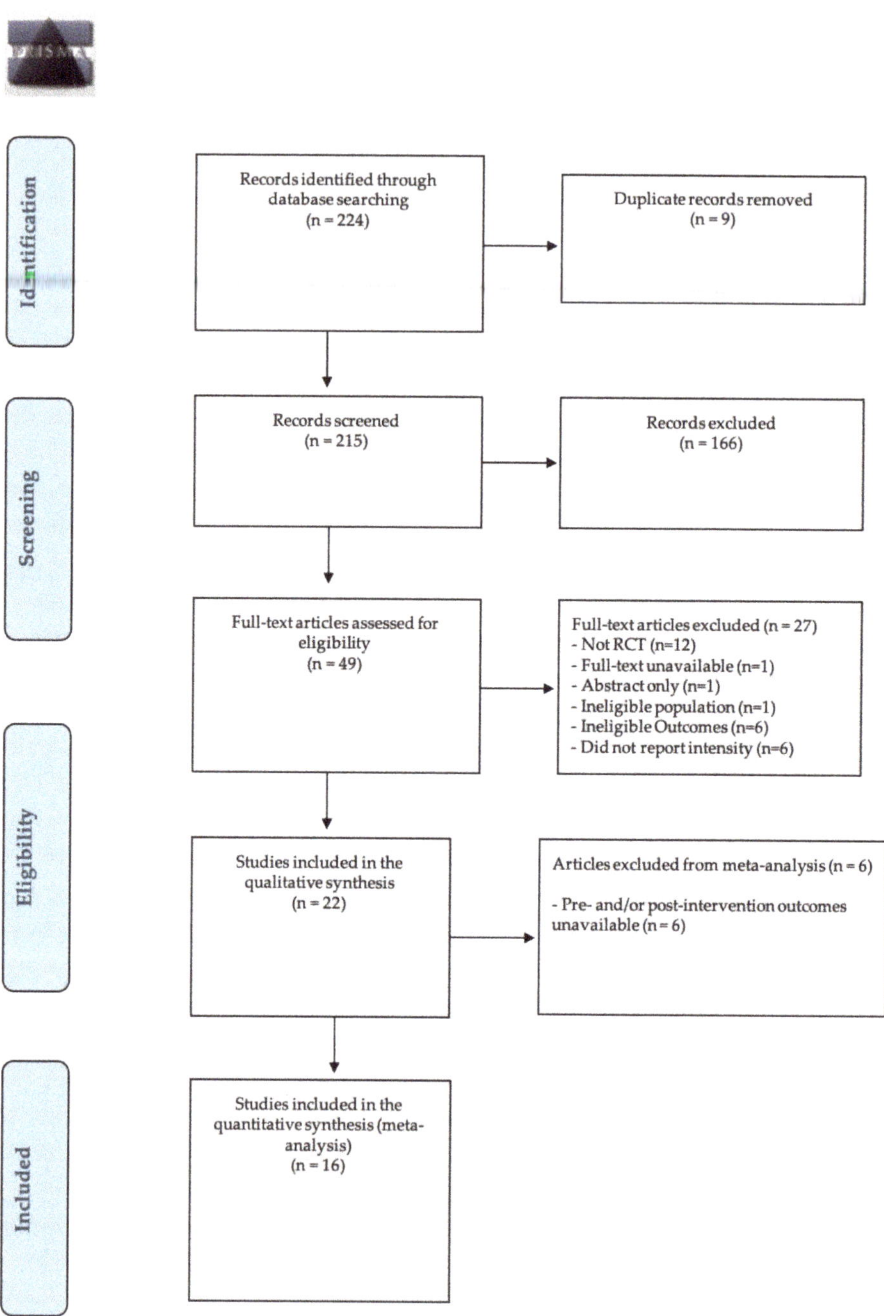

Figure 1. Preferred Reporting Items for Systematic Reviews and Meta-analysis (PRISMA) diagram of literature search strategies.

For each RCT, the author, year of publication, participant characteristics (age, gender, and primary diagnosis), description of the exercise testing protocol and description of the intervention (session frequency and duration, intervention length, exercise modality, resistance training, type of training (interval/continuous), supervision (clinic/home) and intervention type) were extracted. The pre- and post-VO_2peak values and change in VO_2peak were also extracted to assess change in CRF. Outcomes were extracted in relative ($mL\text{-}kg^{-1}\text{-}min^{-1}$) and absolute ($L\text{-}min^{-1}$) terms. Outcomes reported in METs were converted to relative terms (METs $\times$ 3.5 $mL\text{-}kg^{-1}\text{-}min^{-1}$).

2.5. Assessment of Potential Bias

The risk of bias was assessed using the modified Cochrane collaboration tool [25], developed in 2005 to assess and report the risk of bias in RCTs. Bias assessment results from the judgment (high, low, or unclear) of individual elements from seven sources of bias covered six domains: random sequence generation (selection bias), allocation concealment (selection bias), blinding of participants and personnel (performance bias), blinding of outcome assessment (detection bias), incomplete outcome data (attrition bias), selective reporting (reporting bias) and other bias (criteria for selected patients in the studies and the country in which the study was conducted). A detailed description of each source of bias and support for judgement is available elsewhere [25]. The lead reviewer found 16 studies, and discrepancies were discussed and resolved.

2.6. Data Treatment and Analysis

The systematic review was stratified by intensities based on proposed cut-offs [28]. Thereby, each exercise program was ranked as light-, moderate- or vigorous-intensity aerobic exercise (Table 1).

Table 1. Classification of exercise intensity based on physiological and perceived exertion responses.

	$\%VO_{2max}$	$\%HR_{peak}$	$\%HR_{reserve}/$ $\%VO_{2reserve}$	Perceived Exertion *
Light	37–45	57–63	30–39	RPE 9–11
Moderate	46–63	64–76	40–59	RPE 12–13
Vigorous	64–90	77–95	60–89	RPE 14–17
Near maximal to maximal	$\geq$91	$\geq$95	$\geq$90	RPE $\geq$ 18

Table adapted from American College of Sports Medicine (ACSM) [28] and Mitchell et al. [23]. * As per the Borg 6–20 RPE scale. $\%VO_2$max, percentage of maximal oxygen uptake; $\%HR$peak, percentage of peak heart rate; $\%HR$reserve, percentage of heart rate reserve; $\%VO_2$reserve, percentage of oxygen uptake reserve; RPE, rating of perceived exertion.

Studies reporting an intensity that covers the categories of moderate intensity and vigorous intensity (e.g., 60–70% of VO_2peak) were classified as "moderate-to-vigorous" intensity [28]. A separate meta-analysis was performed for each intensity category and length of the trial—e.g., "short-term" (0–6 weeks), "medium-term" (7–12 weeks), and "long-term" (>12 weeks).

The following subgroup analysis was conducted to explore significant heterogeneity: participant characteristics, including (1) age, (2) gender and (3) primary diagnosis; description of the exercise testing protocol and description of the intervention, including (4) session frequency and (5) duration, (6) intervention length, (7) exercise modality, (8) resistance training, (9) type of training (interval/continuous), (10) supervision (clinic/home), (11) intervention type (exercise only/comprehensive); and (12) pre- and postpeak VO_2 values or change in VO_2peak.

Heterogeneity amongst the included studies was first explored qualitatively by comparing the characteristics of the included trials and then by visually inspecting forest plots. It was also assessed quantitatively by the Chi^2 and I^2 statistics. Heterogeneity was considered minimal if I^2 fell between 0–30%, moderate if 30–50%, substantial if 50–90%, and considerable if >90% [25]. I^2 and Chi^2 were considered significant at $p < 0.1$.

Due to the heterogeneity of the protocol, mean differences (MDs) were used, dividing the mean values between different intensities. The differences in means were grouped using the random-effects model. A random-effects model and a standardized means model of averages were used to explain the differences in the methodology of the studies included both in the intensities and length of intervention to ensure a conservative estimate was calculated. A sensitivity analysis was conducted to investigate the possible effects of specific studies on heterogeneity and overall effect.

The dichotomous and continuous variables of the studies were compared with the extracted potential VO_2peak moderator factors. The effect of treatment was calculated for each study for the change in VO_2peak over the intervention using the pooled between-subject SD at both time points. Effects were quantified as trivial (<0.20), small (0.21–0.60), moderate (0.61–1.20), large (1.21–2.00) and very large (>2.00) [29], with the precision of effect size estimates assessed using 95% CI. Pooled SMD was back-transformed using the pooled between-subject SD at baseline within each intensity category. If SD for the mean change in VO_2peak across the intervention was not published [30], it was used for *p*-value entry. If no *p*-values or standard deviations were published, the standard error (SE) of the MD was inputted based on the correlation between preintervention and postintervention outcomes [31]. The imputed SE was then used to calculate the 95% CI for the standardized effect of each study. For outcomes expressed as change in relative VO_2peak ($mL\cdot kg^{-1}\cdot min^{-1}$), a correlation of r = 0.54 from a similar meta-analysis [32] was used. A sensitivity analysis was performed using the estimated correlations of r = 0.30 and 0.70.

Publication bias was analyzed using a funnel plot derived in RevMan5.3 software [30]. The publication bias for the different conditions analyzed (pre- vs. postintervention) was assessed by examining the asymmetry of a funnel plot using Egger's test, and $p \leq 0.05$ was considered to be statistically significant.

3. Results

The initial search resulted in 221 studies. All data were extracted by the principal investigator and their accuracy was assessed by a second author. Search results were entered into EndNote software (Clarivate Analytics, Philadelphia, PA, USA), a reference management tool, and duplicates were removed. After the duplicates were removed, the titles of 212 studies were reviewed. Following a screening of potential records, 49 articles were reviewed for eligibility and their reference lists screened. Twenty-two RCTs met eligibility criteria for the systematic review and meta-analysis. According to our inclusion criteria, sixteen studies [9–14,20–22,33–39] were included in this systematic review (Figure 1).

The main characteristics of the studies and training interventions are described in Tables 2 and 3, respectively.

Table 2. Subgroup analyses assessing potential moderating factors for VO_2peak increase in studies included in the meta-analysis by population characteristics.

Research Studies				Peak VO$_2$			
Group	**N**	**References**	**MD (95% CI)**	**I^2**	***p*^{a}**	***p*-Difference**[b]	
		No. of participants					
<20	4	Ghroubi et al. [20], Tamburus et al. [14], Wu et al. [33], Chuang et al. [34]	2.62 (1.65, 3.58)	88	<0.001		
≥20	12	Abolahrari-Shirazi et al. [9], Blumenthal et al. [21], Giallauria et al. [10–12,36], Kitzman et al. [22], Kraal et al. [36], Kubo et al. [38], Legramante et al. [37], Villelabeitia et al. [13], Zheng et al. [35]	2.75 (2.58, 2.93)	97	<0.001	0.78	

Table 2. *Cont.*

Research Studies			Peak VO$_2$			
Group	N	References	MD (95% CI)	I^2	p[a]	p-Difference [b]
Age, years						
<60	9	Abolahrari-Shirazi et al. [9], Ghroubi et al. [20], Giallauria et al. [10,12,36], Kraal et al. [39], Kubo et al. [38], Tamburus et al. [14], Villelabeitia et al. [13]	4.40 (0.79, 8.01)	97	0.02	
≥60	6	Blumenthal et al. [21], Chuang et al. [34], Giallauria et al. [11], Kitzman et al. [22], Legramante et al. [37], Wu et al. [33]	3.48 (2.09, 4.87)	79	<0.001	0.75
Not reported	1	Zheng et al. [35]	3.10 (2.06, 4.14)	0	<0.001	
Diagnosis						
CAD only	3	Blumenthal et al. [21], Tamburus et al. [14], Villelabeitia et al. [13]	6.41 (−2.70, 15.53)	99	0.17	
CABG only	4	Chuang et al. [34], Ghroubi et al. [20], Legramante et al. [37], Wu et al. [33]	4.27 (1.60, 6.94)	85	0.002	
PCI only	1	Abolahrari-Shirazi et al. [9]	8.20 (4.68, 11.72)	0	<0.001	0.03
CABG/PCI	1	Kraal et al. [39]	3.20 (0.36, 6.04)	0	0.03	
MI	6	Giallauria et al. [10–12,36], Kubo et al. [38], Zheng et al. [35]	2.65 (0.56, 4.74)	91	0.01	
FMD	1	Kitzman et al. [22]	1.60 (−0.13, 3.33)	0	0.07	
Study location						
America	2	Kitzman et al. [22], Tamburus et al. [14]	1.38 (0.39, 2.36)	0	0.006	
Africa	1	Ghroubi et al. [20]	1.70 (−1.07, 4.47)	0	0.23	
Asia	5	Abolahrari-Shirazi et al. [9], Chuang et al. [34], Kubo et al. [38], Wu et al. [33], Zheng et al. [35]	5.33 (2.90, 7.76)	80	<0.001	0.01
Europe	8	Blumenthal et al. [21], Giallauria et al. [10–12,36], Kraal et al. [39], Legramante et al. [37], Villelabeitia et al. [13]	4.23 (1.50, 6.95)	98	0.002	

95% CI, 95% confidence interval. I2, heterogeneity. MD, mean difference. Peak VO2, peak oxygen uptake. Conditions: MI, myocardial infarction. CABG, coronary artery bypass graft. PCI, percutaneous coronary intervention. CAD, coronary artery disease. FMD, endothelial-dependent flow-mediated arterial dilation. Certain enrolled studies were not included because the value used for subgroup analysis was not reported in them. a Test for overall effect. b Test for subgroup differences.

Table 3. Subgroup analyses assessing potential moderating factors for VO_2peak increase in studies included in the meta-analysis by population characteristics.

Research Studies			Peak VO_2			
Group	N	References	MD (95% CI)	I^2	p [a]	p-Difference [b]
Length, weeks						
<6	1	Legramante et al. [37]	2.60 (2.41, 2.79)	0	<0.001	
6–12	9	Abolahrari-Shirazi et al. [9], Chuang et al. [34], Ghroubi et al. [20], Giallauria et al. [10,36], Kraal et al. [39], Kubo et al. [38], Villelabeitia et al. [13], Wu et al. [33]	5.31 (1.24, 9.38)	97	0.01	0.42
>12	6	Blumenthal et al. [21], Giallauria et al. [11,12], Kitzman et al. [22], Tamburus et al. [14], Zheng et al. [35]	2.50 (1.60, 3.41)	52	<0.001	
Frequency, sessions/week						
1–2	2	Chuang et al. [34], Kraal et al. [39]	3.98 (1.96, 6.01)	0	0.001	
3–4	13	Abolahrari-Shirazi et al. [9], Blumenthal et al. [21], Ghroubi et al. [20], Giallauria et al. [10–12,36], Kitzman et al. [22], Kubo et al. [38], Tamburus et al. [14], Villelabeitia et al. [13], Wu et al. [33], Zheng et al. [35]	4.21 (1.82, 6.60)	96	0.006	0.17
5–7	1	Legramante et al. [37]	2.60 (2.41, 2.79)	0	<0.001	
Supervision						
Clinic	12	Blumenthal et al. [21], Chuang et al. [34], Ghroubi et al. [20], Giallauria et al. [10–12], Kitzman et al. [22], Kubo et al. [38], Legramante et al. [37], Tamburus et al. [14], Villelabeitia et al. [13], Zheng et al. [35]	4.01 (2.30, 5.72)	96	<0.001	0.02
Home	1	Wu et al. [33]	8.50 (5.78, 11.22)	0	<0.001	
Mixed	3	Abolahrari-Shirazi et al. [9], Giallauria et al. [36], Kraal et al. [39]	2.99 (−2.89, 8.87)	94	0.32	
Intervention type						
Continuous	13	Abolahrari-Shirazi et al. [9], Blumenthal et al. [21], Chuang et al. [34], Giallauria et al. [11,12,36], Kitzman et al. [22], Kraal et al. [39], Kubo et al. [38], Legramante et al. [37], Wu et al. [33], Zheng et al. [35]	3.27 (2.23, 4.32)	87	<0.001	0.44
Interval	2	Tamburus et al. [14], Villelabeitia et al. [13]	8.67 (−5.86, 23.21)	99	0.24	
Mixed	1	Ghroubi et al. [20]	1.70 (−1.07, 4.47)	0	0.23	

Table 3. *Cont.*

Research Studies			Peak VO$_2$			
Group	N	References	MD (95% CI)	I^2	p [a]	p-Difference [b]
Mode						
Cycle ergometer	7	Ghroubi et al. [20], Giallauria et al. [10–12], Tamburus et al. [14], Villelabeitia et al. [13], Zheng et al. [35]	4.90 [1.52, 8.27)	97	0.005	0.23
Treadmill	1	Chuang et al. [34]	4.80 (1.91, 7.69)	0	0.001	
Walking	1	Blumenthal et al. [21]	1.90 (0.20, 3.60)	0	0.03	
Mixed (treadmill, walking, cycling, calisthenics or/and arm/leg ergometer)	7	Abolahrari-Shirazi et al. [9], Giallauria et al. [36], Kitzman et al. [22], Kraal et al. [39], Kubo et al. [37], Legramante et al. [37], Wu et al. [33]	3.28 (1.17, 5.39)	92	0.002	
Exercise type						
Aerobic	13	Blumenthal et al. [21], Chuang et al. [34], Ghroubi et al. [20], Giallauria et al. [10,12,36], Kitzman et al. [22], Kraal et al. [39], Kubo et al. [38], Tamburus et al. [14], Villelabeitia et al. [13], Wu et al. [33], Zheng et al. [35]	3.94 (1.55, 6.34)	96	0.001	0.86
Aerobic and Resistance	3	Abolahrari-Shirazi et al. [9], Giallauria et al. [11], Legramante et al. [37]	4.24 (1.82, 6.67)	81	0.001	
Intensity						
Moderate	3	Giallauria et al. [10], Kubo et al. [38], Villelabeitia et al. [13]	2.90 (1.64, 4.16)	0	<0.001	0.03
Moderate-to-vigorous	10	Abolahrari-Shirazi et al. [9], Chuang et al. [34], Giallauria et al. [11,12,36], Kitzman et al. [22], Kraal et al. [39], Wu et al. [33], Zheng et al. [35]	5.07 (3.43, 6.72)	92	<0.001	
Vigorous	3	Blumenthal et al. [21], Ghroubi et al. [20], Giallauria et al. [10], Legramante et al. [37], Tamburus et al. [14], Villelabeitia et al. [13]	2.43 (1.33, 3.54)	75	<0.001	

95% CI, 95% confidence interval. I^2, heterogeneity. MD, mean difference. Peak VO2, peak oxygen uptake. Certain enrolled studies were not included because the value used for subgroup analysis was not reported in them. [a] Test for overall effect. [b] Test for subgroup differences.

3.1. Risk of Bias

Sixteen studies were scored by two reviewers, and an absolute agreement (r = 0.94) was obtained from the intraclass correlation coefficient (ICC). Bias was assessed as a judgment (high, low, or unclear) for individual elements from seven sources of bias and the following ICCs for absolute agreement between the two reviewers were obtained: random sequence generation for selection bias (r = 0.90), allocation concealment for selection bias (r = 0.92), blinding of participants and personnel for performance bias (r = 0.98), blinding of outcome assessment for detection bias (r = 0.94), incomplete outcome data for attrition bias (r = 0.79), selective reporting for reporting bias (r = 0.98) and inclusion criteria of patients in the studies and the country in which the study was conducted for other bias (r = 0.88). The risk of bias in the 16 included trials is summarized in Figure 2.

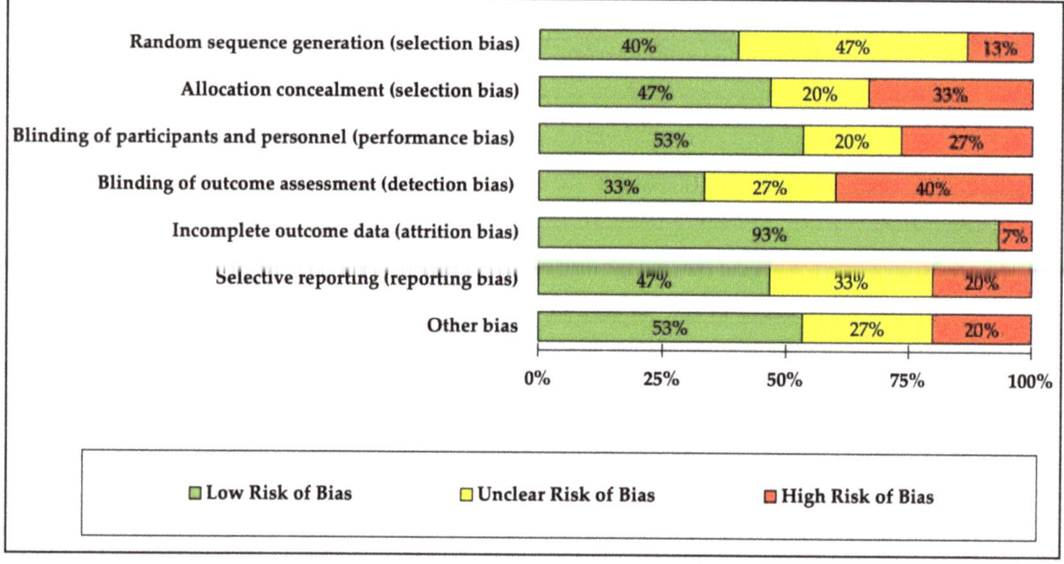

Figure 2. Assessment of risk of bias in included randomized controlled trials.

Of the 16 studies, the risk of bias was low in four or more of the seven sources of bias. Many studies were attributed to high risk in random sequence generation, allocation concealment and blinding of outcome assessment due to the nature of the exercise program. It was high in almost all studies due to the lack of blinding of participants and personnel. However, this issue could not be omitted due to the peculiarity of the intervention (exercise vs. no exercise) and should be taken into consideration.

The most prevalent methodological issues were an inadequate description of randomization (60%), allocation of concealment (50%) and blinding of outcome assessment (70%). Most studies were low risk for incomplete outcome data (90%).

3.2. Study and Participant Characteristics

The total number of CVD participants analyzed across all studies was 969 (267 coronary artery disease (CAD) only, 200 coronary artery bypass graft (CABG) only, 75 percutaneous coronary intervention (PCI) only, 50 CABG/PCI, 310 myocardial infarction (MI), and 63 carotid artery stiffness (CAS)). A summary of study characteristics is shown in the Supplementary Materials.

The number of participants per group ranged between 15 and 48, with four studies reporting <20 participants and twelve studies reporting ≥20 participants, with the majority being males (n = 419). The age range of participants was 52–69 years, with nine studies reporting mean ages <60 years, six studies reporting mean ages ≥60 years, and one that did not report any age information. Individual patient characteristics for each study can be seen in Table 2.

Regarding the characteristics of the patients, the meta-analysis identified statistically significant improvements in VO_2peak in each subgroup of patients with PCI ($p < 0.001$), as well as in patients with MI ($p < 0.01$), CABG ($p < 0.02$) and both CABG/PCI ($p < 0.03$).

3.3. Intervention Characteristics

The included trials tested a variety of interventions to increase VO_2peak (Table 3). In many trials, the interventions were performed with exercise-based clinical supervision [10–14,20–22,34,37,38], a few studies implemented an unsupervised home-based program [33], and some studies performed both programs [9,30,36].

Exercise training was typically continuous [9–12,21,22,33–39], as opposed to interval [13,14], or mixed training [20], and this type of training was shown to be significantly superior in improving VO$_2$peak (3.27 mL-kg^{-1}-min^{-1}; 95% CI = 2.23–4.32; $p < 0.001$; I^2 = 87%).

The frequency of training was typically 3–4 days/week [9–14,20–22,33–35,38], and aerobic training was the most used type of intervention [10,12–14,20–22,33–36,38,39]. Three studies tested aerobic and resistance training together during the intervention [9,11,37]. The meta-analysis identified that cycle-ergometers ($p < 0.05$) and treadmill ($p < 0.01$) significantly favored changes in VO$_2$peak.

Studies were separated into three groups depending upon length (<six, 6–12, and >12 weeks). The intervention length ranged from two to 24 weeks, with one study that reported data for less than six weeks [37], nine studies reported data for 6 to 12 weeks [9,10,13,20,31,33,34,36,37], and six studies reported data for >12 weeks [11,12,14,21,22,35]. The subgroup that included studies of >12 weeks in length was significantly superior in terms of improvements in VO$_2$peak (2.50 mL-kg^{-1}-min^{-1}; 95% CI = 2.23–4.32; $p < 0.001$; I^2 = 52%). Interventions 6 to 12 weeks in length also produced a large increase ($p < 0.01$), demonstrating moderate heterogeneity (5.31 mL-kg^{-1}-min^{-1}; 95% CI = 1.24–9.38; I^2 = 52%).

Based on the American College of Sports Medicine (ACSM) [28] cut-off points, three studies prescribed moderate-intensity exercise (n = 18, 75%) [10,33,38], three prescribed vigorous-intensity exercise (n = 18, 75%) [14,20,21] and ten interventions (n = 62,5%) [9,11,13,22,35–37,39] prescribed a range of intensities that placed them within both the moderate-intensity and vigorous-intensity categories. The meta-analyzed effects found the intervention was beneficial in terms of changing VO$_2$peak in both intensities ($p < 0.001$).

3.4. Subgroup Analyses—Intensity

When interpreting these results (Figure 3), it is essential to consider how exercise intensity was classified. We used a categorical-based approach, in which interventions were categorized according to the prescribed exercise intensity reported in each study, based on the recommendations of the ACSM [28].

The meta-regression analysis displayed in Figure 3 revealed that relative VO$_2$peak was significantly increased in all intensity categories. Moderate-intensity interventions produced a moderate increase in relative VO$_2$peak (0.71 mL-kg^{-1}-min^{-1}; 95% CI = 0.27–1.15; $p = 0.001$) with moderate heterogeneity (I^2 = 45%). Moderate-to-vigorous-intensity and vigorous-intensity interventions produced a large increase in relative VO$_2$peak (1.84 mL-kg^{-1}-min^{-1}; 95% CI = 1.18–2.50; $p < 0.001$ and 1.80 mL-kg^{-1}-min^{-1}; 95% CI = 0.82–2.78; $p = 0.0003$, respectively), and were also highly heterogeneous with I^2 values of 91 and 95% ($p < 0.001$), respectively.

3.5. Subgroup Analyses—Intensity and Length

In the analyses of studies lasting less than six weeks, we evaluated studies that exercised at vigorous intensity. The results (Figure 4) showed a large increase in VO$_2$peak (3.81 mL-kg^{-1}-min^{-1}; 95% CI = 0.16–7.45; $p = 0.04$) and demonstrated significant heterogeneity (I^2 = 98%).

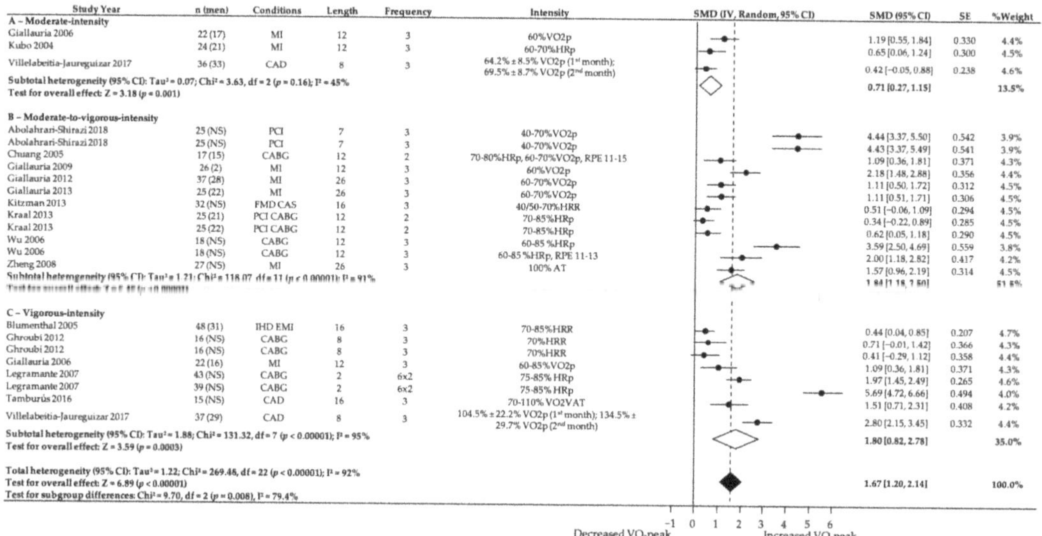

Study Year	n (men)	Conditions	Length	Frequency	Intensity	SMD (95% CI)	SE	%Weight
A – Moderate-intensity								
Giallauria 2006	22 (17)	MI	12	3	60%VO2p	1.19 [0.55, 1.84]	0.330	4.4%
Kubo 2004	24 (21)	MI	12	3	60-70%HRp	0.65 [0.06, 1.24]	0.300	4.5%
Villelabeitia-Jaureguizar 2017	36 (33)	CAD	8	3	64.2% ± 8.5% VO2p (1st month); 69.5% ± 8.7% VO2p (2nd month)	0.42 [−0.05, 0.88]	0.238	4.6%
Subtotal heterogeneity (95% CI): Tau² = 0.07; Chi² = 3.63, df = 2 (p = 0.16); I² = 45%						0.71 [0.27, 1.15]		13.5%
Test for overall effect: Z = 3.18 (p = 0.001)								
B – Moderate-to-vigorous-intensity								
Abolahrari-Shirazi 2018	25 (NS)	PCI	7	3	40-70%VO2p	4.44 [3.37, 5.50]	0.542	3.9%
Abolahrari-Shirazi 2018	25 (NS)	PCI	7	3	40-70%VO2p	4.43 [3.37, 5.49]	0.541	3.9%
Chuang 2005	17 (15)	CABG	12	2	70-80%HRp, 60-70%VO2p, RPE 11-15	1.09 [0.36, 1.81]	0.371	4.3%
Giallauria 2009	26 (2)	MI	12	3	60%VO2p	2.18 [1.48, 2.88]	0.356	4.4%
Giallauria 2012	37 (28)	MI	26	3	60-70%VO2p	1.11 [0.50, 1.72]	0.312	4.5%
Giallauria 2013	25 (22)	MI	26	3	60-70%VO2p	1.11 [0.51, 1.71]	0.306	4.5%
Kitzman 2013	32 (NS)	FMD CAS	16	3	40/50-70%HRR	0.51 [−0.06, 1.09]	0.294	4.5%
Kraal 2013	25 (21)	PCI CABG	12	2	70-85%HRp	0.34 [−0.22, 0.89]	0.285	4.5%
Kraal 2013	25 (22)	PCI CABG	12	2	70-85%HRp	0.62 [0.05, 1.18]	0.290	4.5%
Wu 2006	18 (NS)	CABG	12	3	60-85%HRp	3.59 [2.50, 4.69]	0.559	3.8%
Wu 2006	18 (NS)	CABG	12	3	60-85%HRp, RPE 11-13	2.00 [1.18, 2.82]	0.417	4.2%
Zheng 2008	27 (NS)	MI	26	3	100% AT	1.57 [0.96, 2.19]	0.314	4.5%
Subtotal heterogeneity (95% CI): Tau² = 1.21; Chi² = 118.07, df = 11 (p < 0.00001); I² = 91%						1.84 [1.18, 2.50]		51.5%
Test for overall effect: Z = ... (p < 0.00001)								
C – Vigorous-intensity								
Blumenthal 2005	48 (31)	IHD EMI	16	3	70-85%HRR	0.44 [0.04, 0.85]	0.207	4.7%
Ghroubi 2012	16 (NS)	CABG	8	3	70%HRR	0.71 [−0.01, 1.42]	0.366	4.3%
Ghroubi 2012	16 (NS)	CABG	8	3	70%HRR	0.41 [−0.29, 1.12]	0.358	4.4%
Giallauria 2006	22 (16)	MI	12	3	60-85%VO2p	1.09 [0.36, 1.81]	0.371	4.3%
Legramante 2007	43 (NS)	CABG	2	6x2	75-85%HRp	1.97 [1.45, 2.49]	0.265	4.6%
Legramante 2007	39 (NS)	CABG	2	6x2	75-85%HRp	5.69 [4.72, 6.66]	0.494	4.0%
Tamburús 2016	15 (NS)	CAD	16	3	70-110% VO2VAT	1.51 [0.71, 2.31]	0.408	4.2%
Villelabeitia-Jaureguizar 2017	37 (29)	CAD	8	3	104.5% ± 22.2% VO2p (1st month); 134.5% ± 29.7% VO2p (2nd month)	2.80 [2.15, 3.45]	0.332	4.4%
Subtotal heterogeneity (95% CI): Tau² = 1.88; Chi² = 131.32, df = 7 (p < 0.00001); I² = 95%						1.80 [0.82, 2.78]		35.0%
Test for overall effect: Z = 3.59 (p = 0.0003)								
Total heterogeneity (95% CI): Tau² = 1.22; Chi² = 269.46, df = 22 (p < 0.00001); I² = 92%						1.67 [1.20, 2.14]		100.0%
Test for overall effect: Z = 6.89 (p < 0.00001)								
Test for subgroup differences: Chi² = 9.70, df = 2 (p = 0.008), I² = 79.4%								

Figure 3. Effect of moderate-, moderate-to-vigorous- and vigorous-intensity exercise during exercise programs on change in relative VO$_2$peak (mL-kg^{-1}-min^{-1}). NS, not stated/missing. HRR, heart rate reserve. HRp, heat rate peak. RPE, rate of perceived exertion. AT, anaerobic threshold. VAT, ventilatory anaerobic threshold. 95% CI, 95% confidence interval. SMD, standardized mean difference. IV, Random: a random-effects meta-analysis was applied, with weights based on inverse variances. SE, standard error. Tau^2 and I^2, heterogeneity statistics. df, degree of freedom. Chi^2, the chi-squared test value. Z, Z-value for test of the overall effect. P, p-value. Conditions: MI, myocardial infarction. CABG, coronary artery bypass graft. PCI, percutaneous coronary intervention. CAD, coronary artery disease. IHD, ischsemic heart disease. EMI, exercise-induced myocardial ischemia. FMD, endothelial-dependent flow-mediated arterial dilation. CAS, carotid artery stiffness.

For interventions of 6 to 12 weeks length, moderate-to-vigorous-intensity interventions showed a further increase in VO$_2$peak (2.28 mL-kg^{-1}-min^{-1}; 95% CI = 1.23–3.32; $p < 0.001$; I^2 = 93%) compared to moderate-intensity (0.71 mL-kg^{-1}-min^{-1}; 95% CI = 0.12–1.29; $p = 0.02$; I^2 = 0%) and vigorous-intensity interventions (1.57 mL-kg^{-1}-min^{-1}; 95% CI = 0.12–3.02; $p = 0.02$; I^2 = 0%).

For studies that intervened more than 12 weeks, moderate-to-vigorous-intensity interventions were significantly superior (1.07 mL-kg^{-1}-min^{-1}; 95% CI = 0.64–1.50; $p < 0.001$; I^2 = 51%) to vigorous-intensity interventions (0.92 mL-kg^{-1}-min^{-1}; 95% CI = −0.12–1.96; $p = 0.08$; I^2 = 82%) in improving VO$_2$peak.

3.6. Publication Bias

There was no significant publication bias for studies with moderate-intensity (Egger's test: $\beta = 7.29$; $p = 0.26$) and vigorous-intensity (Egger's test: $\beta = 8.67$; $p = 0.15$) interventions reporting relative VO$_2$peak. However, there was significant publication bias for studies with moderate-to-vigorous-intensity interventions (Egger's test: $\beta = 13.19$; $p = 0.00$). The funnel plot with all studies (Figure 5) showed a significant degree of asymmetry (Egger's test: $p = 0.00$). Nevertheless, false-positive results may occur due to substantial between-study heterogeneity [40], making the disparity in the number of studies included in each intensity category likely to cause significant asymmetry in the funnel plot.

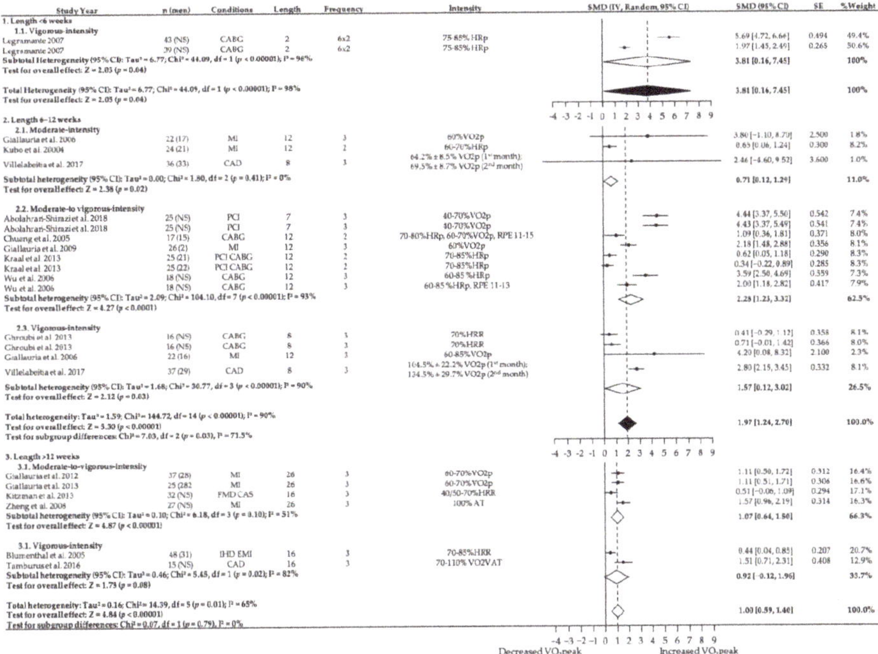

Figure 4. Effect of length in moderate-, moderate-to-vigorous- and vigorous-intensity exercise during exercise programs on change in relative VO₂peak (mL·kg^{-1}·min^{-1}). NS, not stated/missing. HRR, heart rate reserve. HRp, heat rate peak. RPE, rate of perceived exertion. AT, anaerobic threshold. VAT, ventilatory anaerobic threshold. 95% CI, 95% confidence interval. SMD, standardized mean differences. IV, Random: a random-effects meta-analysis is applied, with weights based on inverse variances. SE, standard error. Tau^2 and I^2, heterogeneity statistics. df, degrees of freedom. Chi^2, the chi-squared test value. Z, Z-value for test of the overall effect. P, p-value. Conditions: MI, myocardial infarction. CABG, coronary artery bypass graft. PCI, percutaneous coronary intervention. CAD, coronary artery disease. IHD, ischemic heart disease. EMI, exercise-induced myocardial ischemia. FMD, endothelial-dependent flow-mediated arterial dilation. CAS, carotid artery stiffness.

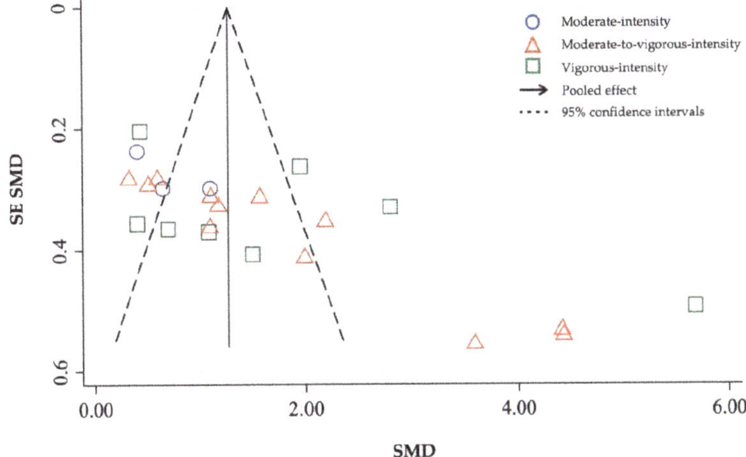

Figure 5. Funnel plot with pseudo 95% confidence intervals for change in relative VO₂peak (mL·kg^{-1}·min^{-1}) by exercise intensity (moderate, moderate-to-vigorous, vigorous). SMD, standardized mean difference. SE SMD, standard error of standardized mean differences.

4. Discussion

The main aim of this systematic review and meta-analysis was to identify the optimal intensity to optimize VO_2peak in patients with CVD following exercise programs. Furthermore, we aimed to gauge whether the length of interventions had an effect on the results.

Our results support the crucial role of physical exercise in patients with CVD. They have shown significant improvements for all cardiac impairments at all ages, regardless of the aerobic exercise mode.

A comparison of the mean effects between intensity classifications showed significant improvements, with moderate-to-vigorous-intensity interventions providing the greatest improvements of VO_2peak. The differences were considered clinically significant ($p = 0.03$) and the retro transformation of the SMD suggested that the difference between the intensities was 3.92 mL-kg^{-1}-min^{-1}. However, when comparing the effects grouped among the intensity classifications, it was found that moderate-to-vigorous-intensity exercises can provide the most significant improvements in VO_2peak. Even so, the differences were not considered clinically significant once the retro transformation of the SMD suggested that the differences between the intensities were, at most, only 1.67 mL-kg^{-1}-min^{-1}. In this regard, our study confirmed the results of previous systematic reviews, pointing out that moderate-to-vigorous- and vigorous-intensity interventions improved CRF to a larger extent than moderate-intensity ones [23].

The difference between moderate-to-vigorous- and vigorous-intensity in our study was 0.4 mL-kg^{-1}-min^{-1} and the difference between moderate- and moderate-to-vigorous-intensity was more significant (1.13 mL-kg^{-1}-min^{-1}). Although these analyses did not yield any consistent findings, they highlighted considerable variability in outcomes for interventions based on VO_2peak that appeared to be consistent across intensities. Although unexpected, this finding is not surprising. Given that VO_2 is not an appropriate variable to regulate intensity during training, in practice, prescriptions are converted to heart rate (HR) estimated to elicit the target VO_2. This approach is confounded in a CR setting by medications (e.g., β-blockers) that alter HR responses, which may cause dissociation of the HR and VO_2 relationship, where a small change in HR may result in varied and disproportionate changes to work rate or VO_2peak [8,17,22].

The first meta-analyses that investigated improvements in CRF following exercise-based CR reported a small improvement in CRF (SMD $\pm$: 95% CI = 0.46 $\pm$ 0.02) [41]. Our study confirmed the results of Mitchell et al. [23] who verified that moderate- and moderate-to-vigorous-intensity interventions were associated with a moderate increase in relative VO_2peak (SMD $\pm$: 95% CI = 0.94 $\pm$ 0.30 and 0.93 $\pm$ 0.17, respectively), and vigorous-intensity exercise with a large increase (SMD $\pm$: 95% CI = 1.10 $\pm$ 0.25), and moderate- and vigorous-intensity interventions were associated with moderate improvements in absolute VO_2peak (SMD $\pm$: 95% CI = 0.63 $\pm$ 0.34 and SMD $\pm$: 95% CI = 0.93 $\pm$ 0.20, respectively), whereas moderate-to-vigorous- intensity interventions elicited a large effect (SMD $\pm$: 95% CI = 1.27 $\pm$ 0.75).

When we subdivided the intensities by length to obtain a more in-depth view of the effect of the different intensities, we found that the vigorous-intensity interventions below six weeks had more significant results in improving the VO_2peak (3.81 mL-kg^{-1}-min^{-1}). Based on the sensitivity analysis, although the results suggest that interventions conducted bidirectionally six times a week resulted in more significant gains of CRF favoring vigorous intensity, the analysis only included two studies and may not be practical to implement. In this sense, not being able to compare with other studies and other intensities within the division by length, the best result obtained was between 6 and 12 weeks in moderate-to-vigorous-intensity exercise, in which there was a significant increase in VO_2peak in relation to the vigorous- and moderate-intensity categories.

Interventions >12 weeks did not show significantly greater gains in CRF compared to other lengths. However, there was a significant improvement in VO_2peak with moderate-to-vigorous intensity. Additionally, there was no significant improvement in VO_2peak

with vigorous-intensity interventions, and there were no studies of moderate-intensity RCTs available for comparison. Furthermore, patients with CVD did not obtain significant VO_2peak improvements when the vigorous-intensity protocol was >12 weeks.

Our results indicate that moderate-to-vigorous-intensity exercise is superior to other intensities in improving aerobic capacity and is likely to be an underestimation of the true differences between groups. This is supported by the methodological decisions favoring the use of a conservative approach in the meta-analysis (by choosing random effects and SMDs) and using the highest calculated SD for studies where no information was published to allow SD calculations.

Thereby, our findings suggest higher benefits from moderate-to-vigorous-intensity exercise lasting 6 to 12 weeks in terms of VO_2peak improvements in patients with CVD. Overall, our findings are in agreement with reports from previous meta-analyses [19,23,32,42,43]. Hannan et al. [24] concluded that HIIT (e.g., of moderate-to-vigorous- and vigorous intensity) is more effective than moderate-intensity exercise in improving CRF in participants of CR (0.34 mL-kg^{-1}-min^{-1}; 95% CI = 0.2–0.48; $p < 0.001$; $I^2 = 28\%$). Still, improvements in CRF were higher in >six-week exercise programs, and the largest improvements in CRF for patients with CAD resulted from programs lasting 7 to 12 weeks, as our study confirmed [24].

Some limitations of this systematic review and meta-analysis should be considered. First, the poor level of reporting within the available RCTs made it difficult to evaluate the most effective doses of intensity on CRF in cardiac patients. Second, the RCTs did not use the same methods to control the exercise intensity and the different variables used to establish exercise intensity added complexity to the analyses. While the variables were based on interrelated physiological constructs (e.g., HR and VO_2), they were not directly comparable. Even in what appears to be the narrow domain of HIIT, there is much heterogeneity in clearly defining what high intensity is.

In our study, each reported intervention was categorized according to the prescribed exercise intensity, based on ACSM recommendations [28]. This approach has two limitations. While some studies reported precise exercise intensities (e.g., 60% VO_2peak), most CR studies prescribed large intervals based on HR responses to exercise (e.g., 40–70% VO_2peak). As these studies often covered several intensity categories, making them difficult to categorize, it was necessary to add an extra intensity category, moderate-to-vigorous intensity. In this category, participants were assumed to have performed similar training interventions, when in fact they may have experienced quite different exercise prescriptions.

We recognized the lack of available data for some intensity analyses when split by program length. For example, in the analysis of subgroups of studies below six weeks, we only had studies that prescribed vigorous-intensity exercise, as well as in the length above 12 weeks, we had no studies that used in their intervention a moderate-intensity program. Furthermore, subgroup analyses for the combined effect of vigorous-intensity programs with lengths below six weeks were based on only two groups of patients, both from the same study that completed the same intervention. As such, we recommend caution when interpreting results where the lack of available data may have limited analyses.

We should consider that medication can influence exercise and therefore should be considered by the therapist when prescribing exercise. Beta-blockers decrease exercise capacity because they create a ceiling effect, meaning the HR will not rise beyond a certain point. Thus, the target HR for monitoring should not be used. Rather, the therapist should use the rate of perceived exertion or calculate the target HR with a graded stress test while the patient is using the medication. Similarly, vasodilators and alpha- and calcium channel blockers may lead to a sudden blood pressure drop while exercising or afterwards.

Therefore, the variables that should be taken into consideration are trainability (result of CRF level, muscular endurance and strength) and risk stratification on the basis of completed medical history. Consequently, these factors may provide options for the optimal type of exercise and intensity level.

Future studies would benefit from being between 6 and 12 weeks in length with an intervention activity carried out at least three times weekly, ensuring that the correct intensity is maintained. For example, appropriate goals for vigorous-intensity exercise include $\geq$85% VO_2peak or $\geq$85% HRR or $\geq$90% HRM and, for moderate intensity, 50–75% VO_2peak or 50–75% HRR or 50–80% HRM. In addition, large ranges of exercise intensities should not be prescribed based on HR responses to exercise. This would allow a more accurate calculation of the exact effects of intensities on CRF and to determine the ideal and most effective "dose" for people with heart problems. Future research should include methods to appropriately describe the compliance of participants with the prescribed exercise intensity and attendance of exercise sessions.

Studies should report standard deviations, conceal allocation, and blind assessors to improve study quality. Moreover, future studies should aim to recruit more women and older participants (<76 years) to ensure vigorous-intensity interventions are more effective than moderate-intensity ones in improving CRF for a broader range of patients with CVD. Finally, further studies that investigate the longer-term benefits of vigorous-intensity interventions and whether these adaptations are maintained would also be beneficial.

5. Conclusions

The most effective doses of exercise intensity to optimize CRF were moderate-to-vigorous and vigorous exercise. Interventions to enhance CRF in patients with CVD are most effective if conducted for 6 to 12 weeks. More research is needed to understand within the moderate-to-vigorous-intensity category which percentage results in increased CRF, assisting in the design of specific prescription protocols.

This review may suggest that countries without guidelines for patients with CVD regarding the intensity of exercise programs, as well as countries with guidelines that recommend lower intensity exercise, should include moderate-to-vigorous intensity and vigorous intensity.

What is already known:

- Cardiovascular diseases are the leading causes of mortality in today's society. They are responsible for up to 30% of all deaths worldwide and 48% of deaths in Europe, and it is expected that these figures will increase in the coming years.
- Exercise programs in patients with cardiovascular disease have several beneficial effects on cardiovascular functional capacity, quality of life, risk factors modification, psychological profile, hospital readmissions, and mortality.
- Exercise-based interventions seem to significantly improve cardiorespiratory fitness in patients following a cardiac event or surgery, but little is known regarding the differential effects of prescribed exercise intensity.

What are the new findings?

- Exercise interventions for patients with cardiovascular disease tend include large ranges of exercise intensities based on heart rate responses to exercise.
- The most effective doses of exercise intensity to optimize cardiorespiratory fitness were moderate-to-vigorous and vigorous-intensity exercises, being more effective when conducted for 6 to 12 weeks.
- More research is needed to understand within the moderate-to-vigorous- and vigorous-intensity categories the percentage that specifically helps to increase cardiorespiratory fitness and the ability to establish specific prescription protocols.

Supplementary Materials: The following are available online at https://www.mdpi.com/article/10.3390/ijerph18073574/s1, Table S1: Complete search strategy for MEDLINE, searched from inception until January 2021; Figure S2: Summary of study characteristics; Figure S3: Outcome of the risk of bias assessment; Figure S4: List of references for included studies.

Author Contributions: Conceptualization, C.G., J.B., A.A. and A.R.; methodology, C.G. and J.B.; software, C.G.; validation, C.G. and J.B.; formal analysis, C.G. and J.B.; investigation, C.G. and

J.B.; resources, C.G. and J.B.; data curation, C.G. and J.B.; writing—original draft preparation, C.G.; writing—review and editing, C.G., J.B., A.A. and A.R.; visualization, C.G., J.B., A.A. and A.R.; supervision, C.G., J.B. and A.R.; project administration, C.G., J.B. and A.R.; funding acquisition, C.G., A.A. and A.R. All authors have read and agreed to the published version of the manuscript.

Funding: This research was funded by Fundação para a Ciência e Tecnologia (Portugal), grant number SFRH/BD/138326/2018.

Institutional Review Board Statement: Not applicable.

Informed Consent Statement: Not applicable.

Data Availability Statement: The data that support the findings of this study are available from the corresponding author, C.G., upon reasonable request.

Acknowledgments: This work was supported by the Fundação para a Ciência e a Tecnologia (Portugal). We thank all authors of the original works cited in the present study, who readily assisted us by sharing their manuscripts for this systematic review with meta-analysis.

Conflicts of Interest: The authors declare no conflict of interest. The funders had no role in the design of the study; in the collection, analyses, or interpretation of data; in the writing of the manuscript, or in the decision to publish the results.

References

1. World Health Organization. *Cardiovascular Disease*; Fact Sheet N 317; WHO: Geneva, Switzerland, 2011.
2. EACPR Committee for Science Guidelines; Corrà, U.; Piepoli, M.F.; Carré, F.; Heuschmann, P.; Hoffmann, U.; Verschuren, M.; Halcox, J.; Giannuzzi, P.; Saner, H.; et al. Secondary prevention through cardiac rehabilitation: Physical activity counselling and exercise training: Key components of the position paper from the Cardiac Rehabilitation Section of the European Association of Cardiovascular Prevention and Rehabilitation. *Eur. Heart J.* **2010**, *31*, 1967–1974. [CrossRef]
3. Mohammed, H.G.; Shabana, A.M. Effect of cardiac rehabilitation on cardiovascular risk factors in chronic heart failure patients. *Egypt Heart J.* **2018**, *70*, 77–82. [CrossRef] [PubMed]
4. Arnett, D.K.; Blumenthal, R.S.; Albert, M.A.; Buroker, A.B.; Goldberger, Z.D.; Hahn, E.J.; Himmelfarb, C.D.; Khera, A.; Lloyd-Jones, D.; McEvoy, J.W.; et al. 2019 ACC/AHA Guideline on the Primary Prevention of Cardiovascular Disease: A Report of the American College of Cardiology/American Heart Association Task Force on Clinical Practice Guidelines. *Circulation* **2019**, *140*, e596–e646. [CrossRef]
5. Anderson, L.; Oldridge, N.; Thompson, D.R.; Zwisler, A.-D.; Rees, K.; Martin, N.; Taylor, R.S. Exercise-Based Cardiac Rehabilitation for Coronary Heart Disease. *J. Am. Coll. Cardiol.* **2016**, *67*, 1–12. [CrossRef] [PubMed]
6. Mezzani, A.; Hamm, L.F.; Jones, A.M.; McBride, P.E.; Moholdt, T.; Stoner, J.A.; Urhausen, A.; Williams, A.M. Aerobic exercise intensity assessment and prescription in cardiac rehabilitation: A joint position statement of the European Association for Cardiovascular Prevention and Rehabilitation, the American Association of Cardiovascular and Pulmonary Rehabilitation and the Canadian Association of Cardiac Rehabilitation. *Eur. J. Prev. Cardiol.* **2013**, *20*, 442–467. [CrossRef]
7. Moholdt, T.; Aamot, I.L.; Granøien, I.; Gjerde, L.; Myklebust, G.; Walderhaug, L.; Brattbakk, L.; Hole, T.; Graven, T.; Stølen, O.T.; et al. Aerobic interval training increases peak oxygen uptake more than usual care exercise training in myocardial infarction patients: A randomized controlled study. *Clin. Rehabil.* **2011**, *26*, 33–44. [CrossRef]
8. Wisloff, U.; Stoylen, A.; Loennechen, J.P.; Bruvold, M.; Rognmo, Ø.; Haram, P.M.; Tjonna, A.E.; Stig, J.H.; Slørdahl, S.A.; Lee, S.J.; et al. Superior cardiovascular effect of aerobic interval training versus moderate continuous training in heart failure patients: A randomized study. *Circulation* **2007**, *115*, 3086–3094. [CrossRef] [PubMed]
9. Rojhani-Shirazi, Z.; Abolahrari-Shirazi, S.; Kojuri, J.; Bagheri, Z. Efficacy of combined endurance-resistance training versus endurance training in patients with heart failure after percutaneous coronary intervention: A randomized controlled trial. *J. Res. Med. Sci.* **2018**, *23*, 12. [CrossRef] [PubMed]
10. Giallauria, F.; Lucci, R.; D'Agostino, M.; Vitelli, A.; Maresca, L.; Mancini, M.; Aurino, M.; Del Forno, D.; Giannuzzi, P.; Vigorito, C. Two-year multicomprehensive secondary prevention program: Favorable effects on cardiovascular functional capacity and coronary risk profile after acute myocardial infarction. *J. Cardiovasc. Med.* **2009**, *10*, 772–780. [CrossRef] [PubMed]
11. Giallauria, F.; Acampa, W.; Ricci, F.; Vitelli, A.; Maresca, L.; Mancini, M.; Grieco, A.; Gallicchio, R.; Xhoxhi, E.; Spinelli, L.; et al. Effects of exercise training started within 2 weeks after acute myocardial infarction on myocardial perfusion and left ventricular function: A gated SPECT imaging study. *Eur. J. Prev. Cardiol.* **2012**, *19*, 1410–1419. [CrossRef] [PubMed]
12. Giallauria, F.; Acampa, W.; Ricci, F.; Vitelli, A.; Torella, G.; Lucci, R.; Del Prete, G.; Zampella, E.; Assante, R.; Rengo, G.; et al. Exercise training early after acute myocardial infarction reduces stress-induced hypoperfusion and improves left ventricular function. *Eur. J. Nucl. Med. Mol. Imaging* **2013**, *40*, 315–324. [CrossRef]
13. Villelabeitia-Jaureguizar, K.; Vicente-Campos, D.; Senen, A.B.; Jiménez, V.H.; Garrido-Lestache, M.E.B.; Chicharro, J.L. Effects of high-intensity interval versus continuous exercise training on post-exercise heart rate recovery in coronary heart-disease patients. *Int. J. Cardiol.* **2017**, *244*, 17–23. [CrossRef]

14. Tamburús, N.Y.; Kunz, V.C.; Salviati, M.R.; Simões, V.C.; Catai, A.M.; Da Silva, E. Interval training based on ventilatory anaerobic threshold improves aerobic functional capacity and metabolic profile: A randomized controlled trial in coronary artery disease patients. *Eur. J. Phys. Rehabil. Med.* **2015**, *52*, 1–11. [PubMed]

15. Beckie, T.M.; Beckstead, J.W.; Kip, K.E.; Fletcher, G. Improvements in Heart Rate Recovery Among Women After Cardiac Rehabilitation Completion. *J. Cardiovasc. Nurs.* **2014**, *29*, 38–47. [CrossRef] [PubMed]

16. Rivera-Brown, A.M.; Frontera, W.R. Principles of exercise physiology: Responses to acute exercise and long-term adaptations to training. *PM R.* **2012**, *4*, 797–804. [CrossRef] [PubMed]

17. Warburton, D.E.; McKenzie, D.C.; Haykowsky, M.J.; Taylor, A.; Shoemaker, P.; Ignaszewski, A.P.; Chan, S.Y. Effectiveness of High-Intensity Interval Training for the Rehabilitation of Patients with Coronary Artery Disease. *Am. J. Cardiol.* **2005**, *95*, 1080–1084. [CrossRef]

18. Rognmo, Ø.; Hetland, E.; Helgerud, J.; Hoff, J.; Slørdahl, S.A. High intensity aerobic interval exercise is superior to moderate intensity exercise for increasing aerobic capacity in patients with coronary artery disease. *Eur. J. Cardiovasc. Prev. Rehabil.* **2004**, *11*, 216–222. [CrossRef]

19. Cornish, A.K.; Broadbent, S.; Cheema, B.S. Interval training for patients with coronary artery disease: A systematic review. *Graefe's Arch. Clin. Exp. Ophthalmol.* **2010**, *111*, 579–589. [CrossRef] [PubMed]

20. Ghroubi, S.; Elleuch, W.; Abid, L.; Kammoun, S.; Elleuch, M.-H. The effects of cardiovascular rehabilitation after coronary stenting, Apport de la readaptation cardiovasculaire dans les suites d'une angioplastie transluminale. *Ann. Phys. Rehabil. Med.* **2012**, *55*, e307–e309. [CrossRef]

21. Blumenthal, J.A.; Sherwood, A.; Babyak, M.A.; Watkins, L.L.; Waugh, R.; Georgiades, A.; Bacon, S.L.; Hayano, J.; Coleman, E.R.; Hinderliter, A. Effects of exercise and stress management training on markers of cardiovascular risk in patients with ischemic heart disease—A randomized controlled trial. *JAMA* **2005**, *293*, 1626–1634. [CrossRef]

22. Kitzman, D.W.; Brubaker, P.H.; Herrington, D.M.; Morgan, T.M.; Stewart, K.P.; Hundley, W.G.; Abdelhamed, A.; Haykowsky, M.J. Effect of endurance exercise training on endothelial function and arterial stiffness in older patients with heart failure and preserved ejection fraction: A randomized, controlled, single-blind trial. *J. Am. Coll. Cardiol.* **2013**, *62*, 584–592. [CrossRef]

23. Mitchell, B.L.; Lock, M.J.; Davison, K.; Parfitt, G.; Buckley, J.P.; Eston, R.G. What is the effect of aerobic exercise intensity on cardiorespiratory fitness in those undergoing cardiac rehabilitation? A systematic review with meta-analysis. *Br. J. Sports Med.* **2018**, *53*, 1341–1351. [CrossRef] [PubMed]

24. Hannan, A.L.; Hing, W.; Simas, V.; Climstein, M.; Coombes, J.S.; Jayasinghe, R.; Byrnes, J.; Furness, J. High-intensity interval training versus moderate-intensity continuous training within cardiac rehabilitation: A systematic review and meta-analysis. *Open Access J. Sports Med.* **2018**, *9*, 1–17. [CrossRef]

25. Higgins, J.; Green, S. *Cochrane Handbook for Systematic Reviews of Interventions*; John Wiley & Sons Ltd.: Chichester, UK, 2011.

26. Leonardo, R. PICO: Model for Clinical Questions. *Evid. Based Med. Pract.* **2018**, *3*, 1–2.

27. Moher, D.; Liberati, A.; Tetzlaff, J.; Altman, D.G.; PRISMA Group. Preferred reporting items for systematic reviews and meta-analyses: The PRISMA statement. *PLoS Med.* **2009**, *6*, e1000097. [CrossRef] [PubMed]

28. American College of Sports Medicine. *ACSM's Guidelines for Exercise Testing and Prescription*, 10th ed.; Lippincott Williams & Wilkins: Baltimore, MD, USA, 2017.

29. Hopkins, W.G.; Marshall, S.W.; Batterham, A.M.; Hanin, J. Progressive Statistics for Studies in Sports Medicine and Exercise Science. *Med. Sci. Sports Exerc.* **2009**, *41*, 3–13. [CrossRef]

30. *Review Manager (RevMan)*; Version 5.3.; The Nordic Cochrane Centre, The Cochrane Collaboration: Copenhagen, Denmark, 2014; Available online: http://community.cochrane.org/tools/review-production-tools/revman-5 (accessed on 21 January 2021).

31. Elbourne, D.R.; Altman, D.G.; Higgins, J.P.T.; Curtin, F.; Worthingtond, H.V.; Vaile, A. Meta-analyses involving cross-over trials: Methodological issues. *Int. J. Epidemiol.* **2002**, *31*, 140–149. [CrossRef]

32. Sandercock, G.; Hurtado, V.; Cardoso, F. Changes in cardiorespiratory fitness in cardiac rehabilitation patients: A meta-analysis. *Int. J. Cardiol.* **2013**, *167*, 894–902. [CrossRef] [PubMed]

33. Wu, S.K.; Lin, Y.W.; Chen, C.L.; Tsai, S.W. Cardiac rehabilitation vs. home exercise after coronary artery bypass graft surgery: A comparison of heart rate recovery. *Am. J. Phys. Med. Rehabil.* **2006**, *85*, 711–717. [CrossRef]

34. Chuang, T.-Y.; Sung, W.-H.; Lin, C.-Y. Application of a Virtual Reality–Enhanced Exercise Protocol in Patients After Coronary Bypass. *Arch. Phys. Med. Rehabil.* **2005**, *86*, 1929–1932. [CrossRef]

35. Zheng, H.; Luo, M.; Shen, Y.; Kang, W. Effects of 6 months exercise training on ventricular remodelling and autonomic tone in patients with acute myocardial infarction and percutaneous coronary intervention. *J. Rehabil. Med.* **2008**, *40*, 776–779. [PubMed]

36. Giallauria, F.; De Lorenzo, A.; Pilerci, F.; Manakos, A.; Lucci, R.; Psaroudaki, M.; D'Agostino, M.; Del Forno, D.; Vigorito, C. Long-term effects of cardiac rehabilitation on end-exercise heart rate recovery after myocardial infarction. *Eur. J. Cardiovasc. Prev. Rehabil.* **2006**, *13*, 544–550. [CrossRef] [PubMed]

37. Legramante, J.M.; Iellamo, F.; Massaro, M.; Sacco, S.; Galante, A. Effects of residential exercise training on heart rate recovery in coronary artery patients. *Am. J. Physiol. Circ. Physiol.* **2007**, *292*, H510–H515. [CrossRef] [PubMed]

38. Kubo, N.; Ohmura, N.; Nakada, I.; Yasu, T.; Katsuki, T.; Fujii, M.; Saito, M. Exercise at ventilatory threshold aggravates left ventricular remodeling in patients with extensive anterior acute myocardial infarction. *Am. Heart J.* **2004**, *147*, 113–120. [CrossRef]

39.   Kraal, J.J.; Peek, N.; Van den Akker-Van Marle, M.E.; Kemps, M.C.H. Effects and costs of home-based training with telemonitoring guidance in low to moderate risk patients entering cardiac rehabilitation: The FIT@Home study. *BMC Cardiovasc. Disord.* **2013**, *13*, 82. [CrossRef]
40.   Harbord, R.M.; Harris, R.J.; Sterne, J.A.C. Updated Tests for Small-study Effects in Meta-analyses. *Stata J. Promot. Commun. Stat. Stata* **2009**, *9*, 197–210. [CrossRef]
41.   Conn, V.S.; Hafdahl, A.R.; Moore, S.M.; Nielsen, P.J.; Brown, L.M. Meta-analysis of interventions to increase physical activity among cardiac sub-jects. *Int. J. Cardiol.* **2009**, *133*, 307–320. [CrossRef]
42.   Vromen, T.; Kraal, J.J.; Kuiper, J.; Spee, R.F.; Peek, N.; Kemps, H.M. The influence of training characteristics on the effect of aerobic exercise training in patients with chronic heart failure: A meta-regression analysis. *Int. J. Cardiol.* **2016**, *208*, 120–127. [CrossRef]
43.   Anderson, L.; Taylor, R.S. Cardiac rehabilitation for people with heart disease: An overview of Cochrane systematic reviews. *Cochrane Database Syst. Rev.* **2014**, *12*, CD011273. [CrossRef]

MDPI

St. Alban-Anlage 66

4052 Basel

Switzerland

Tel. +41 61 683 77 34

Fax +41 61 302 89 18

www.mdpi.com

International Journal of Environmental Research and Public Health Editorial Office

E-mail: ijerph@mdpi.com

www.mdpi.com/journal/ijerph